ORIGIN AND EVOLUTION OF VERTEBRATES

ENCYCLOPAEDIA OF EVOLUTIONARY BIOLOGY SERIES

ORIGIN AND EVOLUTION OF VERTEBRATES

By

Richa Arora

ANMOL PUBLICATIONS PVT. LTD.
NEW DELHI - 110 002 (INDIA)

ANMOL PUBLICATIONS PVT. LTD.
4374/4B, Ansari Road, Daryaganj
New Delhi - 110 002
Phones: 23261597, 23278000
Visit us at: www.anmolpublications.com

Origin and Evolution of Vertebrates

First Edition, 2004

ISBN 81-261-1527-0

PRINTED IN INDIA

Published by J.L. Kumar for Anmol Publications Pvt. Ltd., New Delhi and Printed at Mehra Offset Press, Delhi.

Contents

	Preface	*vii*
1.	Origin of Vertebrates	1
2.	Origin of Land Vertebrates	12
3.	Origin of Reptiles	26
4.	Origin of Dinosaurs	33
5.	Origin of Birds	46
6.	Origin of Mammals	67
7.	Proboscideans	83
8.	Horses	103
9.	Camels	121
10.	South American Mammal Radiation	133
11.	Prosimians	150
12.	The Evolution of Man	192
13.	The Gradual Appearance of Man	219

Preface

All living organisms share a master plan of structural and functional organization because they arose from a common ancestor long ago by the process of evolution. The tremendous work conducted by scientists in this field has increased our knowledge of evolution many folds. In the limited space, it is not possible to go into the details. Hence, a brief and comprehensive account has been produced in the light of recent work and latest information available from various sources.

The present title is an attempt to present an account of what is known or what is postulated about the life, and the subsequent evolution of the main animal groups. Certain basic scientific facts are included, it is true, but the majority of the contents consist of hypothesis which are never likely to be experimentally verified; and the reader should treat the book as a source of facts, ideas and theories on which thought and discussion can be based.

The present title claims to be a complete, thorough and authentic exposition of the topics dealt with. Emphasis has been laid on introducing basic as well as modern concepts. The language of the book is kept simple in lucid style. The subject matter has been interwoven for keeping up interest and curiosity of readers.

Though care has been taken to verify all passages quoted, but it is hardly likely that, in so large a mass of material, all errors shall have been avoided. The author and the publishers would welcome as a favour any suggestions or corrections submitted by interested readers.

Thanks are due to all friends and colleagues whose continuous inspiration have initiated me to bring out the present work.

There can be no claim to originality except in the manners of treatment and much of the information has been obtained from the books and scientific journals available in the different libraries.

—**Author**

Chapter—1

Origin of Vertebrates

We have in the two preceding chapters discussed the evolution of two of the three great lines of descent which the animal kingdom includes. The third line, that of the vertebrates, is to be the subject of our research from now on, to enable us to comprehend in full the place in nature held by mankind.

Definition of a Vertebrate

Notochord

Vertebrates, or to use the more comprehensive term, chordates, have several diagnostic characters which are absolutely distinctive, separating them sharply from all other forms of life. Of these, the first is the possession of a notochord. This is an internal axial stiffening running lengthwise of the trunk and serving to resist the bodily shortening which the contraction of the muscles would otherwise cause. In its most primitive form the notochord is membranous, composed of cellular connective tissue, the cavities of which are so distended with fluid as to render the whole structure turgid, resistant to pressure, but highly elastic. Later the notochord becomes cartilaginous, to be replaced in higher forms by the bony vertebral column consisting of a number of short but often complex vertebræ separated, for mobility, by cartilaginous intervertebral discs.

Perforated Pharynx

The development of apertures known as gill-slits through the walls of the pharynx or throat cavity is the second chordate character. These vary in number from a pair to more than a hundred (*Amphioxus*) and are always present, but by no means always retain their ancient respiratory function, for with ourselves and other mammals, the slits, of which there are several pairs in the embryo, are reduced in number until but one pair is left and these form the eustachian tubes which serve to equalize the air pressure on either side of the ear drum by connecting the middle ear with the cavity of the throat.

Neurocœle

The third diagnostic character is a hollow nerve cord, the so-called spinal cord, which lies immediately above the notochord or the vertebral column. This may be a very simple structure, or again its anterior portion may increase and develop until in its highest expression, the human brain, it has formed what is probably the most intricate thing in nature. In every case, however, simple or complex it may be, the internal canal or neurocœle persists, though exceptions may be said to exist in the tunicates or sea-squirts, which are striking examples of degeneracy resulting from sedentary life. In them the active larva has a nervous system which conforms to our definition, but in the adult it is reduced to a single ganglion, or mass of nerve matter, with no trace of the neurocœle.

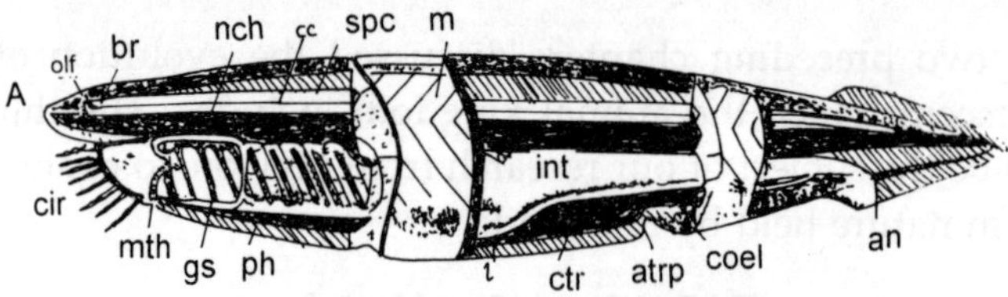

Fig. 1.1: Diagram of *Amphioxus*. A, anterior end, P, posterior end. *an*, anus, *atr*, atrium; *atrp*, atriopore; *br*, brain; *ce*, central canal (neurocœle); *cir*, cirri; *cœl*, cœlome or body cavity; *gs*, gill slits; *int*, intestine; *l*, liver; *m*, muscle bands (myomeres); *mth*, mouth; *nch*, notochord; *olf*, olfactory pit; *ph*, pharynx, *spc*, spinal cord.

Other Characters

Other distinctive features are usually shown by chordates. They are generally segmented, the segmentation showing in the nervous system, gill-slits, vertebral column, ribs and breast-bone, and in the muscles of the trunk. When paired limbs are present, their number never exceeds four, while in the invertebrate there is no such limitation. Finally, there is apparently a reversal of surfaces, for whereas in the invertebrate the bulk of the nervous system lies *below* the gut and the blood system *above*, in the chordate the reverse is true. In the invertebrate the body-wall is equally thick throughout, in the vertebrate there is a remarkable thickening or concentration of the muscles along the dorsal side within which lie the notochord and spinal cord.

It is, therefore, this group of forms, comprising the degenerate tunicates,. the unprogressive *Amphioxus*, the fishes, amphibians, reptiles, birds, mammals, the last of course including man, that we wish to consider, and our immediate problem is to learn, if possible, the time, place, and source of vertebrate beginnings.

Time of Origin

The chordates area very ancient race, dating back probably to the beginning of Paleozoic time, although the tangible record of their existence commences with the fragmentary remains of armoured "fishes" (ostracoderms) found in Middle Ordovician rocks near Canyon City, Colorado, in the Big Horn Mountains of Wyoming, and in the Black Hills of South Dakota. But these relics are those of creatures which had already travelled far along the evolutionary road and, according to most authorities, do not represent the most primitive members of the chordate stem. Hence we may safely say that the time of origin was not later than the beginning of the Ordovidian, and probably long before. There is, however, little chance of finding the geologic record of the ancestral forms, if, as we may suppose, they were soft-bodied, delicate organisms without hard parts for fossilization. An apparent exception lies in the little "fish plate," *Eoichthys howelli,* discovered in the Cambrian shales of St. Albans, Vermont, in 1926. Later study shows this fragment to have been misinterpreted, so the Ordovician record still stands.

Place of Origin

The main contrast between invertebrate and vertebrate animals seems to be that, as a whole, the former are static organisms with little or no power of locomotion, while the latter are essentially dynamic. There are of course exceptions in each group, for the cephalopod molluscs, squid, etc., are splendid locomotor types, and some chordates have become sluggish or even fixed as an outcome of sedentary habits. Whether or not this contrast of type is a result of the physical environment, invertebrates to static marine waters and the vertebrates to dynamic fresh-water streams, as Chamberlin held, is not at all clear. The most primitive chordates existing to-day—tunicates, *Amphioxus,* etc.—are marine, inhabiting for the most part the flatsea, where they lead a wholly or partially sedentary life.

That this is therefore the ancestral habitat seems at first sight plausible, yet within this area there seems to be lacking the necessary physical or external stimulus to impel the evolution of the chordate characteristics, especially segmented body muscles and the notochord. Perhaps the strongest stimuli would be escape and pursuit and the need to remain in the environment. The cephalopod locomotion, acquired in the sea, is largely for the former purposes and is based on an entirely different principle from that of the chordate. There are, however, among the segmented worms, marine forms which swim by a wriggling or undulatory movement comparable to that of the vertebrates, except that it is up and down or dorso-ventral, instead of from side to side or lateral. There is no reason, however, why the sea might not have produced the one as well as the other. Briefly, Chamberlin invoked the dynamic rivers which would give the undulatory movement, extrinsically, to a passive elongated animal temporarily anchored by the mouth. He imagined that the creature might learn to produce the same movements actively, in order to avoid being swept out of the environment into the sea, and thus develop motor organs accordingly, an idea which has had very little general acceptance. Frankly, we do not know the place of origin, nor is there any direct evidence which can be brought to bear upon the problem.

Ancestral Stocks

Several theories have been advanced to set forth the claim of this or that invertebrate group to vertebrate ancestry, but none of them is at this time capable of adequate demonstration. Of these, three are based upon the assumption that a segmented ancestry is necessary to account for the segmentation seen in the vertebrate, an inference which is not necessarily true, as segmentation may readily have arisen anew in the chordate phylum as a response to such conditions as those postulated by Chamberlin.

Annelid Ancestry

The hypothesis of annelid ancestry for the vertebrates derives the primitive chordate from the phylum Annelida, typified by the earth-and marine-worms. In many of the principal organs there is a marked correspondence, with the exception of the general reversal of the relations of the various parts to one another. For, as we have seen, the relative position of blood and nervous systems is diametrically opposite in the vertebrate and invertebrate groups. But by postulating a physiological reversal of the animal—and we know that in the flounder and squid such a change from the morphologically normal posture can take place—the various organs of the worm may be brought into almost complete harmony with those of the vertebrate. Perhaps the greatest difficulty lies in the development of the notochord, but even this seems to have its annelid prototype in "the 'Faserstrang," a bundle of fibres running along the nerve chain and serving as a support. This and the notochord lie in a precisely similar position in relation to the other organs, and in both cases they are enclosed with the nerve cord in a common sheath of connective tissue".

The reversible diagram shows quite clearly this correspondence of parts. In the annelid position we see the mouth at *m*, from which the œsophagus arises, passing through the nervous system and connecting with the long straight gut, *HH*, which terminates at the posterior end of the body at the anus, *a*. The nervous system consists of the large supraœsophageal ganglion or brain from which nerve connectives run, one on either side of the œsophagus, to the ventral nerve chain. The main blood-vessel lies dorsal to the gut and another lies beneath it. These are connected in the anterior region by semi-circular pulsating vessels or "hearts" which cause the blood to flow forward in the dorsal vessel and aft in the ventral one. Reverse our diagram and the form becomes a vertebrate, the blood now flowing forward in the pulsating ventral aorta which serves as a heart, the ancient semicircular "hearts" having

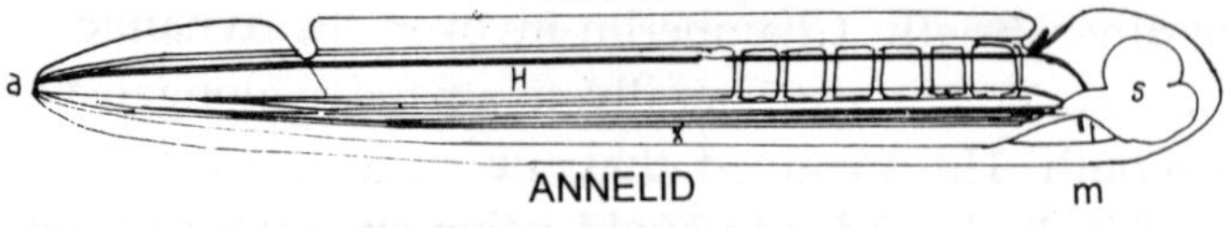

Fig. 1.2: Reversible diagram illustrating the annelid theory of vertebrate origin. Reversible symbols, applying to both forms: *H*, alimentary canal; *S*, brain; *X*, nerve cord. Annelid symbols; *a*, anus; *m*, mouth. Vertebrate symbols; *NT*, notochord; *pr*, proctodæum (anus and rectum); *st*, stomodæum (mouth and pharynx).

relinquished their primal function for that of respiration, since the gill-slits arising between them make them the branchial vessels. As in the annelid, the mouth is again on the ventral side and this can only be brought about through the abandonment of the old and the formation of the new one by an inpushing of the body-wall at *st* until communication with the gut is effected. This stomodæum is balanced at the hinder end of the trunk by the new hind gut, the proctodæum intestine in the tail region being aborted.

The brain and nerve cord are the homologues of the supracesophageal ganglion and ventral nervous chain of the annelid. Indications of the ancient mouth are seen in several structures such as the neuropore in the embryo of *Amphioxus,* which forms in this place a direct communication between the cavity of the nerve cord (neurocœle) and the exterior and is otherwise unaccounted for. Other indications of the early mouth and its œsophagus are the fourth ventricle of the brain, a cavity which lies exactly in the place where in the diagram the annelid œsophagus pierces the nervous system, and also the hypophysis, a structure attached to the lower side of the mid-brain, part of which is pushed up from the alimentary canal, and for which there is as yet no satisfactory explanation.

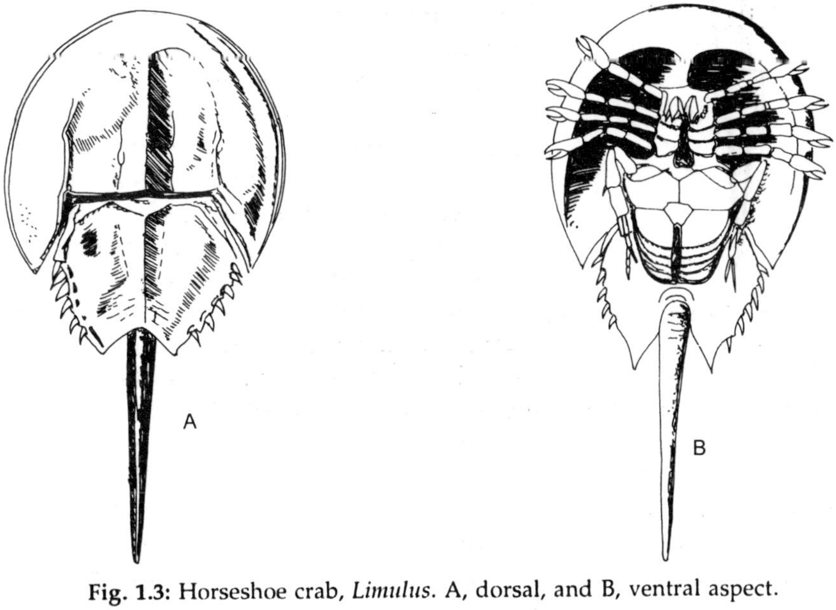

Fig. 1.3: Horseshoe crab, *Limulus.* A, dorsal, and B, ventral aspect.

Add to all this the remarkable correspondence of the kidney tubes or nephridia of the annelids and vertebrates, and the evidence is presented. Wilder says in summation: "Convincing as these comparisons seem when taken by themselves, the influence of later investigation has tended rather away from the annelid hypothesis, and at present, although there are many investigators who seek the ancestor of vertebrates in some worm-like form, there are few who wish to definitely assert that this ancestor was an annelid."

Arthropod Ancestry

In addition to the annelid theory, some authorities have tried to prove vertebrate descent from Arthropoda, especially from the more primitive arachnoids such as are now represented by the scorpion and the horseshoe crab (*Limulus*), and formerly by the extinct Merostomata. By this hypothesis we must set aside as primitive such forms as *Amphioxus* and the cyclostomes and start with the highly specialized ostracoderms which lived in Ordovician and Devonian times and thus were contemporaneous with and in general appearance and probable habits quite similar to the Merostomata. The soft parts of the Merostomata are of course unknown, but it is reasonable to suppose that they were not unlike those of the related scorpion and *Limulus*, and, as Pattern has shown, especially in the brain and cranial nerves of vertebrates and the fused cephalothoracic ganglionic mass found in such arachnoids, there are many points of resemblance. Then, too, the sense organs, especially the eyes, are more or less comparable, and there is in *Limulus* an internal skeletal piece known as the "endocranium" or sternum which serves to protect the central nerve complex, and which in general form and in its relation to other parts resembles the primordial vertebrate skull. Similarities also exist between the heart and arterial systems of each group, and the appendages may be compared. There are, again, the very arthropod-like jaws which Pattern has demonstrated in the ostracoderm *Bothriolepis*, a type which, on the other hand, shows many vertebrate-like characteristics; and the general arrangement of the plates by which the cephalothorax is covered is also very similar in the ostracoderms and contemporary arachnoids, but unfortunately for the argument *Bothriolepis* is a highly specialized end-form from the Upper Devonian. Nevertheless, while the arachnoid theory has been set forth by Gaskell and by Patten, the main thesis has received thus far but little recognition, although the evidence, especially in Professor Patten's book, is based upon an admirably executed piece of research.

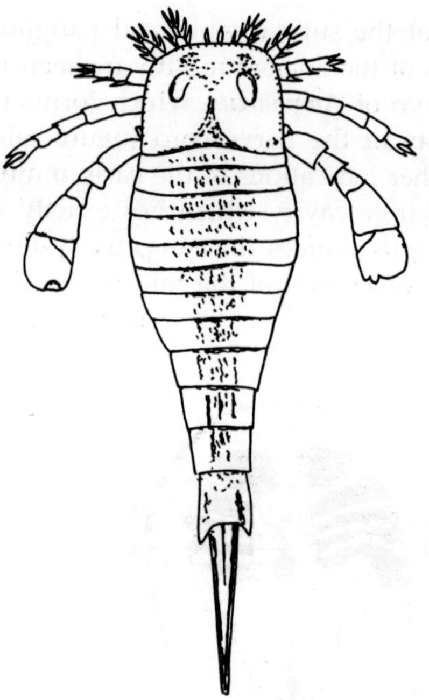

Fig. 1.4: Merostome, *Eurypterus fischeri*, Silurian.

Amphioxus Ancestry

The theory of Amphioxus ancestry places especial emphasis upon the notochord, the gill-slits, and the dorsal position of the central nervous system, and by means of these has traced the line of vertebrate ancestry through a series of transitional forms, externally very unlike one another and each somewhat isolated in its systematic position. Of these, *Amphioxus*, the lancelet, stands nearest the true vertebrates, in fact it is nearest the diagrammatic vertebrate of any living type, although, owing to certain specializations, it can hardly be considered a true stem-form. The lancelet was first described in 1778 as a shell-less snail or slug, and was named *Limax lanceolatus*. It is an inhabitant of the shallow sea, being found off the coasts of all parts of the world. There are sixteen known species most of which are recorded from tropical and *sub-tropical* shores, mainly between latitudes 40° N. and 40° S. In habits they are very sedentary, living for the most part partly buried in the sand or mud in a nearly erect posture, with the anterior end protruding. Aside from their primitive character, their world-wide distribution, coupled with sedentary habits, points to a very great antiquity. As fossils, however, they are thus far entirely unknown and probably always will be.

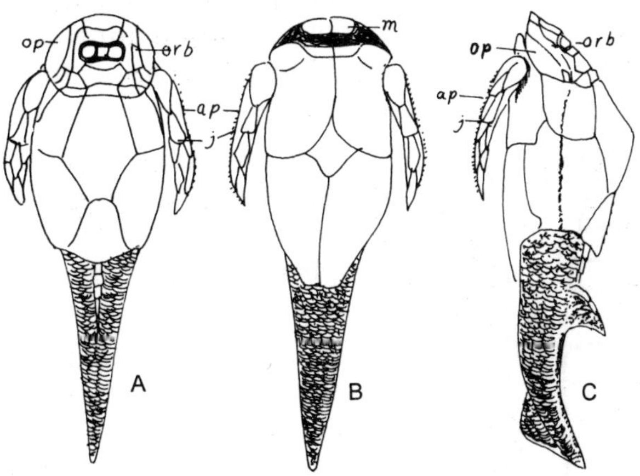

Fig. 1.5: Ostracoderm, *Plerichthys milleri*, Lower Old Red Sandstone, Scotland. A, dorsal, B, ventral, and C, lateral aspects. *ap*, pair of lateral appendages; *j*, joint in appendage; *m*, supposed upper jaw, with notches for narial openings; *op*, operculum; *orb*, orbits.

Description of Amphioxus

The body is lanceolate, compressed, without a distinct head, but with an expansive hood-like structure surrounding the mouth, which is situated beneath the snout. There are no paired fins; but a pair of longitudinal folds, the metapleures, one on either side of the body along the ventral margin, suggest their possible origin. There is a slight median fin running along the back and supported by delicate skeletal elements membranous in character. Around the tail this fin expands into a caudal which extends forward on the ventral side beyond the anal opening, thus displacing the latter to the left of the median line.

A conspicuous feature is the regular segmentation of the muscular system, plainly visible through the transparent skin, the side muscles being divided into a large number (64±) of V-shaped myomeres (Gr.μῦς muscle, and με'ρος, part, each with the apex pointing forward. These do not precisely correspond on the two sides of the animal, and by their successive contraction and relaxation they produce the undulatory movement of the body by means of which locomotion is effected. The notochord has already been described and also the fact of its continuation to the extreme end of the snout, instead of ceasing beneath the mid-brain as in all higher vertebrates (Crainata).

Lying along the dorsal side of this notochord and enclosed with it in a connective tissue sheath is the central nervous system, comparable to that of fishes except that it lacks an expanded brain other than a slight enlargement. The only definite sense organs are an olfactory pit on the left side and a single median pigment spot which serves for the perception of light transmitted through the transparent body.

The alimentary canal lies beneath the notochord, and consists of the mouth and a large pharynx that extends more than half the length of the body and is pierced by numerous gill-slits. The intestine runs directly backward, terminating in the laterally situated anus. There also arises from it ventrally a hollow outpushing known as the liver.

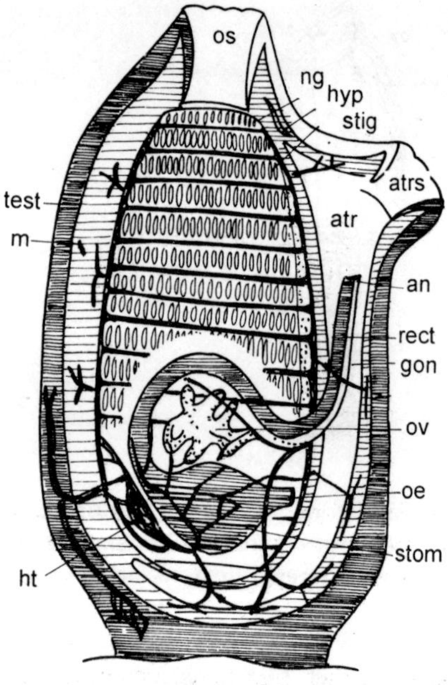

Fig. 1.6: Tunicate, *Ascidia*. Diagram of longitudinal section from left side. Test and mantle removed. *an*, anus; *atr*, atrium; *atrs*, atrial (exhalent) siphon; *gon*, gonoduct; *ht*, heart; *hyp*, hypophysis; *m*, mantle; *ng*, nerve ganglion; *œ*, œsophagus; *os*, oral (inhalent) siphon; *ov*, ovary; *rect*, rectum; *stig*, stigmata (pharyngeal perfora-tions); *stom*, stomach; *test*, test or tunic.

The circulatory system is comparable to that of other vertebrates but is much simpler, and the heart is represented by a pulsatory ventral aorta lying beneath the pharynx. The blood, however, is colourless.

It will thus be seen that *Amphioxus* is a very simple "vertebrate," specialized a little along certain lines, but with several structures of such fundamental importance that they must be borne in mind in our search for yet more primitive forms. These structures are the notochord, the dorsally situated nerve cord, and the pharynx perforated by gill-slits and provided with an endostyle.

The only other living creatures (except *Balanoglossus*), which possess these structures during any part of their career are the *tunicates*, some of the which are plaktonic, others meroplanktonic, in that while they have active larvæ, they soon settle down and become wholly sedentary in their habits. These sedentary forms, curiously enough, have retained more of their primitive characters than those which are plaktonic, as their larvæ are comparatively undifferentiated. The free-swimming forms, on the other hand, are often modified in a remarkable way and may have so complex a life-history that the old-time chordate characteristics have almost entirely disappeared, They always, however, possess the gill-slits except in certain locomotive individuals among the colonial types, which have lost all organs except those of propulsion, that is, the muscles, nerves, and sense organs. The name tunicate comes from the test or tunic which surrounds the entire animal and is comparable to the shell of a mollusc in that it is formed by the body-wall or mantle. This test is unique in being made up of a substance closely comparable to the cellulose of plants.

While the adult tunicate shows certain *Amphioxus*-like characteristics, it is the larva in which these are particularly emphasized, for in this stage the creature, which is tadpole-like, possesses a well developed notochord, segmented muscles, and a prolonged nerve tube with a brain-like vesicle forward which contains a pigment spot and another organ, possibly for balancing. The gill-slits are much fewer in number than in the adult and the endostyle lies in its normal position. The heart also lies ventrally and just behind the œsophagus.

But this comparatively high organization is retained for a very brief time, a few hours only. Then the creature settles down on a pair of adhesive papillæ and undergoes a marked retrogressive metamorphosis during which it loses tail, notochord, and segmental muscles; the nerve tube is reduced to a single ganglion, the sense organs disappear, the gill-slits increase in number, and the animal, after relinquishing practically all of the organs that serve to link it with the vertebrates, degenerates into what is virtually an invertebrate form. It is, however, evident that the tunicates represent a group more or less closely allied to *Amphioxus*, and hence to the other vertebrates, but that since the time of the common ancestor they have taken a divergent road, resulting in a type of degenerate adult whose real affinities are masked by its specializations. Thus, as Wilder says, "The ancestor that we here seek is better seen in the larva than in the adult, and we may believe that there once existed an adult animal with attributes like that of the tunicate larva of the present day, and that this animals was the direct ancestor of that group of which *Amphioxus* is now the only living representative.

Back of the tunicate ancestor there is but one known form which may or may not be near the main ancestral line. This creature is *Balanoglossus*, a marine worm that lives between high and low water marks in fragile tubes of cemented sand.

Balanoglossus is elongated without internal segmentation, but the body is divided into four regions, the burrowing proboscis, a collar with a free anterior edge, a flattened gill region, and a posterior trunk. The mouth is situated just beneath the edge of the collar on the ventral side and receives sand with its contained organic débris. There is a large pharynx communicating with the exterior by two literal rows of gill-slits supported by a skeletal structure comparable to that seen in *Amphioxus*. Owing to the fact that nowhere except among the chordates do such structures occur, naturalists have sought in this animal for the other two chordate characteristics, the notochord and dorsal nervous system. The latter is apparently represented by a very generalized system of nerve fibers and cords somewhat emphasized on the dorsal side, but there

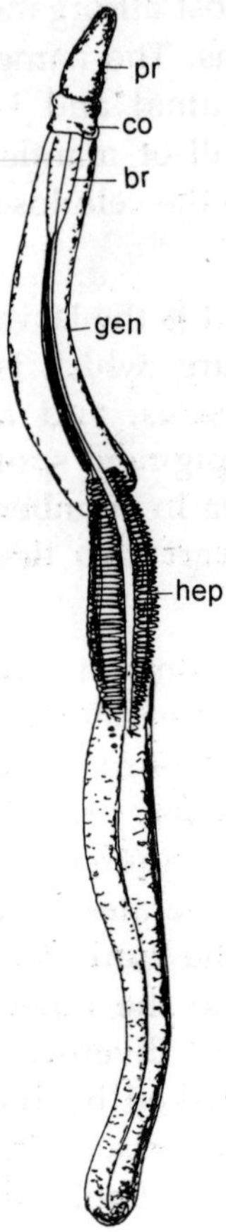

Fig. 1.7: *Balanoglossus. br,* gill region; *co,* collar; *gen,* genital ridges; *hep,* prominences formed by hepatic (liver) cæca; *pr,* proboscis.

is no evidence of the neurocœle. The notochord is supposed to be represented by a small outgrowth arising from the dorsal wall of the pharynx and extending forward into the proboscis. As Wilder says, this supposition has been greatly strengthened through the recent discovery of an allied form belonging to a new genus (*Harrimania*) in which the outgrowth is much larger, and, in its mode of origin, strikingly similar to the true vertebrate notochord, and thus without much doubt homologous with that organ. Wilder further says:

"From the testimony afforded by the structure of *Balanoglossus* and its allied genera (the group Enteropneusta) it may be quite confidently asserted that these forms lie nearly in the line of vertebrate descent, and represent an earlier stage than that of the tunicates. But here the chain seems to end, for *Balanoglossus* is itself unusually isolated and shows no close affinity to any other invertebrate types."

There is but a single rather slender clue to the ancestry of *Balanoglossus,* and that is again afforded by its embryology, for here there is a peculiar ciliated larva, the so-called *Tornaria,* which shows a very marked resemblance to the larvaæ of the echinoderms, and the universal occurrence of this larva within the latter group shows that, whereas they are all to-day, with rare exceptions, either sedentary or vagrant benthos, as their radial symmetry implies, they are descended from a pelagic bilaterally symmetrical ancestry. Thus, according to this belief we may "accept as a very ancient common ancestor of both echinoderms and vertebrates the form which all these larvæ maybe said to copy; a form having the characteristics common to all, including bilaterality, minute size, transparency, locomotion by bands of cilia, and pelagic life. The lineal descendants of this hypothetical ancestor chose two paths, the one leading to the Echinodermata, the other to *Balanoglossus,* the Tunicata, *Amphioxus,* and eventually the Vertebrata".

While this theory is incomplete in many details, it has strength where the other hypothesis are weakest in that it is based not alone upon adult structure but upon ontogeny as well. The weakest link in the chain of evidence is that which binds *Balanoglossus* to the echinoderms—the *Tornaria* larva—because the adult structures are so remote and the echinoderms give not the slightest clue in their bodily make-up to chordate affinities. As Matthew says: The origin of chordates "is still an unsolved problem. We cannot yet point to our ultimate ancestor in the Cambrian fauna. But whether known or unknown, and the latter is by far the more likely hopothesis, we can be pretty sure that we *had* an ancestor at that time, and that his evolutionary status was not above that of the trilobite, and may have been no better than a worm."

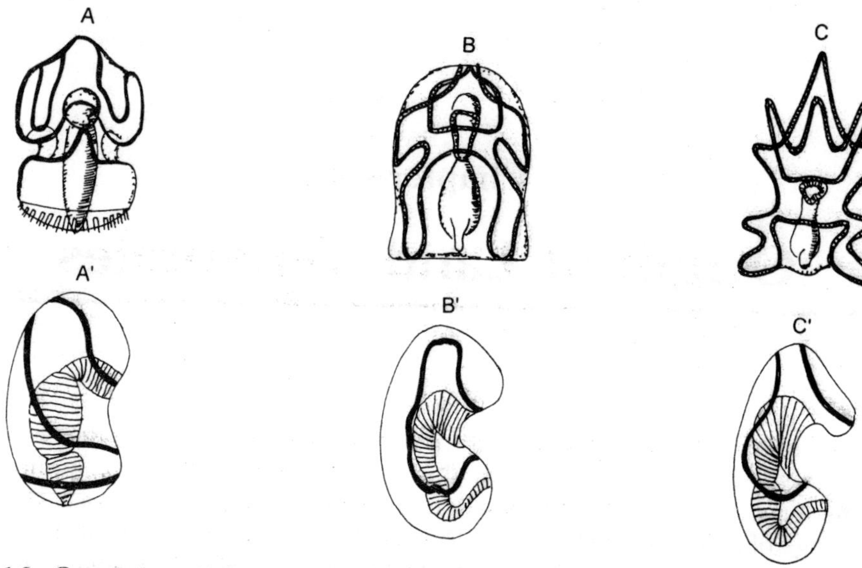

Fig. 1.8: Comparison of *Tornaria* larva with larval echinoderms. Main ciliated bands in black, lesser systems cross-lined. Ventral aspect; A, *Tornaria;* B, *Auricularia* (sea-cucumber); C, *Bipinnaria* (star-fish). Lateral view: A′, *Tornaria;* B′, *Auricularia;* C′, *Bipinnaria.*

Chapter—2

Origin of Land Vertebrates

Next to the origin of the vertebrates from their ancient ancestry the greatest and most dramatic event in all their history is the emergence from the old limiting aquatic environment and the subsequent adaptation to land-dwelling life. The sea is so changeless and the range of its conditions so small that evolution within it is not stimulated as it is on land. Adaptive radiation of marine creatures can accomplish but little; we have seen, on the other hand, what it means on the part of air-breathing forms.

Place of Emergence

The three problems which come to mind are the place, the impelling cause, and the time of this important event, and of these the first has been established, for while certain creatures have forsaken the sea and, crossing the strand, become adapted to terrestrial life, such instances are rare and in no case do they embrace all of the members of a class or phylum, but isolated genera or even species only. Such, for example, are the land crabs, *Birguslatro*, etc., several species of which live in damp woods far from all water and, as they are found on islands which, like the Dry Tortugas, have no permanent terrestrial waters, must have had their initial air-breathing adaptation along the strand. The terrestrial vertebrates, however, apparently did not so emerge, but rather were descendants of inhabitants of land waters, for in such a habitat alone can we find a sufficiently great impelling cause for an event so far-reaching and radical in its ultimate results. Further more, the ancestral habitat could not have been within the limits of the tidal zone but was beyond the influence of the sea.

Impelling Cause

Enemies in the Water

Barrell has discussed the probability of several possible causes which may have led to the emergence of the vertebrates. Of these, the first is enemies in water, which he deems inoperative, for among land-going fishes of to-day those few which crawl on land do not do so to escape their enemies. He also emphasizes the balance which always obtains between carnivorous and herbivorous creatures of a given habitat, and the fact that the amphibia go back to the waters to bring for their young, and the youngest and therefore the most helpless stages are spent in the waters. Add to this the fact that the earliest amphibia which are known from their skeletons, and their ancestors, are, in many instances, powerful armed carnivores themselves, and their forsaking of the ancient habitat for personal safety seems hardly adequate.

Food on the Lands

Food on the lands is also considered an inadequate cause. Here the argument lies in the rarity of the passage of crustaceans, gastropods, and vertebrates from a truly marine to a truly terrestrial mode of life through the ever present path of the tidal zone, which seems to prove that the unused though increasingly abundant food of the land realm cannot operate as a sufficient cause for this change, nor, so far as this factor is concerned, do the river faunas have an advantage over those of the tidal zone.

Lure of Atmospheric Oxygen

That the lure of atmospheric oxygen is also inoperative is proved by the small direct use made of air for respiration by the pelagic marine fishes even when they—the flying fishes, for instance—live an active life near the surface and in frequent contact with the air. It is principally in fresh-water fishes that accessory respiratory organs are employed and their use is in direct relation to the varying impurity of the waters in which they live. This varying impurity, which often means a reduction of respirable oxygen in the water, is literally impossible in the sea. Streams may locally contaminate the adjacent waters by their load of sediment or other impurities, but they cannot seriously lessen the degree of aeration. Then, too, marine fishes are never confined to such localities, but can migrate should conditions become unsuitable; while with fresh-water fishes this may not be true.

Recurrence of Unfavourable Environment

The real cause, therefore, seems to be not the need of safety or food, nor the desire to breathe atmospheric oxygen, but rather an adaptation which has been forced repeatedly to a greater or less degree upon fishes by the recurrence of an unfavourable environment. Hence it appears as though the emergence were compelled by variations in the environment as measured by the amount of dissolved oxygen in the land water, and the question arises whether such variations occur and, if so, under what climatic conditions. The required climate does occur in various parts of the world, but is neither arid nor humid, but semi-arid—conditions which are found in the tropics especially, where, instead of the fourfold change of seasons whose determining factor is temperature, there are alternate seasons of drought and copious rain, occurring in definite cycles. In such a region during the rainy season the streams would be high and fully aërated, but when the rains ceased the waters would gradually slacken their current, until instead of a flowing river of pure water there would be a succession of stagnant water pools, in which the concentrated plant and animal life would die and by its decay charge the waters with impurities and exhaust their free oxygen. Thus we would get a great fluctuation of oxygen content and so a very variable environment in its ability to support water breathing life.

Under such conditions, if life existed, a high premium would be placed upon powers of æstivation, or tormidity induced by summer heat and dryness, or of breathing the atmosphere, or both combined; and a rapid elimination of the unfit, that is, of such as did not possess even the rudiment of this power, would occur.

AIR-BREATHING FISHES

There are among existing fishes a number possessing supplementary respiratory organs which may be one of two structures; either (1) spongy outgrowths of the gills which can retain moisture and, as long as it is retained, utilize the free oxygen of the air for the aeration of the blood; or (2) a modified swim-bladder which may have one of several functions but whose principal purpose is a hydrostatic one—to maintain the fish at a given level in the water.

Teleosts

The first of these structures, that of the accessory organ connected with the gills, is found exclusively in teleostean fishes, a group in which the air-bladder never subserves a respiratory function. Several such teleosts exist, among them the climbing perch, *Anabas scandens* and the mud skippers, *Periophthalmus* and *Boleophthalmus*, but they are generally fishes which voluntarily leave the waters for migration or in pursuit of food, and rarely is their evolution the result of adaptation to the environmental conditions postulated above.

The swim-bladder, on the other hand, is entirely absent in the two groups known as cyclostomes and elasmobranchs and as we have seen, is never of respiratory value in the teleosts, so that its function in this direction lies in the groups between, in other words, among ganoids and dipnoans or true lung-fishes, and these fishes are to-day all denizens of semi-arid tropical climates, living under conditions of varied water aeration arising in the way we have described. These air-breathing fishes are of such importance to our argument, and are so few, and represent so ancient a group or groups that some account of the individual species is worthy of record. It should be borne in mind, however, that these are relic forms, representative of Devonian time when almost all fresh-water fishes belonged to one or the other of these two groups.

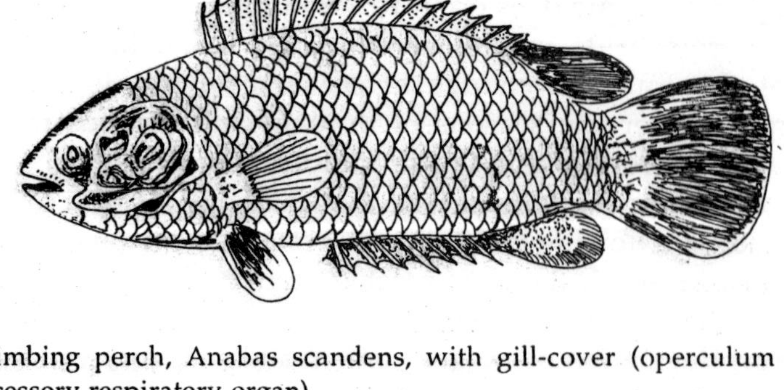

Fig. 2.1: Climbing perch, Anabas scandens, with gill-cover (operculum removed to show accessory respiratory organ).

Ganoids

The ganoids of especial note are specifically of the order Crossopterygii, or the lobe-finned ganoids, and include but two related genera, *Polypterus* and *Calamoichthys*, both African in distribution. Of them the better known is *Polypterus* of which there are several species. *P. bichir* "haunts the deeper holes and depressions of the muddy bed of the Nile, although it is not essentially a bottom-liver or mud-fish. It is most active at night when in search of food, and then it may readily be taken by trawl-lines. The lobate pectoral fins are used for progression, but their primary function is to act as balancers, and they exhibit the characteristic trembling movements so often seen in the balancing fins of teleosts. *Polypterus* does not readily live out of water, rarely longer than three to four hours, and then only when covered with damp grass or weeds. *P. bichir* is said to feed on small teleosts, which it swallows whole, and to these there may be added, in other species, batrachians and crustaceans. The observations of budget show that in captivity *Polypterus* often remains motionless for a long time at the bottom of the water, the anterior part of the body resting upon the tips of the pectoral fins. According to the same observer, the air-bladder is an accessory respiratory organ, supplementary to the gills, rather than a hydrostatic organ" (Bridge). This air-bladder is an out-pushing of the gut, and in the Crossopterygii arises from the ventral side of the gullet and is a paired structure exactly as in the amphibian lung. It is not, however, cellular and is thus a very inefficient respiratory organ.

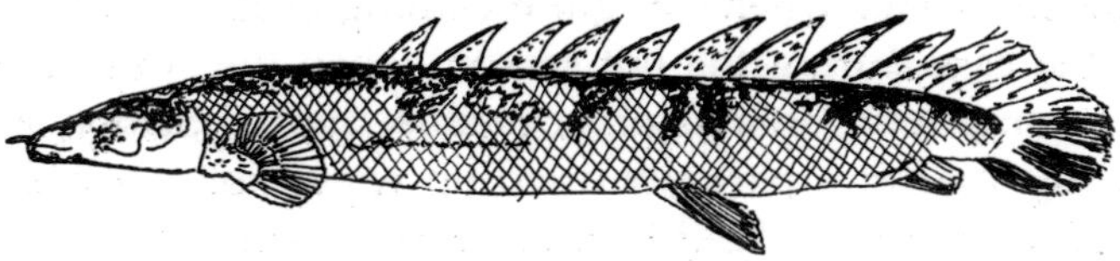

Fig. 2.2: African crossopterygian fish, *Polypterus delhezi*, congo river.

In the genus *Calamoichthys* the body is elongate and eel-like in shape. The pelvic fins are entirely lacking but the pectorals and the series of dorsal finlets are comparable to those of *Polypterus* except that the latter are relatively fewer. *Calamoichthys* has a more restricted distribution than *Polypterus*, being confined to certain rivers in west Africa such as Old Calabar River and those of the delta of the Niger on the coast of Kamerun. It is a very agile fish, swimming like a snake and subsisting on insects and crustaceans. The name signifies palm-fish, from its frequenting the roots of the palm trees.

Neither *Polypterus* nor *Calamoichthys* is known fossil, but the group Crossopterygii to which they belong once included a large number of important fishes. Of these *Holoptychius* of the Devonian is interesting because of the intricate infolded structure of the teeth, which has a striking parallel in those of certain amphibia (labyrinthodonts). *Undina*, another form from the upper Jurassic, exhibits a well developed air-bladder in the fossil specimen.

Dipnoans

Of the dipnoans or true lung-fishes three genera only are extant but they never were as numerous as the Crossopterygii. The living forms are first, the Australian genus *Ceratodus*, or, to be more accurate, *Neoceratodus forsteri*, the barramunda, which is now confined to the Mary and Burnett rivers in Queensland. This form" frequents the comparatively stagnant pools or water-holes which alternate with shallow runs and are usually full of water all the year round. In these pools, filled with a rich growth of aquatic vegetation, and often the favourite haunt of the platypus (*Ornithorhynchus*), the fish is fairly abundant. Inactive and sluggish in its habits, usually lying motionless on the bottom, the fish is easily captured by the natives with hand-nets or baited hooks. *Neoceratodus* lives on fresh-water crustaceans, worms, and molluscs, and to obtain them it crops the luxuriant vegetation of the water-holes much in the same way that a polychæt [worm] or a holothurian [sea-cucumber] swallows sand for the sake of the included nutrient particles, Apparently the air bladder is a functional lung at all times, acting in conjunction with the gills. At irregular intervals the fish rises to the surface and protrudes its snout in order to empty its lung and take in fresh air. While doing so the animal makes a peculiar grunting noise, "spouting; as the local fishermen call it, which may be heard at night for some distance, and is probably caused by the forcible expulsion of air through the mouth. Useful as the lung is as a breathing organ under normal conditions, there can be little doubt that its value as such is much greater whenever gill-breathing becomes difficult or impossible. This seems to be the case during the hot season, when the water becomes foul from the presence of decomposing animal or vegetable matter. Semon records a striking instance of this in the case of a partially dried-up water-hole, in which the water had become so foul that it was full of dead fishes of various kinds. Fatal as these conditions were to ordinary fishes, *Neoceratodus* not only survived, but seemed to be quite healthy and fresh. Such observations are of exceptional interest. Not only do they afford a clue to the conditions of life which, in the course of time, probably led to lung-breathing in *Neoceratodus* but they also suggest the possibility that a similar environment, has been conducive to the evolution of air-breathing vertebrates from gill-breathing and fish-like progenitors.

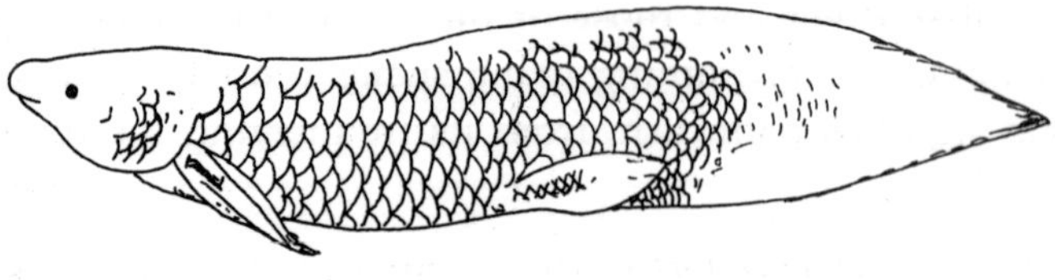

Fig. 2.3: Australian lung-fish (dipnoan), barramunda, *Neoceratodus forsteri*.

"In spite of its pulmonary respiration, *Neoceratodus* more closely resembles the typical fishes in its habits than any other Dipneusti. It lives all the year round in the water. There is no evidence that it ever becomes dried up in the mud, or passes into a summer sleep in a cocoon, and the well developed condition of its gills suggests that these organs play a more important role in breathing than in either *Protopterus* or *Lepidosiren*. The fish is not known to leave the water, and the paired fins, useful no doubt as paddles, are quite incapable of supporting the bulky body on terra firma. In fact when *Neoceratodus* is taken out of its natural element it seems to be more helpless than most other fishes, and in spite of its capacity for lung-breathing, soon dies unless kept moist by artificial means" (Bridge). *Neoceratodus* grows to a length of five to six feet.

Ceratodus, a fossil ally Mesozoic in age was very widespread compared with the limited distribution of its living relative. Its very characteristic crushing teeth occur in the Trias of England, Germany, India and South Africa, and more rarely, in the Upper Jurassic and Lower Cretaceous strata of England and in Colorado and Wyoming (Morrison formation). Its remains are often found associated with those of carnivorous dinosaurs, but the significance of this, if any, is not apparent.

Protopterus and *Lepidosiren*, which represent a separate family of lung-fishes, the Lepidosirenidæ differ from *Neoceratodus* in that the air-bladder is a double organ, while in the latter it is single.

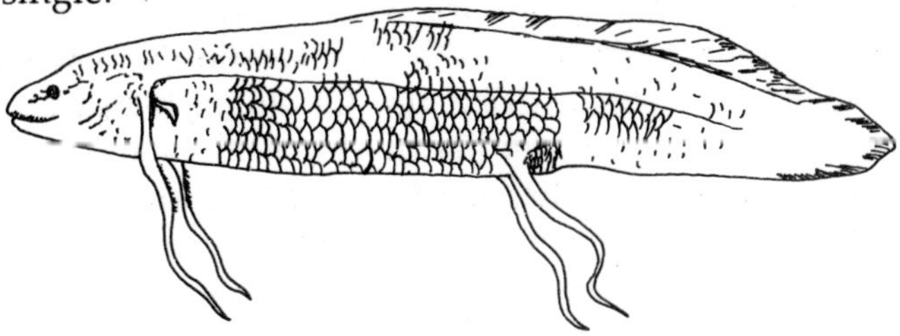

Fig. 2.4: African lung-fish (dipnoan), *Protopterus annectans*.

Protopterus is the African lung-fish and has a wide distribution ranging from the river Senegal and the White Nile on the north to the Congo basin, Lake Tanganyika, and the Zambesi on the south. The three known species live in marshes in the vicinity of rivers. They are carnivorous, their food consisting mainly of frogs, worms, insects, and crustaceans, but when confined together they are very apt to display cannibalistic traits. The tail is the chief locomotor organ and they are remarkably agile and quick in their movements. The limbs are useless for swimming but are used for crawling over the bottom. Then they show a definite elbow-or knee-like flexure at about their mid-length. *Protopterus* ascends to the surface from time to time to breathe air into its lungs. In the dry season, however, it burrows into the mud to a depth of about 18 inches, where it forms a lining to the cavity in which it lies in the form of a capsule of hardened mucus secreted, by skin gland. This capsule has an aperture the margins of which are pulled inward to form a short tube that is inserted between the fish's lips. The fish within the capsule is surrounded by a soft slimy mucus which keeps the skin moist while respiration is effected by drawing the outer air through burrow and tube into the mouth and thence to the lungs. The lungs are, therefore, the sole means of respiration during the period of æstivation, while the body-fat and muscle-tissues of the fish, as in hibernating mammals, supply it with the necessary food. The dry season varies, but lasts in general from August to December, nearly half the year. When, with the advent of the rainy season, the marshes once more become flooded, the capsule is dissolved, *Protopterus* emerges from its burrow, and resuming its active life, very soon begins to provide for its coming young. The larvæ have much the appearance of a young salamander, with four pairs of external cutaneous gills and two pairs of simultaneously developed limbs. It also has chromatophores in the skin whereby its colour may be changed. As in the salamander, the assumption of lung-breathing is marked by a reduction of the cutaneous gills, which takes place about seven weeks after the eggs are deposited. *Protopterus* attains a length of about six feet.

Lepidosiren, the mud-fish, with but a single existing species, is a South American form, occurring along the course of the main Amazon, entering some of its large tributaries and also the Chaco Boreal to the west of the Upper Paraguay River. "The home of the *Lepidosiren* (or `Lolach,' as the natives call the fish), of the Chaco country, is to be found in the wide-spreading marshes and swamps, which for a great part of the year are almost choked by a luxuriant growth of their own peculiar vegetation and covered by a floating carpet of surface weeds, with here and there deeper and clearer water and slow-flowing streams. In the dry season the water gradually shrinks and the swamps eventually become dried up. Of sluggish habits, the fish wriggles slowly about at the bottom of the swamp like an eel, using its hind limbs in an irregular bipedal fashion as it wends its way through the dense network of sub-aqueous plants. *Lepidosiren* is not exclusively carnivorous. The large fresh-water snail *Ampullaria*, which lives in the swamps in enormous numbers, seems to be its favourite food; but masses of confervoid algæ are also eaten, and in its earlier stages it is probable that the fish is more herbivorous than carnivorous "(Bridge). *Lepidosiren* rises to the surface at intervals to breathe. The rate varying with the degree of impurity of the water. It feeds voraciously during the rainy season, storing up a supply of fat against the period of æstivation which is passed in a deep tubular burrow, much as with *Protopterus*. The entrance to the burrow in this instance, however, is closed by a plug of clay perforated by several holes. On the coming of the waters the plug is pushed out and the fish escapes. Development is quite similar to that of *Protopterus* and in each case parallels the amphibia very closely. There are many other parallelisms of structure and habit between the two groups, so many in fact that as Dean said, it is almost impossible to look upon them as of no greater significance than convergences.

Lepidosirenidæ are as yet unknown fossils, but there is reason to believe that their evolution from the older Dipneusti has been in a manner retrogressive. *Dipterus*, the most ancient of lung fishes, may be taken as a starting point and Dollo has selected a remarkable series of genera, *Scaumenacia*, *Phaneropleurom*, *Uronemus*, *Ceratodus* (*Neoceratodus*), *Protopterus*, and *Lepidosiren*, in which the evolutionary sequence agrees perfectly with their succession in time. Back of *Dipterus* lies an unknown ancestry, but one which probably falls within the group of crossopterygian ganoids of which *Polyterus* and *Calamoichthys* are the living representatives. The trend of evolution among the

Dipneusti, if one may judge from *Lepidosiren*, is leading to an elongated eel-like type, which will be both limb-and scale-less, which may be indications of racial senescence. These two interesting groups of air-breathing fishes, the Crossopterygii and Dipneusti, are both on the eve of their racial passing. The lesson which they teach; however, is of great significance and brings us back once more to the theme of our discussion—that of the most momentous emergence in prehistory; and our inquiry now leads us to a consideration of the probable time of emergence.

Time of Emergence

The evidence here is twofold: first, the fossil record, and second the geologic evidence of climatic conditions such as we have assumed.

The fossil evidence, which will be reviewed in greater detail later, points to a time earlier than upper Devonian for it is upon sediment referable to that period that the earliest known footprint of terrestrial vertebrate has been impressed. The time of emergence therefore cannot be later than the age of this footprint, and from the nature of things must somewhat antedate it, although how much we have no means of knowing as it was a time of accelerated evolutionary change.

The climatic evidence points to the same result, for, as Barrell has shown by a careful study of sediments of other phenomena connected with the rocks of Devonian age, these were times of warmth and seasonal rainfall tending toward more marked semi-aridity of climate in the Upper Devonian. There is, moreover, a dominance of dipnoans and crossopterygians in the fish fauna. Of these fishes certain could and probably did adapt themselves after the manner of their living descendants to the increasingly long dry seasons, until the latter became so long that the period of activity was not sufficient for the creature's life needs. Then came the emergence, for instead of æstivation the animal must adopt some other mode of life which would prolong the time of its activity inspite of its climatic restrictions. Thus the more ambitious among the lung-breathers, not content with the limitations imposed upon their lives, emerged from the age-long aquatic home and ventured into the new and untried habitat. Many may have essayed the emergence, but it is probable that relentless nature, weeding out the less fit for so valorous an undertaking, destroyed all but a single sort, for there is no evidence that the ancestry of the amphibia is to be found in more than one evolutionary lineage.

Ancestry

In spite of the many similarities which exist between the Dipneusti and the amphibia, there are few authorities who hold to a possible direct derivation of the one from the other, for very serious anatomical difficulties stand in the way, one of which is the very peculiar and specialized type of limb, the archipterygium, which this group of fishes possess, and out of which it is seemingly impossible to evolve the terrestrial hand or foot. The Crossopterygii, on the other hand, exhibit fewer of these obstacles; in fact there are practically none which evolution cannot overcome. The general opinion therefore, would derive the land-dwelling forms either from the Crossopterygii as such or possibly from some, as yet undiscovered, related group.

Changes upon Emergence

Partial Loss of Armor

The essential changes undergone by the emerging form were, first, partial loss of armour, for while the earliest amphibians, the Stegocephalia (Gr. στεγειν, to cover, and κεφαλη, head) are armoured, the armour is confined mainly to the head as the name signifies, to the breast girdle, and to oblique rows of small scales, chiefly on the under side of the trunk and tail. There is no evidence of their having possessed the heavy enameled scales of the ganoid ancestor.

Loss of Unpaired Fins

The unpaired fins are of course strictly of aquatic use and their loss upon emergence is to be expected. They do, however, recur in forms which have returned to their ancestral habitat. Thus certain salamanders show a rather well developed caudal web of skin which in the male crested newt extends forward along the back, and many aquatic larvae, those of frogs and toads, also have well developed unpaired fins. But these are new structures which have arisen in response to immediate need and bear no genetic relationship to the equivalent fins of fishes, the principal proof there of lying in the fact that there is no trace in the amphibian of supporting fin-rays as all fins show.

Development of Terrestrial Limb

One of the most essential changes upon emergence was the modification of the paired fins of the fish ancestor to support the body on the mud, a function to which they were clearly inadapted in their original condition. The paired fins of the Dipneusti are as we have seen, archipterygia, that is, having a long, jointed bony axis, on one or both sides of which arose a series of parallel rays to support the fin membrane. Such a type of limb, while it may be used as a prop or for slow crawling propulsion in a water-borne form, is in no known instance of suitable strength to support the entire weight of the animal when out of water, nor is it of sufficient surficial area to carry its owner over soft mud, for here a broad member is necessary. Then, too, the skeletal elements are such that one cannot see the slightest prophecy therein of the standard framework of the terrestrial foot.

With the crossopterygians, on the other hand, this is not true, for here the limb is different, having a broad basal lobe containing several bones and a fringe-like expansion so arranged that a much more adequate support is already present, even in the fish stage of evolution. It is particularly in the pectoral fin of the fish *Eusthenopteron*, from the Devonian of Quebec, that the terrestrial limb is foreshadowed, the shoulder-bones corresponding bone for bone, the single proximal bone of the fin to the humerus, the next two to the radius and ulna, and the remainder, or some of them, to the bones of wrist, palm, and digits. Certain bones have naturally been lost and others added, and

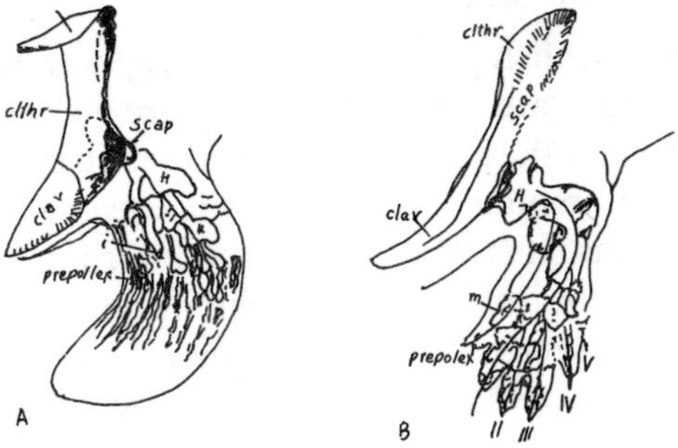

Fig. 2.5: Left pectoral Limb of A, *Eusthenopteron*, and B, *Eryops*, After W. K. Gregory. *H*, humerus; *R*, radius; *Scap*, scapula; *U*, ulna.

the entire fin-rayed portion of the limb abandoned with the relinquishment of the swimming function; but the whole metamorphosis requires no undue stretch of the imagination. The actual transitional limb is as yet unknown to us, but the most ancient footprint, *Thinopus*, is apparently not that of a completely evolved foot and may thus throw light upon the process of evolution. This footprint, while giving no clue to the skeleton of the upper and lower arm and wrist, does give a very adequate idea of the digital structure, which is highly peculiar. There are but two completely formed fingers, probably the first and second, the cleft between them extending deep into the sole of the foot. The phalangeal pads and a rounded, terminal, claw-like portion are already developed and there appears on the outer side of digit II the rudiment of a third, and below this on the outside of the foot the possible *Anlage* of digit IV.

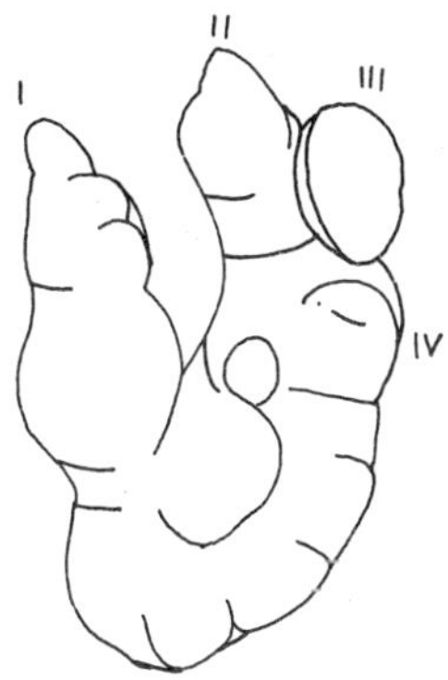

Fig. 2.6: Earliest known fossil footprint, *Thinopus antiquus*, with two fully formed digits, I and II, a partially formed third III, and a possible rudiment of a fourth, IV. Upper Devonian of Pennsylvania. One-half natural size. Original in Yale University Museum.

If this is a normal footprint as we may suppose it to be, it seems to prove that the terrestrial foot, instead of being five-toed from the beginning—and that is certainly the standard undifferentiated type to-day—began as a two-toed organ on the outer side of which the remaining digits arose in orderly succession until the typical number was acquired and the member became standardized.

That this may have been the case is not alone evidenced by the unique footprint which we have discussed but the arrangement of the nerves and muscles and the major and minor axes of the foot and limb are corroborative. The ontogeny of the salamander's foot as figured by Rabl shows the same budding of the lateral digits as the *Thinopus* track implies. So that without having seen the footprint, professor Wilder as a result of his embryological studies postulated an ancestral foot strikingly like that of *Thinopus*.

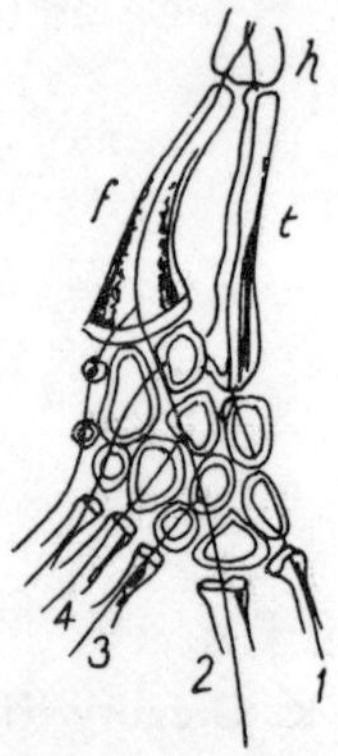

Fig. 2.7: Foot of a reptile, *Ranodom sibericus*. *f*, fibula; *h*, femur; *t*, tibia; 1, 2, 3, 4, digits 1-4.

Loss of Internal Gills

The ancient fish gills, borne on the gill-arches, were also lost upon emergence, for in every instance where permanent gills are seen in living amphibia they are *external* dermal structures of later origin and not strictly homologous with the *internal* gills of the fishes. Some amphibian gills, it is true, *seem* to be internal, as they are occasionally covered by a fold of skin, the operculum, so that they thus come to lie in a gill chamber; but they develop before the gill-clefts open, are restricted to the outer side of the branchial arches, and are always covered by ectoderm, all of which goes to prove them new organs which have assumed the old lost function of aquatic respiration.

Fossil Record

Footprints

The earliest record of a terrestrial vertebrate is the single footprint of *Thinopus antiquus* mentioned above. This is impressed upon a slab of sandstone and is from the upper most Devonian. It was found in 1896 by the late professor Beecher of Yale and by him presented to the Museum where it is now treasured. These same beds contain ripple-marks and mud-cracks, and impressions of rain-drops and land plants also come from the same general horizon. A characteristic *marine* mollusc (*Nuculana*) is preserved on the under side of the footprint slab. The associated strata show dominant delta conditions on the outer margin of which the sea had contributed to the material, for in the wide oscillations of the strand-line characteristic of delta fronts, deposition under shore conditions and deposition under river conditions alternate.

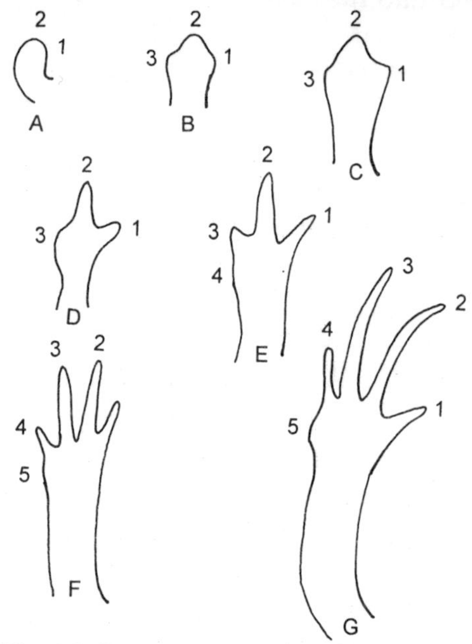

Fig. 2.8: Developmen of the hind foot of a salamander, *Triton tæniatus*. (After Strasser, from Rabl.)

This Devonian is directly overlain by Lower Carboniferous (Mississippian) Coal Measures, represented in Nova Scotia and New Brunswick by the Horton series. These contain the remains of plants and crustaceans and the footprints of amphibians. No bones have been found in these beds, but the footprints indicate, at the beginning of the Carboniferous period and before the deposition of the Lower Carboniferous limestones, the presence of both large and small species similar to those of the coal formation (Dawson). One interesting type, *Hylopus hardingi*, found in the Lower

Carboniferous shales of Parrsboro, Nova Scotia, shows a stride five times the length of the foot, twice the width of the track-way, as though the creature which made it stood high on its legs like an ordinary mammal. This looks very much like a cursorial adaptation; if so, it is the earliest on record.

The next higher level to record the passing feet of these primal terrestrial forms is the Mauch Chunk series of Pennsylvania, assigned by geologists to the upper half of the Lower Carboniferous. Here has been recorded *Palæosauropus primævus*, a five-toed track of considerable size as these early forms run, and more careful search of these same beds at Pottsville has brought to light several other species, some very small and delicately impressed. Other tracks have come from Virginia and are referred to the same general age (Hinton formation).

An amazing series of footprints of Permian age representing a number of genera and species has recently come to light in the Grand Canyon, where they may be seen travelling up the strata until they disappear under the overhanging cliff. Some of these are preserved in the Yale Museum and some in the National Museum at Washington.

First Skeletal Remains

The first amphibian bones were found in the Upper Devonian of Greenland and are comparable in age to that of the *Thinopus* footprint. Other skeletons come from the Edinburgh Coal Measures of Scotland which have been referred to the Lower Carboniferous. These are in no sense transitional forms, but are fully developed amphibians. Above the Lower Carboniferous Coal Measures we have red shales and sandstones in which bones are invariably rare and footprints abundant, and so it is with the Scottish record.

It was during Permian and Carboniferous times especially that the great deployment of amphibia occurred, and we have from various places, notably in Europe and in Nova Scotia and the United States, the remains of a varied assemblage of forms, some small, other huge, heavily armoured types with complex vertebræ, still others with complexly infolded teeth; some with well developed crawling limbs, yet others limbless, elongate, indicating that already the condition which we have called racial old age, or senility, with its attendant degenerative specialization was upon them. One and all were alike in this—they went, presumably, back to the waters to lay their eggs, and their young were therefore aquatic and breathed by means of gills. But there are among them many of which this cannot be proved and some may actually have been transitional, not between amphibians and fishes, but between amphibians and the succeeding class, the Reptilia. Certain of the forms, such as *Cacops*, discovered in the Permian of Texas by professor

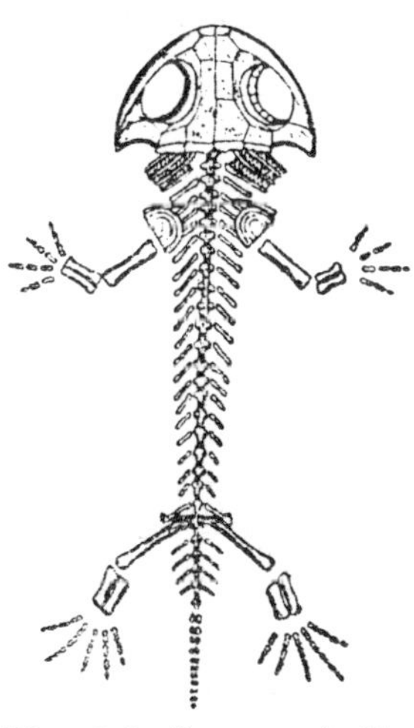

Fig. 2.9: Stegocephalian, *Branchiosaurus amblystomus*, Permian, Germany. Restored larva, showing gill arches.

Williston, show such a combination of characters pertaining both to the amphibia and reptiles that, as the distinguished discoverer says, it may become necessary to revise our definition of the former group.

Summary

"The nature of the geologic record of amphibians indicates that they evolved under climates marked by seasonal dryness, and inhabited river-plains far from the sea. The abruptness of appearance of well developed sustaining legs and feet points to an origin perhaps as far back as the Lower Devonian, but a rapid expansion and evolution in the Upper Devonian. They survived the change to more generally wet conditions in the Lower Mississippian, but showed more convincingly their adaptation to semi-arid continental conditions through the footprint record they have left in the Mauch Chunk shales. The impressions of plants indicate that over the broad river-plains of eastern Pennsylvania there flourished each season an herbaceous vegetation of acrogens following the withdrawal of the river floods, until the advancing seasonal dryness caused it too wither. No traces of an arboreal vegetation have been found, and this taken in conjunction with other facts, suggests that in the dry season the streams completely vanished, or at least were reduced to rivulets and water-holes unable to afford sufficient underground water to support an arboreal vegetation on the banks" (Barrell).

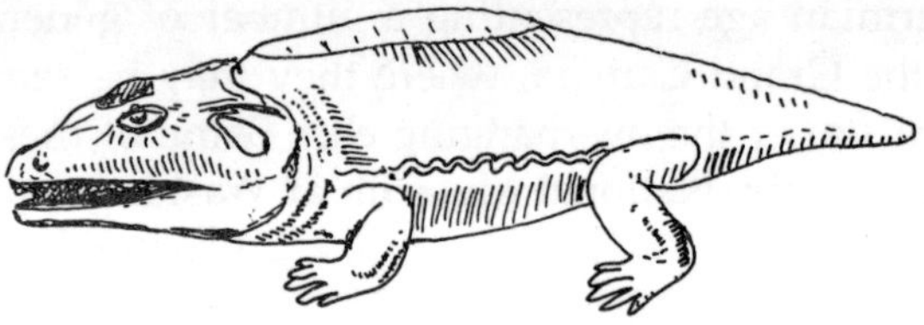

Fig. 2.10: Premian stegocephalian, *Cacops aspidesphorus*, from Texas, allied to the primitive reptiles.

Circumstances such as these were not conducive to piscine life, but were just the conditions under which amphibians would thrive. With further increase in aridity, however, such that no seasonal return of the waters occurred to make aquatic egg-laying possible, came the restriction of amphibia and evolution of reptiles.

REPTILES

Aside from certain anatomical characteristic, which we need not enumerate, two features out sharply in the reptiles in contrast to the amphibians. They are, first the loss of gill-breathing forever, the reptiles and their descendants—the mammals and birds—depending solely upon their lungs for oxygen; and second, the development of certain embryonic envelopes known as the amnion and allantois. The true significance of both loss of gills and gain of allantois is the same-air breathing young.

Embryonic Membranes

The reptilian egg is a complex structure consisting not only of the male and female; germ-plasm but of a considerable amount of nutritive yolk, sufficient to carry the creature well along toward perfection of body and obviating the necessity of a larval stage and a metamorphosis such as so many amphibians possess. This complex egg is surrounded by a protective envelope, the shell, and is invariably laid on land, if laid at all. It is because of this last feature that the amnion and allantois have arisen. The amnion is a two-layered membrane growing out of the ventral wall of the embryo and entirely enveloping it. Between the layers is the amniotic fluid which not only guards the creature against mechanical jars but also serves to resist sudden changes of temperature which might be fatal to the growing young. In other words the amnion is protective in its function. The allantois, on the other hand, is respiratory. It too is a double-layered or sac-like membrane arising in much the same way, an outgrowth in fact of the urinary bladder of the amphibian. It is abundantly supplied with blood-vessels directly continuous with those of the embryo. The allantois lies, in its full development immediately beneath the porous shell, through which oxygen can enter and, passing by osmosis through the allantoic membrane, oxygenate the included blood. Carbonic acid gas is given off at the same time. The blood stream now carries the oxygen to the embryo and brings out more waste and the process is continued. Thus it will be seen that the allantois has a function comparable to a lung and not to a gill, and it is to be doubted whether any reptilian egg could be placed in the water with out drowning the embryo within. At all events, no reptile, bird, or mammal egg, each of which possesses an allantois, is ever laid in the water, but always on land, or else provision is made for its retention within the maternal body as in certain snakes, the ichthyosaurs, and all mammals above the Monotremata.

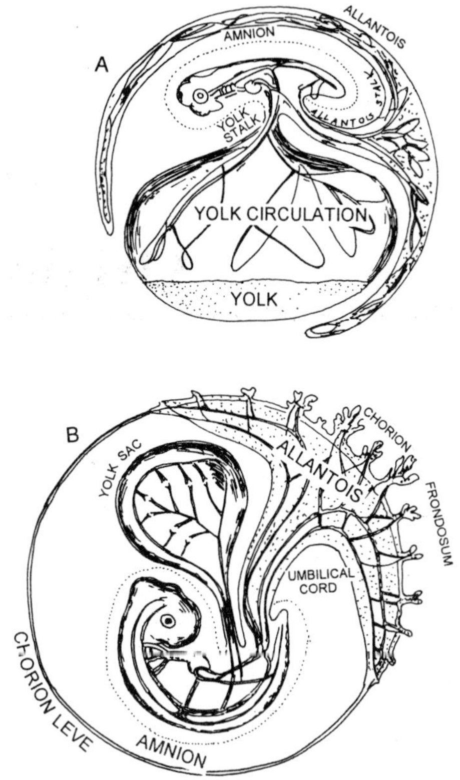

Fig. 2.11: Vertebrate embryos with their membrances. A, reptile or bird; B, placental mammal. In A, the yolk-sac is functional and the allantois respiratory; in B, the yolk-sac is functionless and the allantois becomes the nutritive placenta and umbilical cord.

From this it will be seen that reptiles may survive under conditions of aridity—many are true desert forms—where amphibia might perhaps live as adults but could not pass on their life to future generations. It is logical, therefore, to believe that whereas semi-aridity with seasonally recurring rains impelled amphibian evolution, true aridity with undependable rains or none at all, making amphibian economy impossible, stimulated the evolution of the reptiles.

Chapter—3

Origin of Reptiles

The land dwelling vertebrates evolved from crossopterygian fishes. It was also noticed that the immediate descendants of the latter were the labyrinthodont amphibians that in turn gave rise to the cotylosaur reptiles. All this occured before the beginning of the Mesozoic. Terrestrial vertebrates at that time were probably carnivorous and they could not respond directly to the energy supply offered by terrestrial plants, but they could and apparently did respond to the opportunities presented by the rediation of insects. Probably for the first time in evolutionary history there was an adequate food supply to support fully terrestrial predators. Two forces primarily responsible for the origin and radiation of reptiles were directly related to the exploitation of this new energy source. One was the evolution of a more effective jaw mechanism specifically adapted to feeding on insects and the evolution of more effective locomotion on land. A final change, the evolution of the amniotic egg, was the definitive step that separated reptiles from their amphibian ancestors. The cotylosauria were contemporaneans with numerous primitive amphibians during late Carboniferous and early Permian times, but before the end of the Permian the amphibians had been out-competed and nearly exterminated by the reptiles, and the latter had become dominated.

The best known and most primitive of these is *Seymouria,* a small tetrapod found in the upper portion of the lower permian sediments exposed to the north of the town of *Seymour,* Texas. It is a good example of a persistent structural ancestor. It is also a good example of a "connecting link" of which there are many in the fossil record. It had but one condyle, five digits on the hand (manus), but had, like Amphibia, only one pair of sacral ribs, whereas nearly all reptiles have at least two. The number of

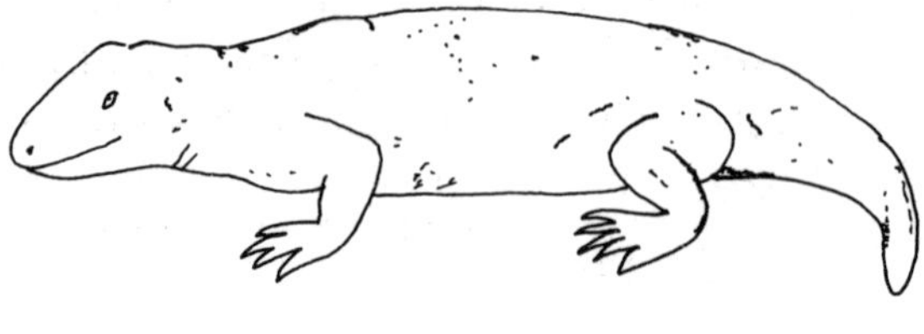

Fig. 3.1: Seymouria.

joints is the fingers is also more than that of Amphibian than that of reptile that is what *Seymouria* may be regarded as incompletely reptilian.

AMBHIBIAN AFFINITIES OF *SEYMOURIA*

Where the *Seymouria* was an amphibian or a reptile, its answer depends on whether *Seymouria* laid an amniote eggs on the land, like modern reptiles, or whether like modern frogs, it returned to the water to deposit its eggs. This problem remained unsolved till *Dr. T.E. White* made a detailed study of *Seymouria*. On the basis of the occurrence of haemal arches he tried to prove the egg laying nature of the animal like large amniotes.

Seymouria also resembled early Amphibia in many features as given below:

(*i*) Flat skull with reduced ossification.

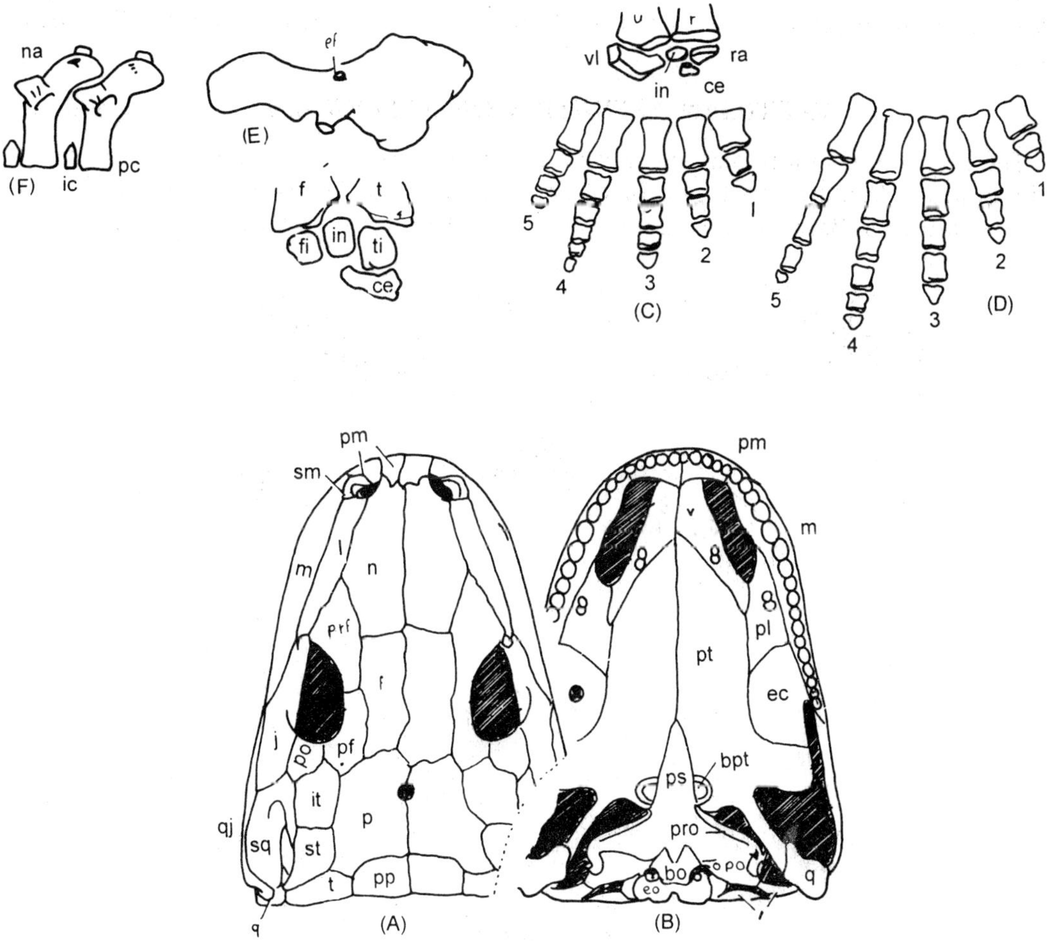

Fig. 3.2: Skull and other skeletal Structure of *Seymouria*.

(*ii*) Presence of inter temporal bone.

(*iii*) Occurrence of primitive palate.

(*iv*) Labyrinthine teeth on vomers and palatines.

(*v*) Position of fenestra ovalis below the basal level of the brain is amphibian.

(*vi*) Traces of lateral line canals in head region.

(*vii*) Presence of one pair of sacral ribs.

(*viii*) Due to the short neck pectoral girdle lies close behind skull.

(*ix*) Skull is without temporal vacuities (an amphibians condition).

These characters of *Seymouria* rather secure on the side of the amphibians. And added to this evidence is the clear indication that some genera closely related to *Seymouria*, notably *Discousouriscus* from Europe, were characterized by gill-bearing larvae.

REPTILIAN AFFINITIES OF *SEYMOURIA*

The mixture of amphibian and reptilian characters seen in *Seymouria* is indicative of the gradual transition that took place between the two classes during the evolution of the vertebrates. Because the change was gradual rather than abrupt it is difficult to

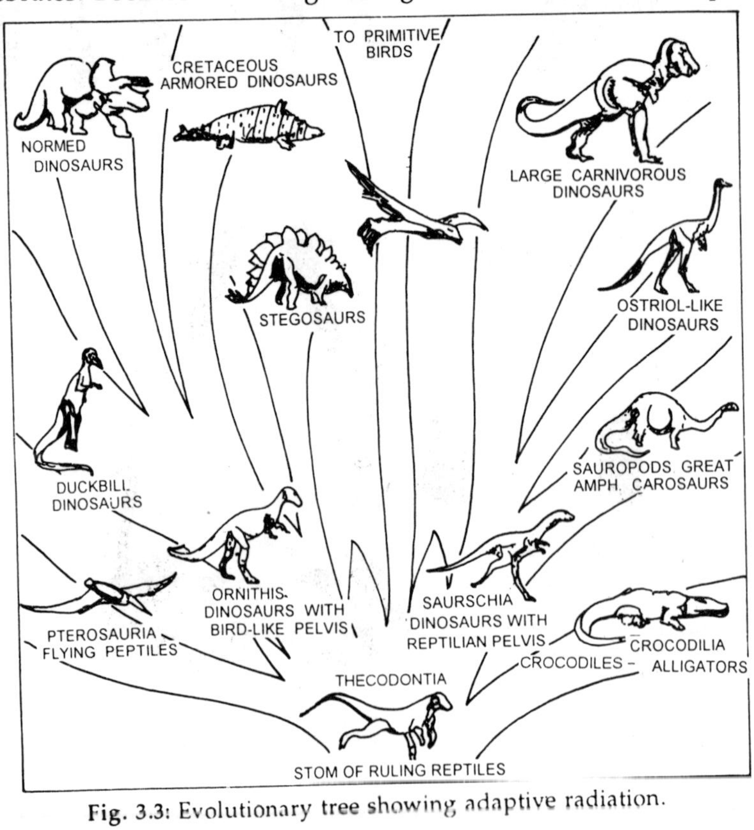

Fig. 3.3: Evolutionary tree showing adaptive radiation.

draw clear cut distinctions between amphibians and reptiles when all the fossil materials are taken into consideration. There are, however, certain character typical of reptiles found in *Seymouria* as given in the following lines:

(*i*) Presence of short, stubby, and muscular limbs.

(*ii*) Monocondylic skull.

(*iii*) Skull is without vacuity as in living chelonians.

(*iv*) Pelvic girdle is attached to the vertebral column by means of sacral vertebrae.

(*v*) Number of phalanges is greater and more reptilian with a basic formula of 2-3-4-5-3.

(*vi*) Presence of a large pineal eye emerging through a hole between the parietal bones.

Due to having mixed characters of amphibians and reptilians, the Symourians are supposed to be the connecting links between amphibians and reptiles. *Romer* treats it as a reptile under a separate heading the *Cotylosauria,* whereas others classify the Seymourians under the head *Seymouriomorpha.*

ADAPTIVE RADIATION

The Mesozoic era has been called the `Age of Reptiles', for while higher forms the birds and mammals, make their appearance during its course. During Permian Period the stem reptiles underwent a considerable radiation, some types becoming pecularly specialized. Food and safety including living space are the basic causes responsible for the entire evolutionary changes. Perhaps the reptiles have shown the greatest evolutionary diversity and adaptive radiations of all vertebrate groups. They were adapted to live in different ecological conditions. Some important reptilian adaptions are taken here under the following heads:

Central Forms

The central form was doubtless a short legged crawling cotylosaur such as *Limnoscelis* found in the lower Permian sediments of New Mexico. It was reptile some five feet or more in length. The skull was solidly roofed, somewhat elongated and rather deep. The eyes were placed laterally. On the skull roof was a well developed pinceal opening. There were sharp teeth on the margins of the jaws, and small teeth on the pterygoid bones of the palate. The body was elongated and the limbs were stout but rather sprawling, probably swamp-dwelling type but potent in evolutionary possibilities.

Fig. 3.4: Limnoscelis paludis.

Arboreal Adaptation

Arboreal habitate is difficult to prove on the part of any Mesozoic or older reptiles, for if any arboreal forms existed, their remains in common with those of other forest-dwelling forms would have had little chance of natural entombment and subsequent preservation. Among living modern reptiles there are numerous arboreal forms like geckos, *Calotes, Chamelaeon* etc. Some of them develop adhesive pads for climbing while others prehensile limbs and tail for the purpose.

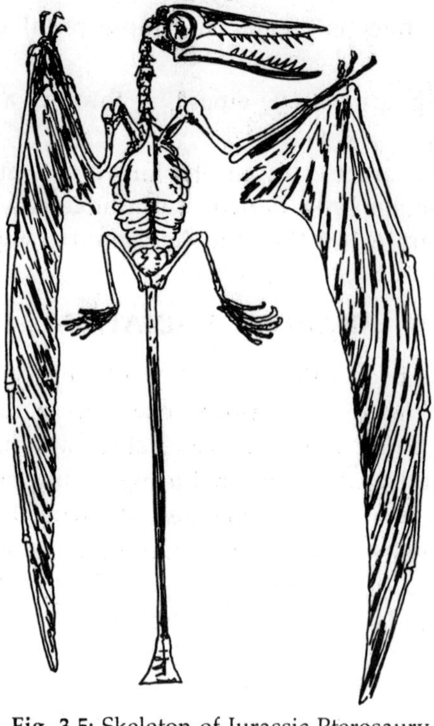

Fig. 3.5: Skeleton of Jurassic Pterosaury.

Aerial Adaptations

Aerial or volant or flying reptiles such as the pterodactyls were finally adapted to their habitat, ranging as they did from the size of a sparrow to the largest of anture's flying mechanism with ample power both of varied and sustained flight, but their origin is lost through the unperfection of the record of Triassic life.

Among living reptiles true reptilian fliers have not yet been reported. Some reptiles like *Draco* simply glides over from branch to branch with the help of its patagi.

Aquatic Adaptations

There are many aquatic forms including both living and extinct, properly adapted for aquatic habitat. Plesiosaurs, Mesosaurs, Ichthyosaurs, Mosasaurs (see lizards) of the among the extinct forms, and turtles, tortoise, terrapins, crocodiles, *Alligators* and

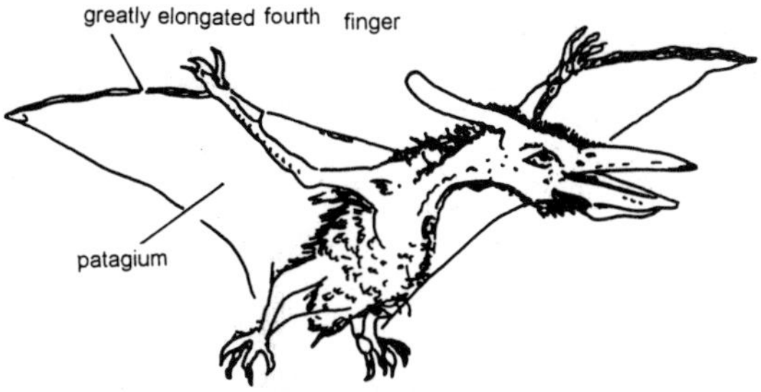

Fig. 3.6: Pteranodon

Gavialis etc. Among the living forms showing aquatic adaptations. For this they have or had paddle like limbs as in *Ichthyosaur, Plesiosaur,* turtles, tortoise etc. The *Hydrophis* or sea snake has a laterally compressed tail.

Fig. 3.7: Ichthyosaurus, Plesiosurus.

Fossorial Adaptation

Fossorial reptiles are rare as fossil for while burial is a prime requisite to fossilization, self burial rarely carries with it the necessary imperviousness to air to insure preservation. Hence we cannot point to a single ancient reptilian group as of extensive fossorial habit, although in the *Cotylosaurs* and *Pelycosaurs* there are certain forms whose powerful implied musculature of arm and leg points to digging powers of no mean degree. Among living reptiles there are many fossorial forms like *Uromastix,* snakes, limbless lizards etc.

and *Pelycosaurs* there are certain forms whose powerful implied musculature of arm and leg points to digging powers of no mean degree. Among living reptiles there are many fossorial forms like *Uromastix*, snakes, limbless lizards etc.

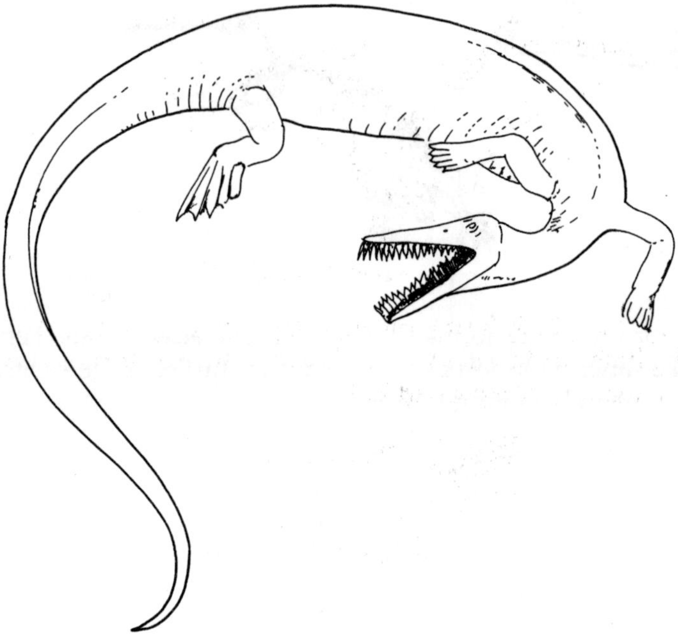

Fig. 3.3: Permian aquatic reptile, Mesosaurus.

Fig. 3.9: Ship–lizard, Edaphosaurus cruciger, Permo-Carboniferous, North America.

Chapter—4

Origin of Dinosaurs

Amphibious Forms

Of the amphibious forms there were many, for increasing humidity—and we have ample evidence of the waxing and waning of moisture—brought with it extensive areas the peopling of which awakened the water-dwelling instinct that had long been dormant in the reptilian blood. Thus we have as partially aquatic forms the turtles ancestral plesiosaurs (Nothosauria), Parasuchia, Crocodilia and many dinosaurs, such as the Sauropoda and Hadrosauria, or duck-bills.

Adaptive Radiation of Squamata

Thus it will be seen that during pre-Tertiary times the reptilian adaptations were ample and varied; with the dawn of the Tertiary, however, came the final extinction of all but four reptilian orders, one of which is represented by but a single relic form (*Sphenodon*) found to-day only in remote New Zealand. This widespread extinction necessarily restricted the range of adaptation, though within the group squamata, which embraces the lizards and snakes, we have a latter-day radiation comparable in a very modest way to the great reptilian radiation of the Mesozoic. For example, we may enumerate:

Ambulatory terrestrial: many lizards, horned toad.

Cursorial: *Chlamydosaurus*, the Australian frilled lizard.

Arboreal: chameleon geckoes.

Aërial: *Draco*, the flying dragon.

Fossorial: *Typhlops, Uromastix*, amphisænians (legless).

Amphibious: many serpents, some of the monitors, *Varanus* which is a living relative of the ancient mosasaurs, the Galapagos sea-iguana *Amblyrhynchus* and others.

Aquatic: the sea-snakes or Hydrophinæ, all of which except one land-locked form in Lake Taal at Luzon, Philippine Islands, are marine and may be found many miles from land.

Today the lizards are kept in their place largely by the mammals. Were the mammals, including man, entirely blotted out so that this control would be removed, it is conceivable that out of the lizard-stock creatures would arise as diverse in habits, size, and prowess as were the reptiles of the Mesozoic, and that another Age of reptiles would be ushered in.

DINOSAURS

The Age of Reptiles may well be called the Age of Dinosaurs, for so far as terrestrial creatures go they were all-important, the other reptiles individually and collectively forming but the supporting cast to these stars in the great drama of medieval life.

Place in Nature

That the dinosaurs were reptiles goes without saying, although their appearance, at any rate in the eyes of those who would restore them in the flesh, was sometimes so very similar to that of certain great mammals of to-day that the uninitiated often confuse relationships and think of *Triceratops*, for instance, as merely a very huge and somewhat better armed sort of rhinoceros. As reptiles they were exclusively lung-breathing and had a large egg which they may or may not have laid before hatching. This is often merely a matter of family convenience as among certain snakes and we know that certain of the dinosaurs were egg-laying. They were scaly or armored, this we know, whether they were cold-blooded or not is debated question. Such activity as they must have shown seems to hint at a possibility of warm blood, and such a belief has met with some support, but in the tropical climate which their known habitat implies there was little more need of a heat-maintaining mechanism than there is in modern cursorial lizards, the Australian frilled lizard *Chlamydosaurus*, for instance. There is no evidence on the part of dinosaurs of a heat-retaining "clothing" such as the birds and mammals possess.

Living Relatives

Of the forms now living, the crocodiles on the one hand and the birds on the other stand nearest the dinosaurs. When one comes to work out a concise technical description of a crocodile and place it beside that of a dinosaur, he will at once see the similarity of the two groups, for the only characters which are *not* common to both orders are the presence in the crocodiles of a false palate, merely a device for eating under water, found in other animals as well and the exclusion of the pubic element of the pelvis from their hip socket.

Ancestral Stock

Dinosaurs, as befits their high estate, were of ancient lineage. One of our best authorities on dinosaurian phylogeny, Doctor Friedrich von Huene, derives them from the primitive cotylosaurian stock which arose in Carboniferous time and continued until the Trias. During Permian time there arose from the cotylosaurs the group Protorosauria, of which *Protorosaurus* is the type, and out of these in turn the Triassic

Parasuchia. While all dinosaurs were derived from the same stock, they are considered diphyletic from their origin; that is, the Ornithischia and Saurischia are unrelated except through a common ancestry.

Duration

The known record of dinosaurs extends from the Middle Triassic (Muschelkalk) to the very close of Cretaceous time. This is particularly true of the saurischian dinosaurs; the plant-feeding Ornithischia, on the other hand, do not appear in the fossil record until late Triassic (Rhætic) but are doubtless older, their subsequent duration being coextensive with that of the others.

Distribution

Dinosaurs were first found in Germany, at Gogolin, Upper Silesia, but this does not necessarily imply that that was their original radiation center. On the contrary the belief has been expressed that one must go farther west, where a great continent is thought by some paleontologists to have extended across what is now the north Atlantic basin, thus connecting Europe and North America, and that somewhere within the confines of this continent the dinosaurs arose and began their world-wide march of conquest. For they extend the world over, across the United states and into Canada; in Brazil, in Patagonia; from England, Belgium, France and Portugal, to Germany and Austria; in far away India, even to Australia; in Africa in its central, eastern, and extreme southern part; and are now recorded in abundance and variety from central Mongolia. Thus they were actually world-wide, except for New Zealand.

Habitat

Their habit was varied, but in all probability their initial evolution, that for speed, took place under stress of semi-aridity of climate, and their main lines were eminently terrestrial forms. With the changing climatic cycles, which came with the passage of time, came dinosaurian adaptation to humid conditions, at least on the part of those we know, so that at least two groups, Sauropoda and Hadrosauridæ, give evidence of an amphibious habitat.

Habits

In habits the dinosaurs were nearly as varied as the mammals are to-day—carnivorous, some small, preying upon such feeble folk as they might overcome, others gigantic, the most terrible terrestrial devourers of flesh the world has ever seen. Again, others were herbivorous—some with feeble dentition, the food being drawn unmasticated down a most capacious throat others with a dental apparatus for the reduction of the most resistant herbage such as would offer little promise of satisfaction to the living herbivores. In certain instances the teeth are greatly reduced, as in *Diplodocus* and *Stegosaurus*, the reduction accompanying other signs of specialization. Yet others like *Struthiomimus* of the North American Cretaceous were utterly bereft of teeth. The dietary of such forms it is difficult to conjecture with any degree of certainty.

Size

The minimum recorded size, that of *Compsognathus*, was about two and one-half feet, with the bulk of domestic cat. The footprints of the Connecticut valley, however, record the existence of feet half as long as those of *Compsognathus*, hardly "terrible lizards" as the term dinosaur implies. The other extreme was reached by *Gigantosaurus* of East Africa; whose over all length has been variously estimate by its Teutonic discoverers to be upward of 120 feet, but by Matthew not to exceed 80 feet. Nevertheless it was an animal of such robust proportions that its weight must have been about 40 tons-greater than that of any living animal except the larger of the modern whales.

CLASSIFICATION

The classification of the dinosaurs is still in a somewhat uncertain state, owing largely to the fragmentary condition of much of the material, especially that from abroad. A somewhat conservative grouping, listing the better known genera, is as follows:

Order Saurischia (Carnivorous Theropoda and Amphibious Sauropoda)

Suborder Cœlurosauria (Carnivorous Dinosaurs)

Podokesaurus—Triassic; Massachusetts.

Hallopus—Triassic; Colorado.

Cœlurus—Jurassic; United States.

Compsognathus—Jurassic; Bavaria.

Ornitholestes—Jurassic; Wyoming.

Ornithomimus—Cretaceous; Wyoming and Alberta.

Deinodonts—Giant Carnivores from Cœlurosaurian Stock:

Gorgosaurus—Cretaceous; Alberta and (?) Wyoming.

Tyrannosaurus—Cretaceous; Wyoming.

Suborder Carnosauria (Megalosaurs, Carnivorous)

Anchisaurus—Triassic; Connecticut valley.

Zanclodon—Triassic; Europe.

Plateosaurus—Triassic; Europe.

Megalosaurus—Jurassic to Cretaceous; Europe.

Allosaurus—Jurassic; Wyoming.

Dryptosaurus—Cretaceous; Wyoming and Alberta.

Suborder Sauropoda (Amphibious Dinosaurs)

Cetiosaurus—Jurassic; Europe.

Brontosaurus—Jurassic; Western United States.

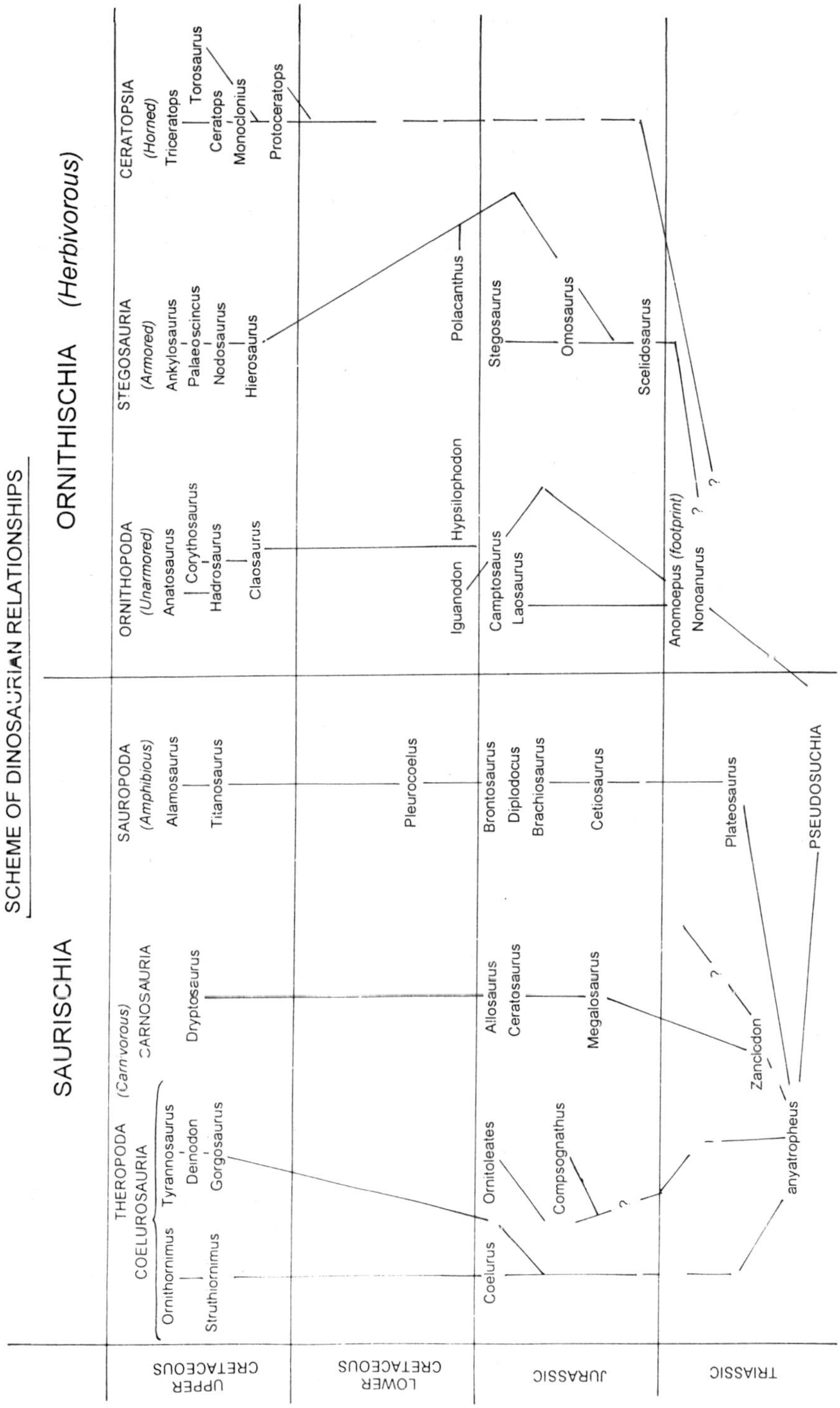
SCHEME OF DINOSAURIAN RELATIONSHIPS
SAURISCHIA
ORNITHISCHIA
(Herbivorous)
THEROPODA (Carnivorous)
COELUROSAURIA
CARNOSAURIA
SAUROPODA
(Amphibious)
ORNITHOPODA
(Unarmored)
STEGOSAURIA
(Armored)
CERATOPSIA
(Horned)
Ornithornimus
Struthiornimus
Tyrannosaurus
Deinodon
Gorgosaurus
Dryptosaurus
Alamosaurus
Titanosaurus
Anatosaurus
Corythosaurus
Hadrosaurus
Claosaurus
Ankylosaurus
Palaeoscincus
Nodosaurus
Hierosaurus
Triceratops
Torosaurus
Ceratops
Monoclonius
Protoceratops
Pleurocoelus
Polacanthus
Coelurus
Ornitoleates
Compsognathus
Allosaurus
Ceratosaurus
Megalosaurus
Brontosaurus
Diplodocus
Brachiosaurus
Cetiosaurus
Iguanodon
Hypsilophodon
Camptosaurus
Laosaurus
Stegosaurus
Omosaurus
Scelidosaurus
Zanclodon
anyatropheus
Plateosaurus
Anomoepus (footprint)
Nonoanurus
PSEUDOSUCHIA
UPPER CRETACEOUS
LOWER CRETACEOUS
JURASSIC
TRIASSIC

Diplodocus—Jurassic; Western United States.

Brachiosaurus—Jurassic; Wyoming.

Pleurocœlus—Jurassic; United States and Europe.

Titanosaurus—Cretaceous; Europe, Africa, and South America.

Order Ornithischia (Predentata, Beaked Dinosaurs)
Suborder Ornithopoda (Unarmoured Herbivorous Dinosaurs)

Nanosaurus—Triassic; Colorado.

Anomœpus (footprints)—Triassic; Eastern United States.

Camptosaurus—Jurassic; Western United States and ? Europe.

Iguanodon—Lower Cretaceous; Europe.

Hypsilophodon—Lower Cretaceous; Europe.

Corythosaurus—Cretaceous; North America.

Anatosaurus—Cretaceous; North America.

Suborder Stegosauria (Armoured Herbivorous Dinosaurs)

Scelidosaurus—Jurassic; Europe.

Polacanthus—Lower Cretaceous; Europe.

Stegosaurus—Jurassic; Western United states and Europe.

Palæosincus—Cretaceous; Western North America.

Ankylosaurus—Cretaceous; Western North America.

Suborder Ceratopsia (Horned Herbivorous Dinosaurs)

Protoceratops—Cretaceous; Mongolia.

Monoclonius and *Ceratops*—Cretaceous; Western North America.

Triceratops—Cretaceous; Western United States.

Torosaurus—Cretaceous; Western United States.

Contrast of Phyla

The two main phyla of dinosaurs to which the rather unwieldy names of Saurishia (Gr. σαυρος, lizard, and ισχιον, hip-joint) and Ornithischia (Gr. ορνις, bird) have been given, while undergoing in many ways a remarkable parallelism of evolutionary change, show notwithstanding some very constant contrasting features. These two orders may be thus briefly defined:

Saurischia: Generally carnivorous, except the sauropoda, with teeth in the anterior portion of the mouth; compressed, slightly curved crowns with serrated margins, or spoon-or-pencil-shaped teeth. Without predentary bone connecting the two halves of the lower jaw; with powerful sharp-pointed, curved claws, and dense hollow bones with well finished articulation except in the Sauropoda. Pelvis triradiate, with a hip-bone or ilium elongated fore and aft, a simple pubis directed downward and forward and an ischium directed downward and backward. Includes the bipedal carnivores and the quadrupedal, amphibious, herbivorous sauropods.

Ornithischia (Predentata): Teeth in rear of jaws only, sometimes forming a wonderful magazine of successional teeth, and a toothless predentary bone in front which, with rare exceptions, opposed the equally toothless premaxillary bones above; premaxillary and predentary sheathed with horny skin or turtle-like beak for the prehension of food. Limb-bones less compact and less well finished at their extremities than in the carnivores; claws depressed, sometimes hoof-like. Pelvis tetraradiate in that the pubis normally consists of two branches, one of which, the pre-pubis, extends downward and forward while the post-pubis lies parallel with the ischium. The ornithischian pelvis and the ossified tendons are quite suggestive of those of bird; the saurischian, on the other hand, is more crocodilian.

SAURISCHIA THEROPODA

The earliest known dinosaurs belong to the Saurischia, and they exist with conservative changes to the close of the Mesozoic. The principal evolutionary changes which they show are a gradual increase in size of body and a proportionate decrease in that of the fore limbs, the function of which in bipedal Theropoda is not locomotion, but prehension. These creatures therefore walked or ran entirely on the hind legs, the anterior part of the body being balanced by weight of the long, slender tail when they ran the limbs were well under the body and the stride was alternate like that of an ostrich or bipedal lizard, as their numerous bird-like footprints impressed upon the sands of the Connecticut valley imply. Some of the better known carnivorous dinosaurs are:

Cœlurosauria

These are in general the smaller and more agile of the carnivorous dinosaurs which, Matthew thinks, were largely upland dwellers. As a rule their bones were hollow and the entire skeleton light and bird-like. Their range is from the Triassic to the close of the Cretaceous, and they are widespread, being found in Mongolia, as well as in North America. During the Cretaceous they diverge into two races, the ornithomimes, or bird mimics, represented by *Struthiomimus* and *Ornithomimus,* some of which at any rate were toothless and doubtless of omnivorous habits; and a group of huge Carnivora, the deinodonts, which converged toward the megalosaurs and include among their number some of the most terrible devourers of flesh the world has ever seen. The culminating member of this subrace is *Tyrannosaurus,* described below.

Representative of this group are: *Podokesaurus,* an extremely slender, agile carnivorous dinosaur from the Triassic sandstones of the Connecticut valley (South

Hadley, Massachusetts). The body, tail and limbs were preserved in the one specimen known, unfortunately destroyed by fire, and indicate an animal about four feet in length of which the tail includes more than half. Many footprints described under name of *Grallator* (he who walks on stilts) were undoubtedly made by creatures similar to this.

Cœlurus, with bones so delicate that the walls of the vertebræ for instance, are of paper-like thinness. It was a small form from The Morrison formation of Wyoming and the Potomac formation of Maryland and is incompletely known.

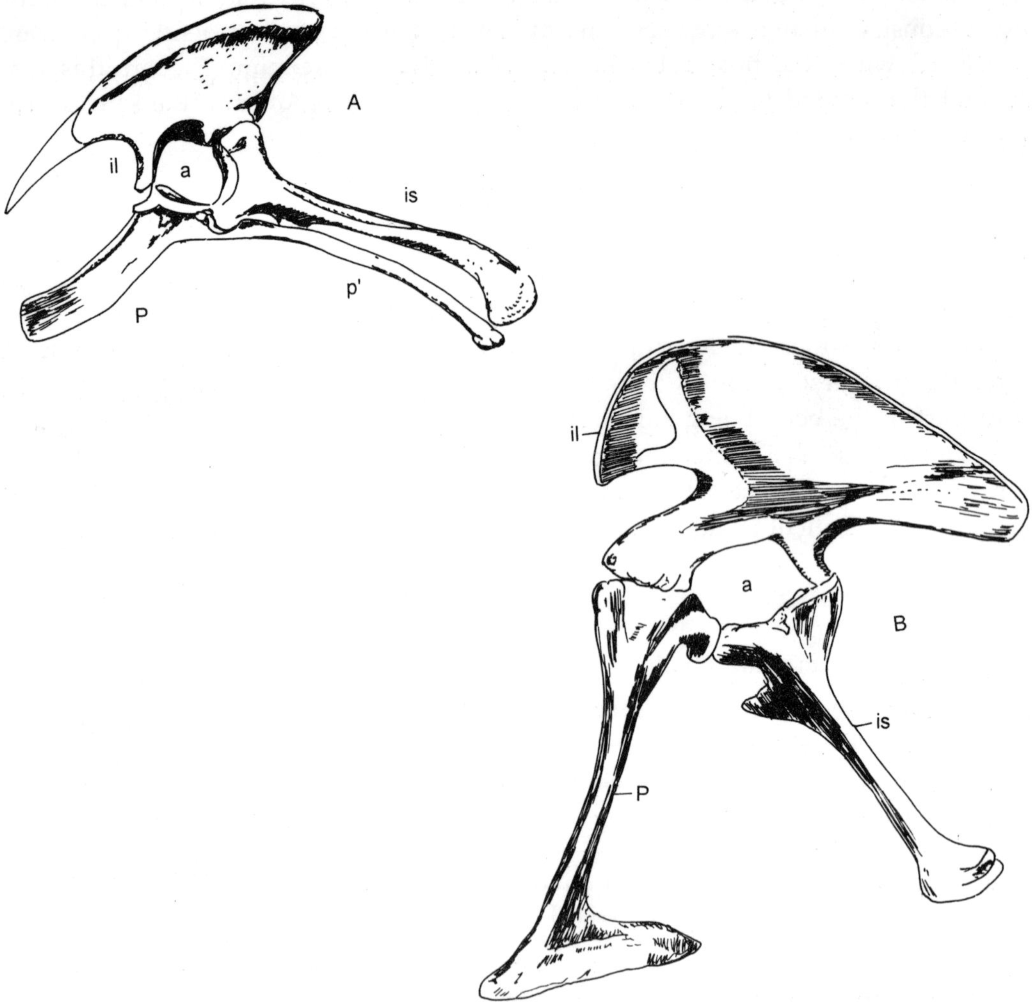

Fig. 4.1: Pelvis of (A) predentate dinosaur (Ornithischia), *Camptosaurus*; and (B) carnivorous dinosaur (Saurischia), *Allosaurus*. One-twelfth natural size. *a*, acetabulum or hip socket; *il*, ilium; *is*, ischium; *p'*, pubis; *p'*, post-piubis. (After Marsh.)

Fig. 4.2: Restoration of *Podokesaurus holyokensis,* Upper Trias, Connecticut Valley. Length, nearly four feet. (After a model by Lull.)

Compsognathus, known from a very perfect skeleton from the Jurassic of Eichstädt, Bavaria and coming from the famous Solenhofen quarry which also produced *Archæopteryx* the earliest known bird. *Compsognathus,* as has already been stated, is the smallest recorded dinosaur.

Out of the famous Bone Cabin quarry in eastern Wyoming where the author was initiated into the mysteries of bone-digging and which has produced *Allosaurus* (described below), comes *Ornitholestes,* an extremely slender from whose total length was not more than 7 feet and whose bulk could not have exceeded that of a setter dog. This form had long, slender fingers none of which was armed with the cruel curved claws of the megalosaurs. This suggested the idea that perhaps it may have preyed upon contemporary birds, so the name *Ornitholestes,* the "bird robber" was given to it. Another suggestion was that it may have preyed upon fish, and its association with amphibious dinosaurs lends colour to the proposition. Be that as it may, the contrast between the marked agility of *Ornitholestes* and the more ponderous character of *Allosaurus* must have been reflected in the prey. A successor to the Jurassic *Ornitholestes* was *Ornithomimus* of late cretaceous time, a form long known from its slender, very bird-like feet and a few other elements of its frame. The entire skeleton of an intermediate form, *Struthiomimus,* was discovered in Alberta in 1914. In its general proportions it is what one would be led to expect from the character of feet, but the surprise came in

Fig. 4.3: Restoration of the small carnivorous dinosaur, *Ornitholestes,* based upon a specimen in the American Museum of National History. Length seven feet. Jurassic, Wyoming. (After Lull, from Schuchert's *Historical Geology.*)

Fig. 4.4: Restoration of *Tyrannosaurus*, based upon a specimen in the American Museum of Natural History. Length 47 feet. Cretaceous, western North America. (After Lull, Schuchert's *Historical Geology*.)

the fact that its jaws are entirely toothless, the skull reminding one quite forcibly of that of large cursorial bird.

Apparently out of cœlurosaurian stock came an amazing, race of giant carnivores, the deinodonts of the late Cretaceous, which in many ways paralleled the megalosaurs yet to be described. The more notable of these were *Gorgosaurus* and *Deinodon*, but especially *Tyrannosaurus*, the climax of evolution of the great flesh-eating dinosaurs.

Tyrannosaurus reached a length of 47 feet and in bulk must have exceeded the largest of elephants. The massive hind limbs supported the weight of body and, in a standing position. The animal was 18 to 20 feet high. The head was 4 feet in length and the powerful jaws bore teeth from 3 to 6 inches long. To this armament were added the great eagle-like claws of the hind feet and probably of the fore feet, 6 to 8 inches in length. The fore limbs were relatively very small compared with the huge size of the animal but were probably constructed much as in *Allosaurus*, with two or three large curved claws, the inner opposing the others. As the hands could not reach the mouth, it is difficult to conjecture their use other than in combat and possibly in mating.

Carnosauria

Another group of carnivorous dinosaurs were the megalosaurs which ranged from the Triassic onward. Notable genera are *Anchisaurus* and *Yaleasaurus* known from more or less complete skeletons in the Connecticut Trias. The three most perfect come from the town of Manchester and range in size from perhaps 5 to 8 feet. The hand was large compared with that of later forms and bore five fingers with one large and two smaller grasping claws. These skeletons are preserved at Yale.

Zonclodon and *Plateosaurus* were larger Old world forms; out of the latter (Pateosaurus) probably arose not only the later megalosaurs but the amphibious dinosaurs (Sauropoda) as well.

Fig. 4.5: Restoration of *Allosaurus*, based upon the specimen in the American Museum of Natural History. Length, 34 feet, Jurassic, Wyoming, Colorado, and Maryland.

Among other genera are: *Allosaurus* one of the best known of carnivorous dinosaurs, for there is in the America Museum of Natural History a practically complete skeleton mounted in a most lifelike pose. This creature is of great size, being 34 feet 2 inches in length by 8 feet 3 inches high in its present almost horizontal posture. It was collected from Como Bluff, near Medicine Bow, Wyoming whence came so many of the wonderful dinosaurian and contemporaneous mammalian specimens which the Yale museum possesses. *Allosaurus* had a comparatively inflexible though deep body and a tail which could undergo little movement as compared with that of a modern lizard or a snake, since its principal use, that of a counterpoise, was best subserved by its being held out rather stiffly behind. The jaws were loosely hung and could evidently be opened very widely, and there is evidence of some movement of the upper jaw upon the cranium as though the chunks of the prey which the creature tore off were at times of considerable size or possibly the victim was swallowed whole. The teeth are compressed, recurved, and admirably adapted to the owner's implied habits, Both hands and feet were armed with powerful curved claws, doubtless sheathed with talons like those of a hung eagle.

Ceratosaurus, which is known from a single well-preserved skeleton in the United State National Museum, was a contemporary of *Allosaurus*, though not so large. The remarkable thing about *Ceratosaurus*, as the name implies (Gr. *κερας*, horn), is the presence of a compressed, horn like process upon the nose which must have borne a horny Sheath. This is one of the extremely rare associations of the possibly defensive horns and carnivorous habits, for among the mammals, and indeed the later dinosaurs as well, it is the herbivores only which are thus endowed.

In Europe and elsewhere than in North America the remains of carnivorous dinosaurs are far less complete, and as a consequence, except for some very well preserved Triassic types, we know but little of them. The generic name of *Megalosaurus*, which gives its title to the group, is applied to most of the later forms from the Jurassic to the final extinction, but doubtless covers as varied an assemblage of larger dinosaurs as lived in the New World.

Sauropoda

To this group various names have been given. They were called Cetiosauria by Seeley in allusion to their whale-like bulk, and Opisthocœlia by Owen because their neck vertebrae, which have a ball-and socket articulation, have the concave facet behind and the convex one in front. According to the law of priority this latter term takes precedence over the others, but professor Marsh's term sauropoda is the one in most common use.

These creatures were all quadrupedal, although there is reason to believe that when water-borne some may have reared upon the hind feet. Their backbone is a marvel of complexity and has been described in some detail in Chapter XII. The greatest economy of material is manifest in its structure, giving maximum strength with a minimum of bony substance. The limb-bones, on the other hand, are extremely massive, with very rugose ends, as though the joints were formed very largely of cartilage, in sharp contrast with their bony perfection in the carnivorous forms. This imperfection of the joints in the Sauropoda admits of but one interpretation a semi-aquatic life, when the weight, largely water-borne, did not subject the ends of the limb-bones to the mechanical impact as it would were the animal wholly terrestrial. A line drawn from shoulder to hip separates the lighter portion of the animal's frame from the weightier as though it represented the water-line. Extreme lightness, especially of the neck, is necessary that it be not unwieldy in the creature's search for food, while weighty limbs were equally necessary to enable their owner to wade into comparatively deep water, for these forms were doubtless more into comparatively deep water, for these forms were doubtless more wading than swimming in their straight habits. In order to support their huge weight, the limbs had become more or less pillar-like as the straight, limb-bones imply, and the sprawling gait of most living reptiles is unthinkable with so ponderous a form. In 1936 footprints of huge Sauropoda have been found in the bed of the Paluxy River, Texas. These deeply, impressed tracks show conclusively that the animals walked as the mounted specimens imply. The absence of a tail trace shows that the tail was buoyed up by the water and did not drag on the ground.

The teeth of the Sauropoda are clearly derived from those of carnivores as they occupy the same place and arise in a similar way. They have, however, lost their sharp point and serrated margins and have become more or less spoon-shaped. As a rule they are large but in *Diplodocus* they are reduced to the size of a lead pencil and are confined to the extreme anterior part of the jaws. Claws also are clearly derived from those of carnivores, as they are laterally compressed, but not so curved and give no evidence of grasping powers. The foot bore at least three such claws, while the hand evidently possessed but one. The food must have consisted of some abundant and easily obtainable aquatic plants which were probably dislodged by the claws and rake-like teeth and swallowed without mastication. The occasional presence within the ribs of highly polished siliceous pebbles of a material foreign to the matrix in which the specimens were found points to some sort of a muscular gizzard-like structure which, aided by the stones, could reduce the otherwise inert mass of food to a proper condition for subsequent digestion. Such a thing is known in some existing animals such as certain birds and seals.

Among the notable sauropod genera is *Brontosaurus*, of which the very complete original specimen is preserved at Yale. Another specimen of similar proportions in the American Museum of Natural History measures 66 feet 8 inches long and in an estimated living weight of 38 tons. This specimen is from the Jurassic near Medicine bow, Wyoming and the Yale specimen came from Como Bluff, half a dozen miles away.

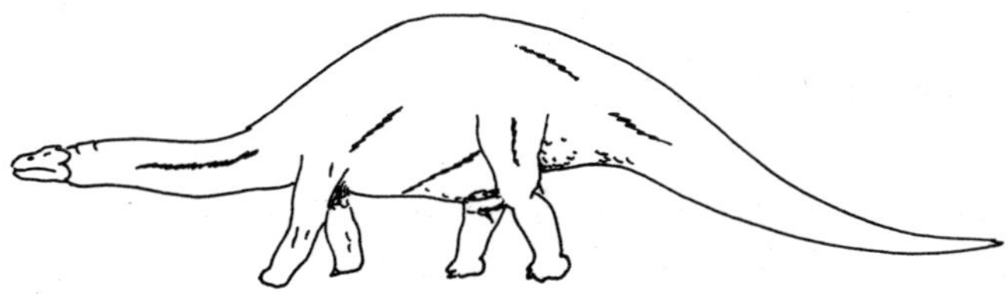

Fig. 4.6: Restoration of *Brontosaurus* based upon a specimen in the American Museum of Natural History. Length about 67 feet. Jurassic, western North America. (After Lull, from Schuchoret's *Historical Geology*.)

Diplodocus another well-known sauropod, differs from *Brontosaurus* in the more slender form, so that even with a length of 87 feet it was by no means so weighty as the latter. All of these creatures had most of their length in the extremely slender neck and tail the body being comparatively short and compact, quite elephantine in fact, especially when viewed in connection with the limbs. In *Diplodocus* the terminal ten feet of the tail was like a whiplash, as the contained vertebrae did decrease further in size. This may have proved a very efficient weapon of defence if its use was analogous to that of certain modern lizards in which the tail has much of effectiveness of a black-snake whip. Aside from this caudal whiplash the Sauropoda were apparently unarmoured and weaponless, relying entirely upon their huge bulk.

Fig. 4.7: Restoration of the sauropod dinosaur, *Diplodocus*, based upon the mounted specimen in the Carnegie Museum. Length, 87 feet. Jurassic Wyoming. (After Lull, from Schuchert's *Historical Geology*.)

Chapter—5

Origin of Birds

ORNITHISCHIA OR BEAKED DINOSAURS

The general characters of the Ornithischia, the predentate or beaked dinosaurs, have already been given. They were derived in common with the Saurischia from the parasuchian stock, and the Ornithopoda, the more conservative of them, underwent an evolution which very closely paralleled that of the carnivores. Like the Saurischia, they too gave rise to aberrant races, which however, did not as with the Sauropoda emphasize bodily bulk so much as arms and armour, and among the later of them were those whose grotesque bizarrerie exceeded that of any known terrestrial forms. None of the Ornithischia was huge compared even with some carnivores, and their bulk was vastly less than that of *Brontosaurus*.

First Record

The first known record of Ornithischia is that of their fossil tracks upon the Connecticut valley's late Triassic rocks, for with some of the footprints are seen the impressions of smaller hands whose five fingers were armed with rounded claws, like those of known predentates but totally dissimilar to the grasping claws of a carnivore. And in contemporaneous rocks from far-off Colorado has been found *Nanosaurus* which has been recognized as pertaining to this order. Doubtless the Ornithischia antedated the close of Triassic time, but there is little evidence that their antiquity is as great as that of the Saurischia.

Ornithopoda

As we have seen from the classification, three suborders of Ornithischia are recognized. Of these the first is the Ornithopoda, which includes bird-footed forms, unarmoured, and bipedal in gait, though unlike the carnivores they occasionally assumed the quadrupedal posture while resting or feeding. Two of the more notable genera are the Connecticut valley *Anomœpus* and *Sauropus*—known only from the footprints and with an estimated length varying from two to eight feet. The Colorado *Nanosaurus* mentioned above is another small form, as yet imperfectly known. These are all Triassic. Jurassic Ornithopoda are unknown in America until near the end of the period, though they presumably lived in some part of the continent, the absence of a fossil record corresponding to the absence of strata suitable for their preservation.

The late Jurassic (Morrison formation) is characterized by two American genera, *Laosaurus*, a slender type not exceeding six and one-half feet in length, and *Camptosaurus* several more or less complete skeletons of which are known from Wyoming. The latter are conservation in character and in size, ranging from 7 to 17 feet.

A related form is *Iguanodon* form Europe, known from no fewer than seventeen remarkably preserved skeletons found in a coal mine at Bernissart in Belgium. These creatures had evidently fallen into an open fissure in the ancient Carboniferous strata and there, unable to extricate themselves, they died, were buried, and were subsequently preserved together with the remains of other reptiles. Their skeletons, ten of which are mounted erect, the others prone on the rock, are to be seen in the Brussels Museum of Natural History. *Iguanodon* was about 34 feet in length and bore upon the hand by way of weapon a peculiar spike-like thumb. Remains of the genus have also been found on the Isle of Wight, whence it was described by Sir Richard Owen long before the fortunate Belgian discovery.

Fig. 5.1: Restoration of the predentate dinosaur, *Camptosaurus*. Average length, about 10 feet. Late Jurassic, North America.

The English Wealden besides yielding *Iguanodon* has produced a smaller form, *Hypsilophodon*, peculiar in having teeth in the anterior part of the mouth, a characteristic which distinguishes it from every other beaked dinosaurs except *Protoceratops* of which our knowledge is sufficiently complete to make a comparison, and in this regard it must be a persistently primitive form. Possibly the Triassic forms also possessed these teeth and this may yet be proved. Both *Camptosaurus* and *Iguanodon* are known from the European Middle Jurassic on.

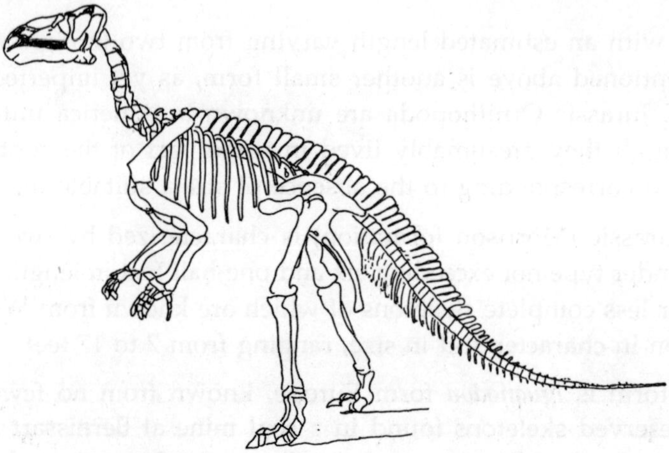

Fig. 5.2: Skeleton of *Iguanodon bernissartensis.*

Fig. 5.3: Restoration of *Iguanodon.* Cretaceous of Belgium and the Isle of Wight.

The Cretaceous of Tendaguru, Tanganyika, whence came the *Gigantosaurus,* has also produced an *Iguanodon* like form, but the details of its structure are not yet announced from Berlin.

Cretaceous time produced several genera of Ornithopoda, which may collectively be called duck-billed dinosaurs from the peculiar character of the mouth, the anterior part of which was often broad as the name implies while the hinder portion of the jaws contained the wonderful dental battery of which we have spoken. No creature of whatever sort is known to have possessed more teeth than *Anatosaurus* the terminal member of the race.

Fig. 5.4: Restoration of *Anatosaurus*, based upon the mounted skeleton in the Yale Peaboby Museum. Length, 29 feet. Cretaceous of North America.

Anatosaurus is the best known genus, as mounted specimens from Wyoming and Montana may be seen at the Yale, American, and United States National Museums. Its length did not exceed 30 feet but it was a fine cursorial type. It also possessed a powerful tail which, judging from the vertical expansion implied by the bones, was admirably adapted for swimming; for *Anatosaurus* was the contemporary of *Tyrannosaurus* and the swimming tail must have been must effectively used when, hard pressed, it took to the water for safety. Several mummied specimens of *Anatosaurus* have come to light, one of which from Wyoming, preserved in the American Museum, is truly marvellous in the degree of its perfection, for not only is the skeleton entirely articulated except for its hind feet and tail, but the skin, shrunken down on the bones by the heat of the sun immediately after death, is also perfectly preserved, together with traces of muscles and tendons. The skin was utterly without defensive armour, for as now preserved it is very thin and is covered with small tubercle-like scales. The hands, which possessed four fingers, were webbed, as were probably also the feet.

Anatosaurus comes from the close of the Cretaceous, the so-called Lance formation. In rocks of a preceding age, Belly River and Edmonton, there have been discovered some curiously grotesque allied forms, *Saurolophus* with a backward extended crest on the skull, and *Corythosaurus* with a most remarkable helmet-like heightening of the cranium. Both of these are known from articulated specimens collected in the Red Deer region of Alberta, Canada, by Mr. Barnum Brown, who by his discoveries has brought us to a clearer understanding of so many of the formerly ill-known Cretaceous forms. These creatures, however, were not confined to the West, for the New Jersey Cretaceous marl beds have produced *Hadrosaurus*, an ally of *Anatosaurus* and a form which has been known to science for many years. Hadrosaurs are also known from Mongolia.

Stegosauria

The Stegosauria or armoured dinosaurs were all quadrupedal, doubtless owing to the great weight of their armament. The armour took the form of high crested plates or spines, or of massive armour plates sometimes welded into a broad cuirass over the hips, evidently the most vulnerable portion of the creature's anatomy. They may well have evolved from the iguanodont-like dinosaurs, developing the armour and consequent modifications of the frame as a defence against their carnivorous enemies. The degree of perfection of the armour in some of the later forms is such as to render them practically invulnerable to any form of animal attack.

The oldest known stegosaurian is *Scelidosaurus* from the English Lias (Lower Jurassic) formation, where the single known specimen was entombed, strangely enough, in marine strata, due probably to its floating carcass having drifted out to sea from some ancient river beside which it lived. As preserved in the British Museum of Natural History, *Scelidosaurus* is only 12½ feet long and its armour consists of two rows of oval bony scutes each with a low fore-and-aft keel not unlike similar elements in a modern crocodile. In addition to these there are a pair of large spines on the shoulders. In late Jurassic and early Cretaceous (Wealden) time we have recorded several armoured

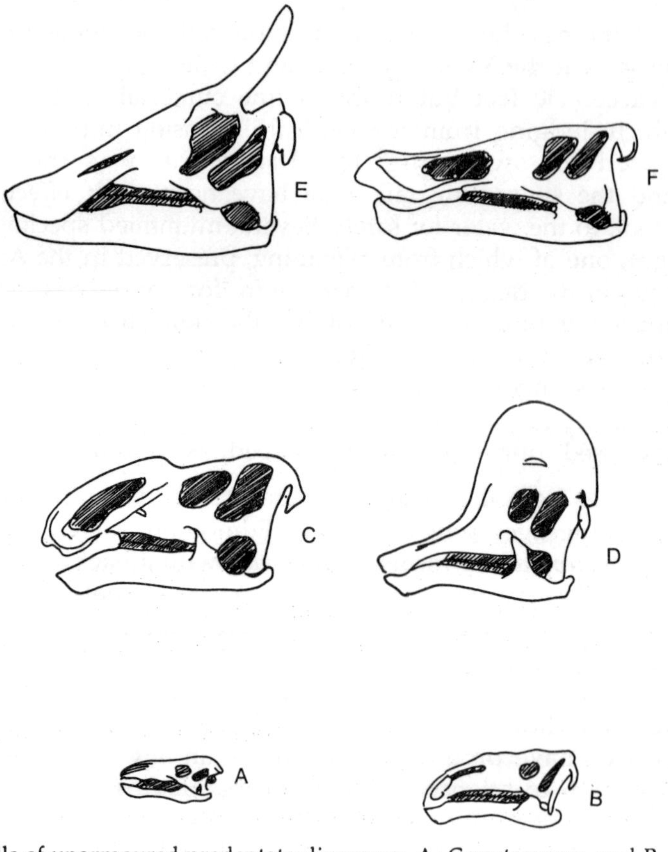

Fig. 5.5: Skulls of unarmoured predentate dinosaurs. A, *Camptosaurus,* and B, *Iguanodon,* Jurassic and Lower Cretaceous; C, *Kritosaurus,* and D, *Corythosaurus,* Mid-Cretaceous (Belly River); E, *Saurolophus,* late Cretaceous (Edmonton); F, *Anatosaurus,* latest Cretaceous (Lance). All one twenty-fifth natural size.

dinosaurs in England and the adjacent continent, known as *Polacanthus* and *Omosaurus*. Of these the former is small, not over 12 feet in length by 3 in height, with a broad rump-shield formed of coalesced plates, and sundry spine-like plates which have been arranged by the restorer in two rows along the neck, back, and tail. Their precise position however, is not assured. *Omosaurus* is another heavily armoured form, some species of which were very large; it is as yet incompletely known.

Fig. 5.6: Restoration of the armoured dinosaur, *Stegosaurus*, based upon the mounted skeleton in the Yale Peabody Museum. Length, about 20 feet. Jurassic, Wayoming and Colorado.

Stegosaurus was a late Jurassic type, but in spite of the fact that it has gives its name to group of armoured dinosaurs, it was not typical, being a highly spinescent senile side branch which died out without issue. In many ways this was the most grotesque of dinosaurs, an awkward angular brute, very high at the rump and low at the withers, with the back ornamented by two rows of huge upstanding plates and the end of the tail armed with fearful horn-like spines 25 inches or more in length. The huge plates which ran along the back culminated in the ones over the pelvis or the base of the tail, beyond which they diminished. A study of the skeleton mounted at Yale and another articulated specimen at Washington betrays great muscular power especially in the tail and hind limbs. The inference is therefore that the plates served for passive, the spines for active defence possibly offence. But the nervous system of *Stegosaurus* is the most remarkable feature, for its brain was very

Fig. 5.7: Brain space (A) and sacral enlargement (B) of the spinal canal of *Stegosaurus*. Drawn to scale, one-fourth natural size.

diminutive, the entire cranial cavity having a volume of but 56 c.c. Thus the estimated weight of the brain could not have exceeded two and one-half ounces, while the total weight of the animal must have been greater than that of the largest of living elephants whose brain averages at least 8 pounds and may run to nearly 11, 80 times that of the dinosaur. In comparing the relative potential intelligence of the two, one has also to bear in mind the great preponderance of the cerebrum, the seat of intellect, over the other parts of the elephantine brain while in *Stegosaurus* the cerebrum constituted hardly more than a third of the entire brain weight. Or, as Professor Williston has expressed it, the seat of a stegosaur's intelligence is no greater in volume than that of a three-weeks-old kitten! In contrast with the diminutive brain-case, however, the neural canal in the sacrum is of startling dimensions, for a cast thereof displaces no less than 1200 c.c. of water, thus giving it a mass more than twenty times that of the brain. This was the seat of the reflex and coördinating control of the huge hind limb and caudal muscles and is further evidence of their very frequent and effective use. The life of *Stegosaurus* was not psychological, but essentially physiological—an animated, largely automatic machine!

The Cretaceous rocks of North America have produced other more conservative armoured dinosaurs, *Nodosaurus*, *Ankylosaurus*, and *Palæscincus*, of which the last is the best known. *Ankylosaurus* was a contemporary of *Tyrannosaurus* and the duck-billed dinosaurs, and was not unlike a huge horned toad. The head was broadly triangular with a horny beak and very feeble teeth, while the trunk was covered with a veritable cuirass of keeled and spined plates which formed an invulnerable armour. The tail was also enclosed in a bony sheath which probably bore spines, like the battle mace of a mediæval knight, and was the offensive weapon. Even *Tyrannosaurus* could not effectively attack one of these creatures when it settled down and tucked in its short legs beneath its body.

Ankylosaurus was a remarkable reptilian prototype of the armoured mammalian glyptodonts of the Pleistocene, the similarity even extending to a club-like fusion of armour on the tail, especially in the genus *Dœdicurus*. In *Dœedicurus*' day it was the great saber-tooth cat, *Smilodon* against whose attacks the creature may have guarded itself in much the same way as the ankylosaur met those of *Tyrannosaurus*.

Ceratopsia

The horned dinosaurs, or Ceratopsia are except for Mongolia largely North American, and more than that, their remains come almost entirely from the eastern uplift of the rocky Mountain region from Alberta to New Mexico. That they lived only within these narrow limits, however, is hardly to be supposed. Their geologic range is also brief in extent as they are confined exclusively to the Upper Cretaceous period. The degree of their evolution when they first appear is, however, indicative of an origin not later than the lower Cretaceous.

Triceratops is a late Cretaceous member of the group, but is the best known and may be taken as typical. It was a huge creature of rhinocerine aspect, from 20 to 25 feet in length, of which from one-quarter to one-third consisted of the great head, for whereas in all other dinosaurs the head is small, here the reverse is true. This is due partly to the great expansion of the cranial roof for the support of the horns, but particularly to the backward extension of the rear of the skull into a widely expanded frill or crest for the protection of the neck, and also for the attachment of the powerful muscles of the back of the neck, giving this group of animals tremendous prowess, correlated with their armament and the great bulk and power of the enemies which they were called upon to withstand. One *Triceratops* skull in the Yale collection measures 8 feet 4 inches over-all, and that of the allied *Torosaurus* 8 feet 7 inches, making the latter the largest known skull of a land animal.

The skeleton of *Triceratops* gives evidence of enormous strength especially in the great fore limbs and shoulders. But the armament was the most striking thing, for the mouth was armed anteriorly with a sharp cutting beak like that of a turtle, and the nose and portion of the skull above the eyes bore huge horns, three in number, hence the name *Triceratops* (three-horned face). In the earliest American forms the nasal horn was always the dominant one, straight or curved either forward or backward, while the frontal horns above the eyes ranged from mere rudiments to fairly well developed organs. The principal Belly River genera were *Monoclonius* and *Ceratops*. Another was *Styracosaurus*, which was remarkably spinescent, having a huge straight nasal horn and at least eight more horn-like processes around the margin of the frill. The crest in all of these earlier types was incomplete in that it was penetrated by two large apertures one on either side. In that latest Lance forms the frontal horns were predominant, and were very long in the later types, while the nasal tended to reduce, becoming entirely obsolete in one genus, *Diceratops*. *Triceratops* and *Diceratops* both had a completely bony

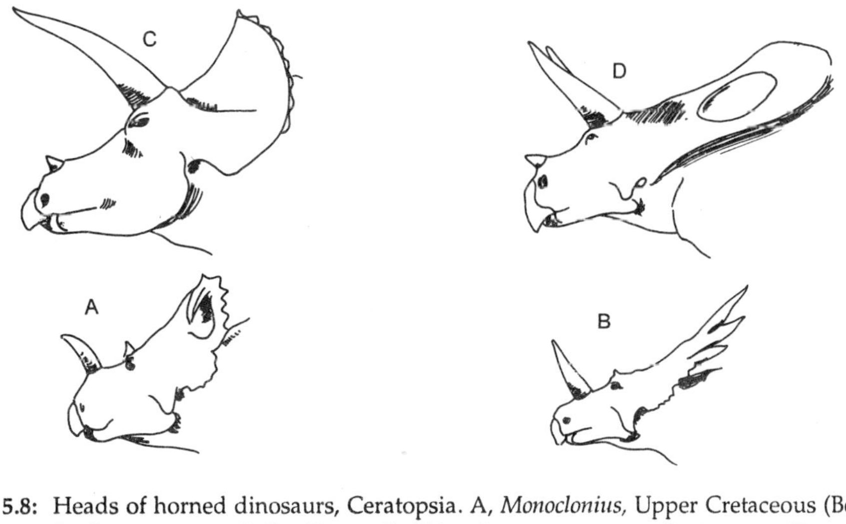

Fig. 5.8: Heads of horned dinosaurs, Ceratopsia. A, *Monoclonius*, Upper Cretaceous (Belly River); B, *Styracosaurus*, Belly River; C, *Triceratops*, uppermost Cretaceous (Lance); and D, *Torosaurus*, Lance. Drawn to scale. *Torosaurus* skull 8½ feet long.

frill with no trace of the ancestral apertures. In *Torosaurus*, on the other hand, while the horns were like those of *Triceratops*, the two apertures still persisted in the immensely expanded crest. That these horns and the defensive frill were put to actual use is highly evident, for broken and healed horns, broken jaws, and punctured crests are unusual with these skulls. And these are the deep and grievous wounds, doubtless few compared with the many superficial injuries which the creatures must have suffered in the combats of rival males or in defense against their arch enemy, the tyrant saurian. The Mongolian *Protoceratops* at least structurally ancestral, possessed the bony frill but was hornless. This animal laid some of the dinosaur eggs found by the Andrews Expedition to Mongolia, and not only were the eggs found, but 72 skull representing the entire ontogeny from the unhatched young to the full-grown animal, which is believed to be without parallel in vertebrate paleontology.

Fig.5.9: Restoration of the horned dinosaur *Triceratops*. Length 20-25 feet. Upper Cretaceous (Lance), western North America.

SUMMARY

As in the history of nations the members of various human races intermingle, so the dinosaurs are interwoven not alone with the various types of their own stock, but with the kinds of animals and plants which together with physical conditions go to make up the environing complex. A summary of the changing life conditions with their influence on the successive dinosaurian societies is necessary to an understanding of their evolution.

The first relics of dinosaurian life are found in the early Mesozoic (Mid-Triassic) rocks of Central Europe, though as they appear shortly after in those of North America one is led to infer that the initial evolution occurred in what may have been an annectant land mass lying across the North Atlantic. Ancestrally, the dinosaurs were quadrupeds remotely related to the crocodiles; but the increasing aridity of climate, clearly indicated by the character of the geologic sediment, placed a high premium on ability to travel rapidly and far in search of food and drink, and may well have been the impelling force that raised these creatures erect from the prone gait and posture gait and posture of their progenitors and stimulated their rapid evolution into the several types.

The so-called continental rocks include such as are formed by stream-borne and lake-borne sediment or by sands and volcanic ash carried by the winds. In contrast to marine deposits these are extremely scarce, and yet they alone, with rare exceptions, contain the remains of terrestrial animal. We are given, however, aside from scattered records, three or four vivid pictures of the environment, both physical and organic, wherein the dinosaurs dwelt. Fortunately for us these glimpses are given in the early stages, in the middle, and at the close of the dinosaurian career, showing the race in the period of its youth, its full maturity, and in its extreme old age.

Triassic

The first of these pictures is of great interest, for the scene is laid in what is now the Connecticut valley; the time, late Triassic. Here one must imagine a broad valley rimmed by environing uplands of older rocks, in its climate and sparse vegetation similar to the conditions to be seen to-day by the traveller through the semi-arid regions of the great Southwest. The plant life is somber: huge scouring rushes, ferns, and pines, with no flowering plant to break the monotony of the dull dust-covered greens. Here and there are dry stream-beds carrying at times great floods, while occasionally there are formed extensive bodies of water with characteristic insects and other invertebrates, fishes, and crocodile-like reptiles. In the Connecticut valley deposits, skeletons are extremely rare and the few which have been found are largely those of dinosaurs, and in but three or four localities. The footprints, however, for which this region is justly famous, are numberless, and indicate hosts of creatures of more than a hundred kinds, as full a record of vertebrate life as in any other time and place in the geologic past. The dinosaurs are mainly carnivorous though impressions of the feet of herbivores give indubitable evidence of their presence at this time.

Jurassic

During the long Jurassic period, the second division of the reptilian age, the known fossil-bearing rocks are so largely of marine origin that, except for rare instances where the animal has accidentally been carried into the sea, the record of terrestrial life is almost a blank. At the close of the Jurassic however, we have another glimpse of well-nigh the entire assemblage of dinosaurian life. The scenes are laid along the eastern United States from Maryland southward, but more notably in Wyoming, Colorado, and Utah, and in the Old World in southern England and northern France and Belgium. In contrast with the aridity of the former scene the climate is now moist, and the eye ranges over an extensive, low-lying country with misty bayous in which the huge Sauropoda, now in the flower of their evolution, find retreat. In Europe and western America the vegetation, though far more luxuriant, is of the same monotonous type as that of the Trias; but in central and eastern America appear representatives of the flowering plants, the dominant flora of to-day. Here are predentate dinosaurs, armoured, and unarmoured, and while the former are still relatively few, *Stegosaurus* already shows the senile grotesqueness which heralds its extinction. Carnivores small and large, now in their millennium of numbers and differentiation, ceaselessly seek their prey.

Late Cretaceous time ushers in the third and last scene, of much the same geographic extent as before, but with sharp contrast of an essentially modernized flora, for the forests now contain many familiar trees and plants though often in unfamiliar combinations. Again our record is of low-lying country—stretches of everglade and swamp lands with higher areas between—but the denizens have changed; for while the carnivores are now in their maximum of size, the Sauropoda have run their course and, except for a few remote survivors, are largely extinct. Unarmoured dinosaurs are in their prime, only in rare cases showing indications of degeneracy, while the relatively few heavily armoured types are represented by the most impregnable of their race. The horned dinosaurs begin and end their brief evolutionary career.

Thus it will be seen that the law of the balance of nature was as operative during the Mesozoic as it is to-day, and that as the carnivores grew and waxed mighty, the herbivores were forced to meet the menace of their aggression in several ways; either by increase in numbers, for the remains of carnivores are very rare compared with those of the herbivorous orders, or by speed, or by increase of bulk, while also meant a partial forsaking of the terrestrial habitat as with the Sauropoda and Hadrosauridæ. Or it meant the development of armament, either of defensive armour as in the stegosaurs, or of aggressive armour and weapons as in the ceratopsians. The lack of brain power placed a premium upon brutality, and never, perhaps before nor since has the animal world felt to so great an extent the burden of preparedness.

Extinction

One of the most inexplicable of events is the dramatic extinction of this mighty race, for in the rocks of undoubted Tertiary age not a single trace of them remains. One student has argued internecine warfare amongst the dinosaurs themselves; another, the destructive slaughter, not of adults but of the young, possibly while yet in the egg, by small bloodthirsty mammals; yet another change of climate, either by the diminution of the necessary heat without which no reptilian race may thrive, or of the moisture with an accompanying change of vegetation. These are all conjectural causes of extinction; but this we know, that with the extensive changes in the elevation of land areas which marked the close of the Mesozoic came the draining of the great inland Cretaceous seas along the low-lying shores of which the known dinosaurs had their home, and with consequent restriction of old haunts came the blotting out of a heroic race their career was not a brief one, for the duration of their recorded evolution was thrice that of the entire mammalian age. They do not represent a futile attempt on the part of nature to people the world with creatures of insignificant moment, but are comparable in majestic rise, slow culmination, and dramatic fall to the greatest nations of antiquity.

THE ORIGIN OF BIRDS

Of all the animals the birds are the best known and most early recognised. The study of comparative anatomy and of the fossil evidence of ancient life has convinced zoologist that birds are unquestionably descended from reptiles. The bird skeleton shows so many reptilian features that *Huxley* said that, "birds are glorified reptiles" and many zoologist have preferred to combine birds and reptiles in a single class the *Sauropsida* which means reptile like animals. It is also well known that the evolutionary potentialities of archosaurs expressed themselves not only in a variety of adaptations to terrestrial and amphibian life but also in two experiments in flight represented by pterosaurs and ancestral birds.

The birds are far more reptilians in their organization than the mammals. The evolution of feather presumably from scales of reptilians type confered obvious advantages for the avian flight. The *Archæopteryx* and *Archaeornis* though bridge the reptiles and birds, but still the early fossil remains of birds are scare. Therefore, there is a still demand by the paelaentologists to have a transitional fossil to record exactly what evolution of birds actually means. Thus in few of sufficient transitional type between reptiles and aves, discussion of birds origin has taken the form of theories as to how flight got started among the birds. Several theories exist to account for this. Thus the origin of birds in current times in the absence of exact fossil link is generally dealt on the anatomical peculiarities of the *Archaeopteryx* and the theories of hypothetical proaves.

Affinities of Birds with Archosauria

(1) The birds are far more reptilian in their organization than mammals. The structure of dianosaurs and of pterosaurs, and the birds shows many similarities which point the common origin of all three groups from the primitive thecodont stock.

(2) The establishment of bipedalism (walking on two legs) among early thecodont left the fore limbs free for conversion into wings, by the ancestors of pterosaurs and the birds. But as the contrasting wing structure in the two groups clearly shows that they must have arisen independently and evolved along the different line from the very first.

(3) Resemblances such as the pneumatisation of the bones, the tendency to develop toothless beak which was present among some dinosaurs the enlargement of the sternum and the form of brain are doubtless showing parallel evolution.

(4) Birds are like the ornithichians dinosaurs in the structure of their pelvic but are more like sauroschian dinosaurs in their structure of both arms and legs and this fact misled early writers who believed that birds were descendent from the bird like dinosaurs. But on the contrary the above dinosaurs lack a clavicle and they differed widely from birds in other skeletal peculiarities.

Fig. 5.10: Restoration of probable fore-runners of birds.

Affinities of Birds With Living Reptiles

Birds show many resemblances to the recent existing reptiles in the anatomy, embryology, physiology, palaentology and these are enumerated below:

ANATOMY

Exoskeleton

In both the birds and reptiles there is an exoskeleton formed from the epidermis, the feathers of birds are nothing but, the modified epidermal skeleton of reptiles.

Endoskeleton

(*i*) *Skull*. In both the cases the skull articulates with the vertebral column by means of single occipital condyle. The diapsid condition of skull is formed in both reptiles and birds.

(*ii*) The lower jaw in similar in both the groups.

(*iii*) The suspensorium is biquadrate and is autostylic.

(*iv*) The epiphysis are absent in both the cases.

(*v*) The ribs, as in *Rhyncocephalia*, have uncinate processes.

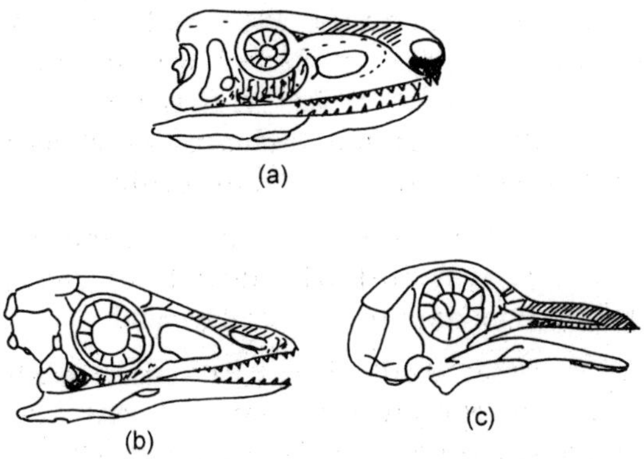

Fig. 5.11: Comparison of reptile and bird skull to show evolutionary stages: (a) The pseudosuchian reptile. (b) Archaeopteryx. (c) The common pigeon.

Visceral Organs

(*i*) Respiration in both is pulmonary and there is no gill in the adult. The air sacs which formed in the birds are also present in the lung of *Chameleon*.

(*ii*) The heart is similar to crocodiles.

(*iii*) The red blood corpuscles are oval and nucleated.

(*iv*) There are 12 pairs of cranial nerves in both the groups.

(*v*) The eye is surrounded by sclerotic bones as formed in some dinosaurs.

Embryology

(*i*) Birds are oviparous as reptiles are, with certain exceptions.

(*ii*) The ova are very large with much yolk usually with calcareous shells.

(*iii*) Cleavage is *meroblastic*.

(*iv*) The early development amnion and allatois are very similar in both the groups.

Physiology

Experiment on the physiology made in the recent years show a close resemblance between birds and reptiles. The precipitation method of blood test shows a degree of genetic relationship between the two groups. They further indicate that the birds are more near to the turtle-crocodile series than the lizard-snake series.

Archeopteryx as Palaeontological Evidence

Relative of Reptiles and Birds. The oldest fossil bird *Archeopteryx* which is more reptilian than birds furnishes a good example of reptilian ancestory of the birds. The earliest known birds appear rather late on the scene than the pterosaurs towards the end of the Jurassic. Only two specimens and one isolated feature are known to the palaeontologist. These were discovered in State as in a quarry located near Pappenhiem, Bavaria. The first skeleton was found in 1861 and was named as *Archeopteryx lithographica*. The second skeleton (now in Berlin) was found in 1877 and has been called *Archiuornis siemensi*. The skeleton of British museum of natural history has been restudied by *De beer* 1954. By the use of new techniques such as the examination under ultraviolet rays that these two skeleton's were conspecific.

The skeleton of *Archeopteryx* is more reptilian than the avians and infact *Lowey* (1944) denied that *Archeopteryx* was bird. The despite of *Lowey* assertions, however, the avian affinities of *Archeopteryx* are now generally held.

These two were the Jurassic birds. The Cretaceous birds found as fossil are *Ichthyornis* and the other *Hesperonis* both were marine forms. The *Hesperonis* was found in the marine Niobarara Cretaceous chalk pads of Cansas. This was described by *Marsh* (1880) and these birds have been spoken as the birds of Cretaceous or the toothless Cretaceous birds. It is generally agreed now that both birds and dinosaurs are descendant from *primitive unspecialized* thecodant reptiles belonging to sub-order *Pseudosauchia*. There were small carnivors possessing many teeth in sockets in the jaws. Their hind limbs were larger than their front limbs and in some forms the 5th hind toe was reduced in length.

Theories of Flight

Nopcas theory or Cursorial origin of flight. Cursorial means fast running forms. The theory of the cursorial origin of flight was advanced by *Nopcas*, a Hungarian palaeontologist. *Nopcas* considers that there is a very fundamental distinction, between flight based on membranous plumes like those in bats and pterodactylus. The plumage is made up of feathers. For the former involves marked adaptation of the hind limbs, whereas the latter involves only the fore limbs and leaves the hind limb unchanged. The hind limbs of birds are essentially homologous with those of cursorial dinosaurs. It is, therefore, argued that the origin of flight involves changes in fore limbs only and that beginning of flight occurred while running efficiency was at its height. The conclusion is that the first birds arose from some long tailed reptiles.

A hypothetical *pro-avies* that spread over the earth on its strong hind legs, stretched out its fore limbs for the sake of maintaining balance and probably flapped. To aid the speed of running these flapping fore limbs or pro-wings developed more surface, partly by flattening our and partly by the backward growth of scales of the posterior margin. Similar large scales are supposed to have developed laterally on the tail. Thus evolution of specialized flight scales into feathers is thought to have been a more matter of a continued increase in size and numbers, accompanied by regional specialization. According to this theory, "A feather is regarded as more or less a specialized scale". The gradual modification of the remaining body scales into feathers would be the logical sequence of events to long list of flight adaptations would appear as correlated variation.

The first step in flying would be prolonged leaps aided by flapping pro-wings then short souring flight, would be made followed by larger flight, accomplished by energetic flapping of the wings alternating with periods of soaring. While rather accepted some ways the theory of the cursorial origin of flight has not gained general acceptance.

Fig. 5.12: Restoration of a hypothetical pro-aves.

Theory of the Arborial Origin of Flight

The theory of the arborial origin of flight has met with more widespread approval. Two phases of this general theory have been advanced the pair wing theory and the four wing theory.

Pair Wing Theory

The pair wing theory is derived directly from a study of characters of *Archaeopteryx*. The long clawed, prehensile, probably climbing wing fingers of ancestral bird point toward an arboreal habitat. *Archaeopteryx* is believed to have not been a true flier, but merely a soarer or gliders, capable of only short passages from twig to twig or from tree to tree. The lack of any foundation for a strong flight musculature argues against the possibility that the creature could have been taken any long flight in which propulsion by means of wings would be necessary.

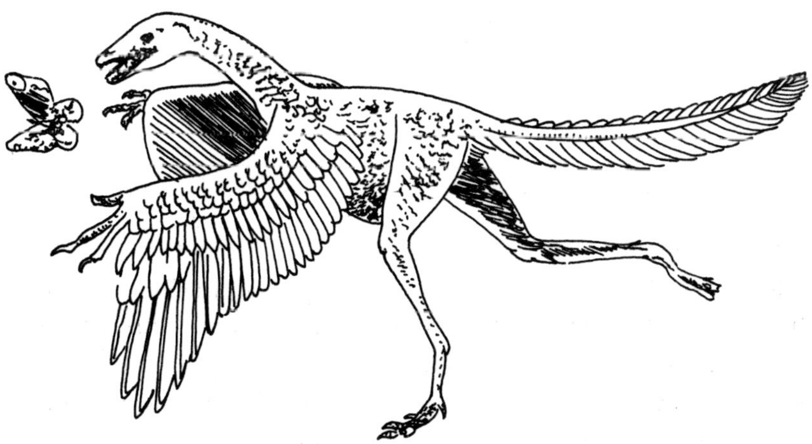

Fig. 5.13: Restoration of Archaeopteryx.

The Four Wing Theory

The four wing theory is of the *Beebe,* is the most recent theory, dealing with the origin of flight. This author made the remarkable discovery that vestigial flight feathers occur on the thigh of a number of species of modern birds. Traces of similar feathers were found on the thigh of *Archaeopteryx*. These discoveries led to the conclusion that the first flier had wings on both arms and legs and used both sets in gliding from trees to the ground and from trees to trees. Later the origin of the legs degenerated as the tail feathers took up the duty of acting as a posterior plane and effectiveness as motor organs.

Fig.5.14: A—The arboreal Pro-aves, B—Arboreal *Tetrapteryx*.

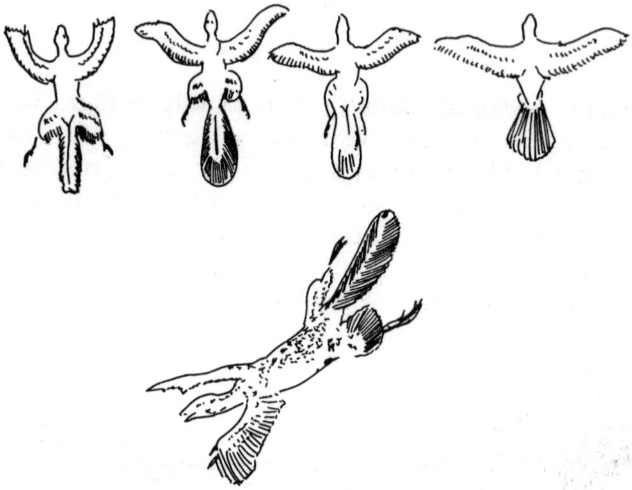

Fig. 5.15. Four wing theory of origin of flight.

Gregory's Compromise Theory

Gregory's compromise theory of the origin of flight is perhaps more nearly acceptable than any of those hitherto given, and is here presented in his own words.

The pro-aves were surely quick runners, both on the ground and on the trees, but it is not clear weather the upright position was first attained upon the ground or on the trees. They very early acquired the habits of searching upright on the branches as shown by the consolidated in step bones, grasping first digit and strong claws of *Archaeopteryx*. Their slender arms ended in three long fingers provided with large claws which were at first doubtless used in climbing. These active pro-aves contrasted widely in habit with their sluggish remote reptilian fore bears. In pursuit of their prey they jumped lightly from branch to branch and finally from tree to tree, partly sustained by the folds of skin on their arms and legs and later by the long scaly feathers of the pectoral and pelvic wings and tail. That they held the arms and legs perfectly still through out the gliding leap appears doubtful, for all recent animals that do that have never attained true flight. It can not avoid the impression that a vigorous downward, flap of the arms even before they became efficient wings would assist in the take off for the leap and that another flap just before landing would check the speed and assist in true landing.

Fig. 5.16: Reconstruction of arboreal *Archaeopteryx* compared with modern birds.

Gregory in 1916 stated that the birds had a two fold or dual origin, some form cursorial (fast running form) and other from arboreal forms. They have conclusively proved that the birds have descended from the reptiles.

Diving Origin of Flight

According to this view the pro-aves used the fore limbs together with their membranous and elongated scales or plumes to aid in diving. The value of such accessories obvious, the dive being more definitely directed the descent mode flatter so as to carry the diver farther out from shore and the force of the plumage eased up sufficiently to avoid shock. If the wings were flapped more or less, a longer glide out over the water could be made and possible circling movements could be taken over the water while searching for fish. It would appear, therefore, that the prowings as placed in diving would serve as useful a function as in running or leaping. We would then have to suppose that some of the arctics diving birds such as the *Punguins* underwent a specialization of the primitive wings using them for underwater flying.

The strong flying sea birds would then be derived from ancestral diving types that had gradually perfected their flight, while land birds of all sorts would be derivatives of the sea birds. There are infact many evidences that the sea is the ancestral home of the birds and that they have invaded the land in comparatively recent time.

Archaeopteryx of course seems to indicate against the diving origin of flight for it is assumed to be a climbing arboreal bird but might not climb the equally appropriate as an aid in sealing puffs after diving and swimming in the water. On the whole the existence of *Archaeopteryx* is no more a barrier to acceptance of the diving than of the arborial origin of flight when other considerations appear to make the former more probable than the latter.

Fig 5.17: Restoration of a hypothetical proavis, or ancestor of the birds.

GEOLOGIC RECORD

Upper Jurassic

The earliest recorded birds are known from two well-preserved specimens, one headless, now in the British Museum in London, (*Archæopteryx*) the other, which bears a head in Berlin (*Archæornis*). They are both from the lithographic quarry at solenhofen, Bavaria and were contemporaries with *Compsognathus*, not as we have seen, the earliest but the smallest known dinosaur.

Archæornis was about the size of a crow, feathered, and with fair powers of flight. There were however, several characteristics wherein it was more reptile-like than are modern birds. These are the presence of teeth in both jaws, the free, clawed fingers of the hand which were not yet fused into the form of the modern wing, the feebly developed sternum, and especially the possession of a long tail on either side of which the rectrices or steering feathers were arranged. In all subsequent birds the tail is shortened and the feathers are disposed fan-wise.

Cretaceous

The Cretaceous chalk (Niobrara) of Kansas has produced the next recorded avian remains in geologic time. These strata are marine, for besides invertebrates and sharks and other sea-fishes, they contain mosasaurs sea-turtles plesiosaurs, and the fish-eating pterodactyls *Nyciosaurus* and *Pteranodon*. The birds belong to two main sorts, both of which were doubtless aquatic, but the large of them *Hesperornis* was especially so, since it had lost the power of flight. The other, Ichthyornis, was a small tern-or gull-like bird well endowed with flying powers and essentially modern except that, like Hesperornis its jaws still bore teeth. It is interesting to note, however, that in common with the predentate dinosaurs (except one or two) the teeth in both of these genera were confined to the maxillary and dentary bones, the pre-maxillary, which forms the forward part of the upper jaw, being toothless.

Hesperornis was a splendid bird measuring over four and one-half feet in length, with powerful hind limbs which while rendering the bird awkward on land must have been very adequate swimming organs. Loss of flight is indicated by the reduction of the shoulder-girdle and especially of the wing which is represented by along slender humerus the fore arm and hand being entirely lacking. The breast bone also is devoid of a keel for muscular attachment resembling that of an ostrich. On the whole Hesperornis finds its recent analogy in the loons or great divers, and except for its flightless condition may have had quite similar habits of life. Other cretaceous genera are known, though very imperfectly, but, with one possible exception they all agreed in the possession of teeth, although in other respects they were essentially modernized. Two original mounted skeletons of Hesperonis and two of Ichthyornis, the latter unique are preserved at Yale.

Tertiary birds leave but little to the imagination as they are similar to those of to-day. It is interesting to note however, that the loss of flight occurred apparently several times among Tertiary forms, for even from the Eocene formation in most parts of the world numerous big Ratites (*i.e.*, cursorial birds), such as *Diatryma*, are known, which can only have originated from badly flying ground-birds, whereas in more modern times the ratites are apparently vanishing from the earth's surface.

Fig.5.18: *Diatryma*, a giant Eocene bird contemporary with the four toed horse Height about 7 feet.

The birds as a class are a very compact group and do not begin to show the range of size and adaptation of the reptiles as a whole; in fact, in this respect they hardly rank with the dinosaurs. With them it is perfection and multiplication of detail, and the most essentially modern among them are the small tree or perching birds of the order Passeres.

Chapter—6

Origin of Mammals

Mammals may be defined as warm-blooded creatures whose body is more or less clothed with hair, whose young are produced alive, except in the egg-laying monotremes, and are nourished for a while after birth by the secretions of milk (mammary) glands. The skeleton shows several important distinctions from most reptiles and birds, having a double occipital condyle—the articular facets which unite the skull with the first cervical vertebra—and having a simple lower jaw. In reptiles and birds, on the other hand, the condyle is generally single and the jaw is a bony complex. There is also an intervening bone, the quadrate, between the jaw and the skull, which is lacking in the mammal. A further mammalian distinction lies in the fact that the vertebræ and limb-bones ossify from three separate bone-forming centers: the body or shaft, as the case may be, and the articular ends or epiphyses. The mammalian dentition is peculiar in the local differentiation of the teeth (heterodonty) in to incisors, canines, premolars, and molars, and also in that there are but two sets in series, the milk or lacteal teeth and the permanent ones; never does one see anything comparable to the amazing successional teeth of the predentate dinosaurs, for instance. Secondarily acquired simplicity of the teeth may occur as in the toothed whales, and the number of teeth may be increased or reduced.

The mammals are certainly the highest class of vertebrates from many standpoints; in some ways, however, this place may be disputed by the birds, but the latter represent the culmination of one line of ascent and the mammals another.

ORIGIN OF MAMMALS

Stock

At least two views have been held as to the origin of mammals. The older one, that advocated by Huxley in 1880, would derive them from Amphibia. Much in Huxley's theory is undoubtedly correct, except that authorities do not now believe that the Amphibia represent the *next* lower stage, but that there was an intervening condition, one in which gill-breathing had been lost but truly mammalian characters had not yet appeared, although some of them were already foreshadowed.

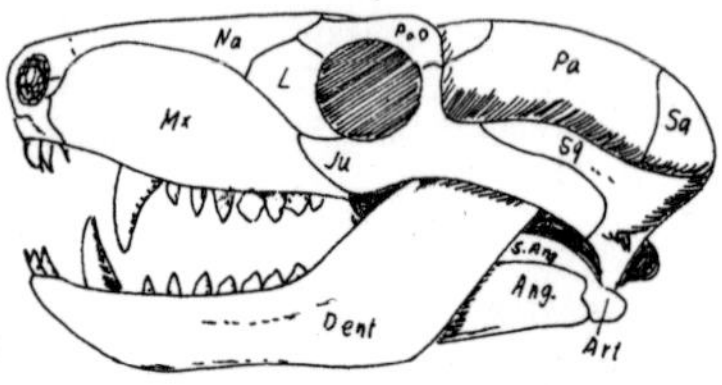

Fig. 6.1: Skull of cynodont reptile, *Nythosaurus larvatus*, Trias, South Africa. Note the mammal-like tooth differentiation, but complex reptilian jaw. *Ang*, angular; *Art*, articular; *Dent*, dentary; *Ju*, jugal; L, lachrymal: *Mx*, maxillary; *Na*, nasal; *Pa*, parietal; *Pmx*. premaxillary; *PoO*, postorbital; *Pr. F.*, prefrontal; *S. Ang*, surangular; *Sq*, squamosal.

There are found in Triassic rocks in South Africa a group of reptiles known as the Cynodontia, in allusion to their dog-like teeth, and there seems to be a large body of evidence in favour of the view that out of this group, although from no known member of it, the mammals have been derived. The cynodonts resemble the mammals in the possession of a heterodont dentition, in that the teeth are clearly divisible into incisors, canines, and molars, and in the paired occipital condyles, in addition to which there are similarities of construction. Structurally the cynodonts bridge the gap between reptiles and mammals because while the dentary the single bone of the mammalian lower jaw, is large and important, the jaw is nevertheless complex in that it possesses the several bones typical of the reptile. There are other reptilian characters as well, all of which may reasonably be expected in the remote ancestors of the Mammalia. These cynodont reptiles are low in the reptilian scale, far removed from the dinosaurs and birds with we have been concerned, although capable of as high degree of specialization along other lines.

Place of Origin

While we find members of this order in both North America and Africa, although they must have been much more widely diffused, it was possibly in the latter place that the mammals arose, probably due not so much to the potentiality possessed by the reptiles of one place over the other as to happy combination in Africa of a potent stock and an impelling cause.

Cause

A cause for the origin of the mammals has been mentioned under the caption "Significance of cursorial adaptation." Therein is emphasized the statement made by Broom that all of the characters wherein a mammal differs from a reptile are the result of increased activity, for he says that when the cynodont took to walking with feet underneath and body off the ground, it first became possible for it to become a warm-blooded animal. But back of this lay impelling geologic cause which Broom does not even hint at. These were, first increasing aridity of climate, which from Permian time well into the Jurassic seems to have been characteristic of all lands, as the extensive series of red sediments may imply. And aridity has been found to be a great stimulative to speed. Add to this the evidence, especially in the southern hemisphere between latitude 20° and 35°, of extensive glaciation—greater even than in the more familiar

Pleistocene Glacial period—and we have a high incentive to the retention of bodily heat. For it is well known that cold more than any other factor limits the activity of reptiles and effectively prevents their distribution into the higher latitudes. For a while, perhaps, there were recurring warm seasons, sufficiently long and frequent so that a normal reptilian life could still be led, the creature hibernating when the weather became too severe. But, as in the case of the æstivating lung-breathing fishes, sufficient time must still be had for the active portion of the creature's career, so that, although in the origin of terrestrial forms premium would be placed upon the capability of atmospheric respiration, here it would be upon ability to withstand the cold and yet remain active, and the acquisition of warm blood and a heat-retentive clothing is the only possible means to this end. Hence as immediate geologic causes of mammalian evolution we have, first, aridity the incentive for speed, rendering *possible* the development of warm blood, and second, the increasing cold to place a premium upon such as did develop it and to eliminate those which did not.

Time

The time of mammalian evolution can only be fixed within certain limits. Geological evidence, if we have read the cause a right, points to its inception in Permian time for "there is now undeniable evidence that in middle Permian times most of the lands in the southern hemisphere were subjected to as severe a glacial climate as that of the present polar areas, and that this Permian glaciation was likewise interrupted by times of warmth" (Schuchert).

On the other hand, the geologic record seems to point to the Triassic at any rate as the time of the *culmination* of the evolutionary movement, for here are found for the first the cynodont reptiles and the actual relics of the mammals themselves. But the cause must always precede the effect and it may be that the known cynodonts were persistent reptilian survivors of the group out of which the mammals actually sprang, and that the earliest known mammals themselves had already a long transitional period. It seems reasonable to suppose therefore, that the time of mammalian origin was not later than Middle Triassic nor earlier than Lower Permian.

MESOZOIC MAMMALS

Deployment

As with the dinosaurs three important vistas are open to our scientific vision, so it is with Mesozoic mammals. We see them in late Triassic time in Germany, England, and South Africa in Jurassic in eastern Wyoming, and in late Cretaceous in the same general region. Again as in the dinosaurs, there are other occurrences of less importance, but the three mentioned above are so placed in time that their interest is thereby greatly increased, for they may be considered roughly to mark the stage of evolution of this all-important class at the close of each of the three great periods of the Mesozoic—the Triassic, the Jurassic, and the Cretaceous.

General Characteristics

The general characteristics of Mesozoic mammals are, first, their small size, for the average among them could hardly have exceeded the stature of a rat, although some attained that of a fox terrier. Their habits were doubtless varied. Some had sharp-pointed teeth comparable to those of living insectivores, and like them adapted to a varied animal diet—insects, worms, young birds, and reptiles—in other words such creatures as they overcame, for in all probability the gratification of their appetite was limited largely by their lack of prowess. Others had teeth better fitted for an herbivorous than an animal diet, with sharp cutting incisors, almost like those of a rodent in front, shearing premolars and many-cusped (multi-tuberculate), broad-crowned, grinding teeth behind. In certain instances their teeth are quite suggestive in general form of those of the rat-kangaroos of Australia and Tasmania, all of which are small animals, hardly exceeding a rabbit in size, nocturnal and feeding on the leaves of various kinds of grasses and other plants as well as roots and bulbs which they dig up with their fore-paws. There is little doubt of the herbivorous character of these Mesozoic forms, although in trying to fix upon a precise dietary from analogy the possibilities of the contemporaneous vegetation must always be borne in mind.

Habitat

Matthew has discussed in some detail the implied habitat of the Mesozoic mammals and has come to the conclusion that they were largely arboreal, his conviction being based chiefly upon the skeletal characteristic displayed by their descendants. He says: "The Cretaceous ancestors of the Tertiary mammals were small arboreal animals of very uniform skeletal characters, but probably somewhat differentiated in dentition according to fruit, seeds and nuts, or insects formed the staple of their diet. At the beginning of the Mesozoic the available modes of life for land vertebrates were chiefly the amphibious-aquatic the arboreal, and the aërial, the terrestrial habitat being subordinate because the upland flora was largely undeveloped and inedible as compared with its present condition. The three available provinces were occupied by reptiles mammals, and birds respectively. In the later Cretaceous the spread of a great and varied upland flora vastly extended the terrestrial province, and opened a new and constantly widening field for the expansion of the Mammalia.... The little that is known of the Mesozoic Mammalia fits in with our hypothesis of their arboreal habitat but adds little to the evidence in its favour. Practically nothing is known of their skeletal structure; they are all of small or minute size, with teeth of insectivorous or

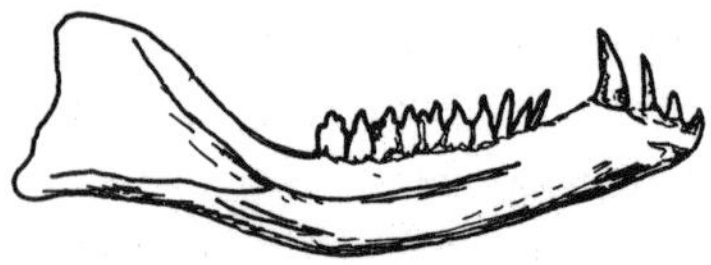

Fig 6.2: Cynodont reptile, *Dromatherium sylvestra*, Trias, North Carolina. Twice natural size.

granivorous type... Their minute size, and association, in strata of fresh or brackish water origin, with large amphibious and aquatic reptiles and with abundance of fossil wood, suggest that the deposits in which they occur were laid down in extensive forest-clad river deltas and coastal swamps, and that the minute Mammalia represent the arboreal fauna of these forests."

Classification

The order "Protodonta," which apparently has no validity, included two small jaws called respectively *Microconodon* and *Dromatherium*. They were discovered in a coal mine at Egypt, North Carolina, and are of Upper Triassic age, contemporaneous with the far-famed Connecticut valley beds. The dentition is that of the cynodont reptiles rather than mammalian, and the principal evidence of mammalian affinity lies in the apparently simple jaw, which, as preserved, consists of but a single bone. Recent restudy by Dr. George G. Simpson has shown, however, that the specimens are defective and that, in all probability the jaw was complex. This necessitates their reference to the cynodont reptiles rather than to the mammalian class.

In the order *Triconodonta*, the teeth are more perfectly formed, but the molars are characterized by having but three cusps arranged in a single longitudinal row; of these the middle cusp is usually dominant, the others being much smaller. At times how ever, the latter may equal the median cusp in height. Their relationships are doubtful.

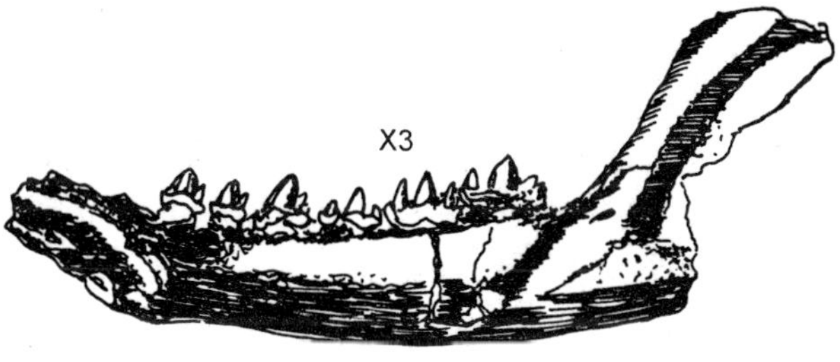

Fig. 6.3: Triconodont mammal, *Priacodon ferox, Jurassic*, Wyoming. Three times natural size.

Triconodonts first appear in the Lower Jurassic (Stonesfield slate) of England, in the genus *Amphilestes*. *Triconodon* itself comes from the Upper Jurassic of England, while a related, possibly equivalent genus comes from the Jurassic of Como Bluff, Wyoming, which is the upper limit of their range.

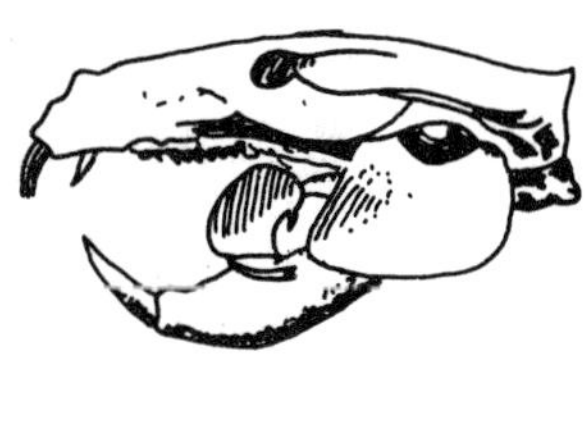

Fig. 6.4: Multituberculate mammal (allothere), *Ptilodus gracilis*, Paleocene (Ft. Union), Wyoming. About natural size.

The *Allotheria* or *Multituberculata* are among the oldest of mammals, for their characteristic molar teeth are found in Upper Triassic rocks of Germany and they range into Lower Eocene time. Geographically they are very widespread, as they been reported from Europe, Asia, and Africa and from

North America. Little is known of the skeleton, but jaws and teeth are more or less abundant and several skulls have been found. They had a single pair of rodent-like incisors above and below, while the molars bore two or three longitudinal rows of tubercles, hence the name Multi-tuberculato; the premolars were either simple-cusped or compressed, sharp-edged, cutting teeth.

The most notable forms are *Tritylodon,* a large type from the Karroo beds (Lower Jurassic) of South Africa; *Thomasia* of the Upper Trias of Europe; *Plagiaulax* from the Jurassic of Europe and North America; *Ptilodus* and *Taenoilobis* from the Paleocene of North America. The Allotheria have been included under the Marsupialia, but there is not the least likelihood that any existing mammals were derived from them.

The *Symmetrodonta* are a little-known group confined to the Upper Jurassic of England and North America. Their molars have three main cusps arranged in a triangle *Spalacotherium* from England and *Tinodon* in North America are the best known representatives. They were apparently an isolated order, distantly related to the Pantotheria.

In the next order, the *Pantotheria,* the dentition, while simpler, suggests that of the insectivores to such an extent that they have been considered by good authority as the actual forerunners of that group. The molars have four or more cusps, but instead of being arranged in lineal series as in the triconodonts, the three principal cusps are in the form of a triangle or trigon, the main one being on the inner side of the teeth in the lower jaw and on the outer side in the upper jaw. In habits the pantothers were probably insectivorous and they have been classed with the placental mammals as primitive Insectivora. Their time range is throughout the Jurassic, and although they then became extinct as an order they may still survive in their descendants—possibly the true insectivores among other order.

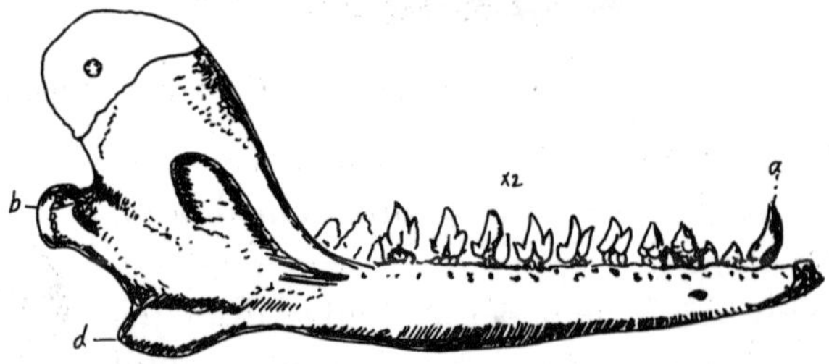

Fig. 6.5: Trituberculate mammal (pantothere), *Dicrocynodon victor,* Jurassic, Wyoming. *a*, canine tooth; *b*, condyle; *c*, coronoid; *d*, angle. Twice natural size.

Amphitherium, one of the oldest genera, is from the Lower Jurassic (Stonesfield slate) of England, *Amblotherium* is from the English Purbeck (Upper Jurassic), while the more familiar American forms are *Dryolestes* and *Dicrocynodon* from the Jurassic of Como, Wyoming.

Scheme of Relationships of the Mesozoic Mammals. After Simpson

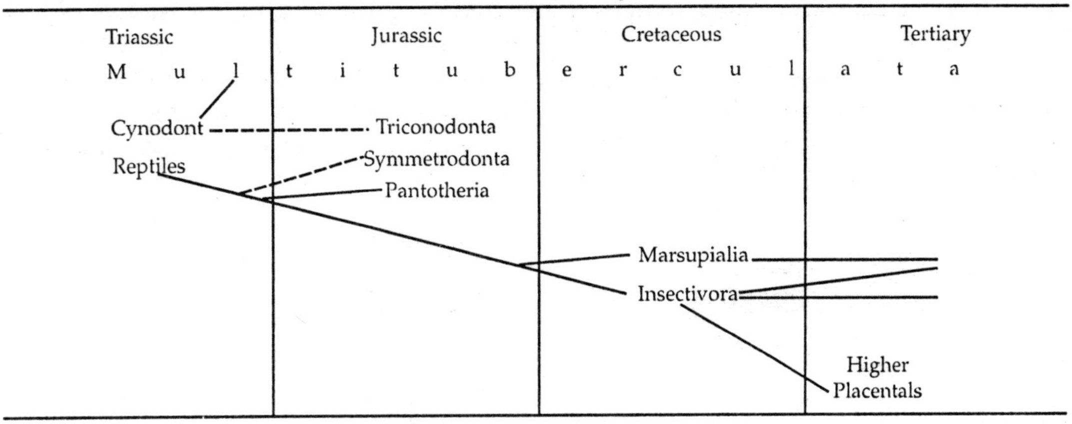

Check on Mesozoic Mammalian Life

Perhaps the most remarkable thing about the Mesozoic mammals is their apparent conservatism as regards evolution, for we find so little recorded change, compared with that of the reptiles during their long-drawn-out career, that we look instinctively for some inhibiting cause. After the establishment of the mammals in late Triassic time, there are no great geologic or climatic changes of a revolutionary character to quicken their evolution until the close of the Cretaceous and, while reptilian dynasties wax and wane, the trend of their evolution seems pretty well established after the early Jurassic, the remarkable types which appear later being largely the florescence which characterizes racial old age. But with the close of the Age of Reptiles came a most momentous change in mammalian evolution, when the sluggish stream of their existence was quickened into life and their remarkable radiation began. This may have been due in part to the expansion of the upland flora which, as Matthew believes, was either restricted in its development or of a sort not suitable to mammalian dietary during the Mesozoic. But it seems reasonable to suppose that the inhibition of mammalian evolution was due not so much to lack of a suitable physical and floral environment as to an overwhelming check against which these small creatures could not contend.

There is in the Yale Museum a remarkable series of mammal jaws and teeth from what is known as "Quarry 9" at Como Bluff, Wyoming, which lies in strata of Jurassic age. Associated with them, among other reptilian remains, was a single tooth of a carnivorous dinosaurs, perhaps *Allosaurus*, a tooth keen-pointed and terrible, like a curved dagger with serrated margins and many, many times the bulk, not only of the teeth but of the entire jaws of the associated "higher" forms. This tooth, suspended like the sword of Damocles over the head of these actually associated mammals, brings before the mind's eye broad vistas of low-lying, well-watered woodland with ever alert furry forms taking such refuge as the tree or shrubbery or occasional hiding holes could offer, in the midst of stalking terrors such as the world never saw before. That the mammals managed to maintain themselves is not surprising, for there is a teeming horde of small mammalian folk in the tiger-haunted jungles of India to-day; and that they did not dispute with the dinosaurs the realms of greater opportunity is but a logical assumption.

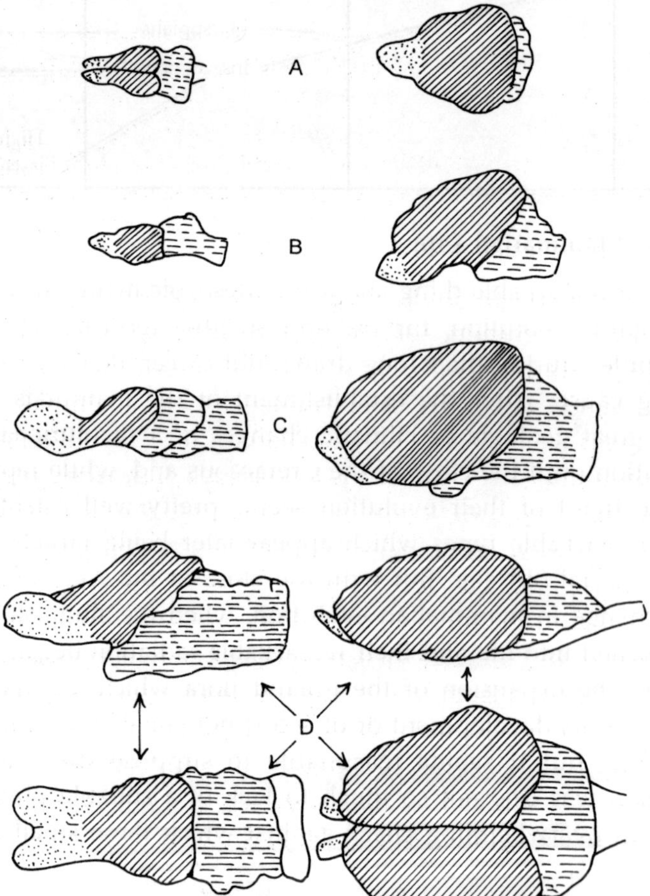

Fig. 6.6: Brain proportions in archaic (left) and modern (right) mammals of similar size. Olfactory lobes (dots), cerebral hemispheres (oblique lines), cerebellum and medulla (dashes).

Arctocyon	A *Canis*
Phenacodus	B *Sus* (domestic)
Coryphodon	C *Rhinoceros*
Uintatherium	D *Hippopotamus*

The Release

If our premise be true, the great Tertiary expansion of mammals therefore is only in part the direct outcome of changing geologic conditions, the primal incentive being the removal of the reptilian check.

ARCHAIC MAMMALS

This is the name given to the creatures which, in early Tertiary time, supplanted the great reptiles in their vacated habitats. They constitute the first great mammalian adaptive radiation, but one of short duration for they were soon to be displaced in their turn by mammals of a higher sort, the so-called modernized mammals. For a while these archaic types served very well, and doubtless had it not been for a competition which they could not meet, they might have survived for a longer period; but there was written over against them the memorable indictment—"Thou art weighed in the balances and found wanting."

Defects

There are two prime essentials to every creature's adaptation to its environment—it must have safety and food. Hence two principal structures are of paramount importance: locomotor organs, that it may flee from its enemy or overtake its prey, and efficient teeth that it may utilize such food as is available. In other words, two organs whose contact with the environment is most intimate are the feet and teeth, and these are seen to suffer the most profound changes with passage of time and the consequent changing of the environment. Not only must these organs be adapted to immediate need, but adaptable to the inevitable changes of conditions which time will bring. Thus it is that by a study of feet and teeth so much of an animal's life conditions and consequent habits can be deduced. Add to this a structure of which the dinosaurs made but little, but which in mammalian evolution became increasingly important—the brain—and the tale of the requisites for future evolutionary success is complete.

It was specifically in these three things that the archaic mammals were deficient, for while size, strength, physical prowess, arms and armament were theirs in full measure, their feet and grinding teeth were conservative, inelastic, and incapable of meeting new conditions as they arose. Their brain too was singularly old-fashioned, generally small, but always relatively undeveloped in comparison with that of modernized mammals of equivalent bulk, especially in the part wherein the intelligence lay. Hence it is not surprising that the cares of these forms was brief and that with rare exception they have suffered racial death and vanished as utterly as did the dinosaurs before them.

Classification

"Nature" as Osborn says, "deals in transitions rather than in sharp lines. We can not circumscribe the archaic mammals sharply, nor be sure as yet that some of them did not give direct descent to certain of the modernized mammals. Yet the mammals of the basal Eocene [Paleocene] of both Europe and North America are altogether of very ancient type; they exhibit many primitive characters, such as extremely small brains, simple, triangular teeth, five digits on the hands and feet, prevailing plantigradism. They are to be collectively regarded as the first grand attempts of nature to establish insectivorous, carnivorous, and herbivorous groups, or unguiculates [clawed forms] and ungulates [hoofed forms]. The ancestors or centers of these adaptive radiations date far back in the Age of Reptiles. At the beginning of the Eocene we find the lines all separated from each other but not as yet very highly specialized. The specialization and divergence of these archaic mammals continue through the Eocene period and reach a climax near the top, although many branches of this archaic stock become extinct in the Lower Eocene. The orders which may be provisionally placed in this archaic group are the following:

"Marsuipialia

Placentalia

Insectivora

Tæniodonta. Edentates with enamel [-banded] teeth

Creodonta. Archaic families of carnivores

Condylarthra. Primitive light-limbed cursorial ungulates

Amblypoda. Archaic, typically heavy-limbed, slow moving ungulates.

"This group is full of analogies, but is without ancestral affinities to the higher placentals and marsupials. There are forms imitating in one or more features the modern Tasmanian 'wolf' (*Thylacynus*), the bears, cats, hyaenas, civets, and rodents of to-day, but no true members of the orders Primates, Rodentia, Carnivora, Perissodactyla, Artiodactyla have been discovered."

Of the archaic mammals we will turn our attention to three groups, of which the first is the *Creodonta*. These forms resemble in many details the hoofed Condylarthra next to be described, but differ from them chiefly in the skull and teeth, in that they have more the aspect of a true carnivore than the condylarths which were largely of vegetarian diet. The terminal phalanges (unguals) are also more claw-like, although there are exception to this rule, notably in *Dromocyon*. The skull of a creodont differs from that of a carnivore, for while it is always large for the size of the animal, there is a much smaller brain-case, thus necessitating a high crest of bone along the mid-line of the cranium to obtain sufficient surface for muscular attachment. There are expanded temporal or zygomatic arches for the same purpose. The teeth also differ in not being so perfectly adapted for a flesh diet as in the true carnivores. In the latter, certain cheek teeth are almost always enlarged and modified to form a wonderful shearing device, and these so-called carnassial teeth (Lat. *caro*, flesh) are, when present, *invariably* the fourth upper premolar and first lower molar, expressed thus: p^4/m^1 With the creodonts the carnassials may not be developed at all, and if they are, are variable

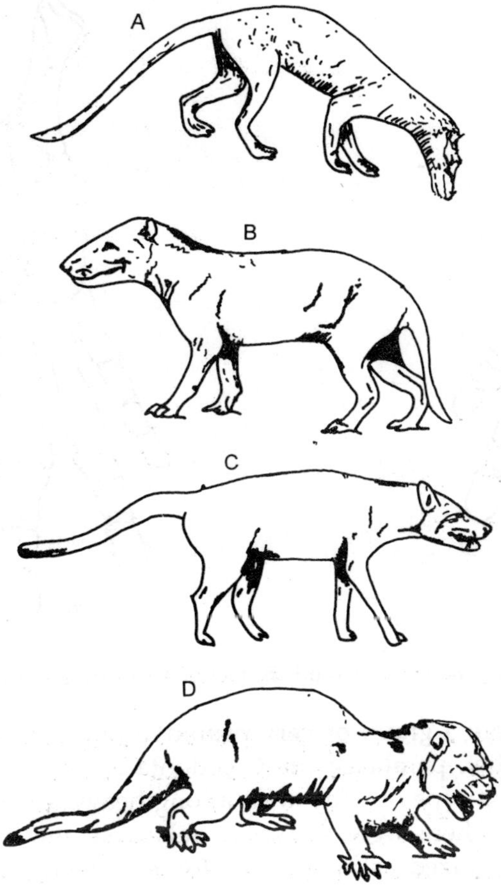

Fig. 6.7: Creodonts. A, *Tritemnodon*, a primitive hyænodont, Middle Eocene, North America. B, *Hyænodon*, the last survivor of the archaic carnivores, Lower Oligocene, North America and the Old World C, the dog-like *Dromocyon*, Middle Eocene, North America. D, *Patriofelis*, Middle Eocene, North America.

and not necessarily, indeed rarely, p^4/m^1' and in addition they are rarely confined to a single pair of teeth but are two or more in number.

The creodonts have been divided into at least six distinct families, of which but one probably gave rise to true carnivores, the rest dying out one after another until by Upper Oligocene time none were in existence. The creodonts foreshadow the true carnivores in a number of ways, in that certain of them were bear-like (*Arctocyon*), others dog-like (*Dromocyon*) or otter-like (*Oxyæna, Patriofelis*), some like the minks (*Sinopa*), other cat-like (*Dissacus*) or resembling hyænas (*Hyænodon*). The last genus is of especial interest because together with its Old World ally *Pterodon* it is the last creodont survivor, existing until the Middle Oligocene. One form, *Andrewsarcus*, from Mongolia attained gigantic size, the skull alone measuring nearly a yard in length.

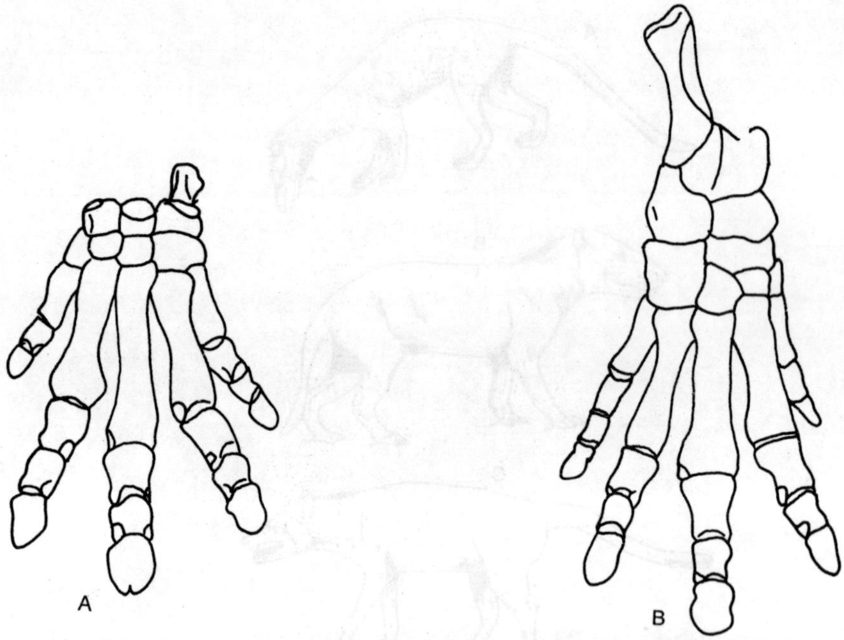

Fig. 6.8: Fore (A) and hind (B) feet of *Phenacodus primævus*.

The *Condylarthra* were a group of very primitive ungulates which, aside from the implied differences in diet, paralleled the Creodonta closely, for in both groups there was the same generalized type of body with a long heavy tail and rather stocky more or less cursorial limbs. There were, however, relatively few of the condylarths but four families being recognized as against six for the creodonts. They range in time Paleocene and Lower Eocene, but very little is known as yet of their geographical extent. Scott says of them: "They may be looked upon as the connecting link between the clawed and hoofed mammals and therefore ancestral to most, or perhaps all of the ungulate orders." One of the condylarths, *Phenacodus*, was hailed by its discoverer, Professor Cope, as the five-toed ancestor of the horse, but this is now known to impossible as it is too large and too highly specialized in certain directions, although very primitive in others and also too late in time to be the founder of the great equine lineage. This genus, from the Wasatch beds (Lower Eocene), ranged in size from a fox to a small sheep. While the canines were tusk-like, they were not large, and the grinding teeth were low-crowned and of simple pattern, suited undoubtedly to a rather succulent herbage. The skull was long and low, with a well

Fig. 6.9: Cursorial archaic mammal, condylarth, *Phenacodus primævus*, Lower Eocene, North America.

developed sagittal crest and, while that portion of the cranium behind the orbits was relatively long as with most primitive skulls, the brain-case was of very small capacity. The feet are five-toed, semi-plantigrade, and built on a very primitive plan. *Phenacodus* and the earlier *Euprotogonia,* represent the family Phenacodontidæ while the other family, Meniscotheriidæ, embraces but a single known genus, *Meniscotherium.* These forms, while contemporaneous with the phenacodonts, were more advanced in tooth structure, for the cusps of the grinders have begun to assume a crescent shape such as one often finds in the higher odd- and even-toed ungulated. The body and tail were long and the limbs, while long, resemble so much those of the Hyracoidea of Africa as to cause the inclusion of *Meniscotherium* in that group by certain authorities. Others have considered the Hyracoidea to be surviving condylarths. There are, however, no very good grounds for such an assumption. The Condylarths are of interest, however, in that they represent or were very similar to what was probably a very widespread group of primitive ungulates out of which, possibly, all of the other orders of ungulates arose. The genera which we know could not have been the direct ancestors, but they show us the nature of the ungulate ancestry.

The *Amblypoda,* or short footed ungulates, are another group of hoofed forms, among which were some that attained a huge, almost elephantine size and in spite of a basic primitiveness developed a superficial specialization of a very remarkable sort. Their geologic range is throughout the Eocene period, when they in their turn suffered extinction. Four families are recognized, of which the two most primitive are both Paleocene in distribution. *Pantolambda,* the type of the second family, while undoubtedly an ungulate, shows many points of similarity with the creodonts. It is described as having a head and body somewhat smaller than those of a sheep, and much shorter legs. The body and tail had somewhat the proportion of the large cats and the skull as with the condylarths, was long and low, with small brain capacity and prominent sagittal crest the limbs were very short and relatively heavy with five spreading toes on each foot.

Fig. 6.10: *Coryphodon,* a swamp-dwelling amblypod, Lower Eocene, North America.

Coryphodon represents the third family and is in many ways a remarkable beast. The different species vary in size from a tapir to an ox, and thus are the largest forms we have so far considered. They were heavy, unwieldy animals whose short powerful limbs and spreading feet point to swamp dwelling if not aquatic habits. The skull was large and flattened in such a way that no median crest is visible, nor are there any indications of horns such as the next genus possessed. The canine teeth were developed into huge flaring tusks suggesting those of the swine. Altogether it was a heavy sluggish brute whose very small brain gives evidence of great stupidity.

Uintherium (*Dinoceras*) represents the last family of amblypods and in many ways—size, up to seven feet in height, dentition, and horns—was by far the most specialized, in fact grotesquely so. Its limbs were graviportal, quite like those of the Proboscidea and like them an adaptation to carry the creature's great weight. The elephant-like characteristics extended also to the body, but there the resemblance ceased, for the skull was totally dissimilar in that it was extended upward into a series of horn-like

Fig. 6.11: Horned amblypod, *Uintatherium*, the culmination of its race, Upper Eocene, Wyoming (After Osborn).

prominences. These consisted of a pair upon the nose, which from their appearance may have borne dermal horns like those of rhinoceroses. The second pair were higher, with bluntly rounded ends, and were probably not sheathed with horn but covered with skin as in the giraffe. There was also a third pair, massive structures 8 to 10 inches high, which again could not have borne horny sheaths. There was a high transverse occipital crest at the hinder end of the skull connecting the posterior pair of horns and giving, together with the prominences, a unique basin-shaped character to the top of the skull. Another remarkable feature lay in the greatly developed canine teeth, which were curved sabers in some genera and spear-shaped in others, and were doubtless important weapons. Both the tusks and horn prominences were apparently better developed in the male than in the female, for their variation constitutes about the only difference seen in certain skulls. There is no indication of a proboscis, as the nasal bones, which are long and prominent in *Uintatherium*, are invariably shortened whenever that useful organ develops. The molar teeth *Uintatherium* were very conservative, for while one may trace a very marked evolution in the skull and tusks, these important organs hardly change at all. The brain also was absurdly small for so large a creature. The armament of *Dinoceras* may have served a useful purpose but one is constrained to believe that together with the relatively great size, it indicates racial senility—the extreme of over-specialization attained by a primitive stock.

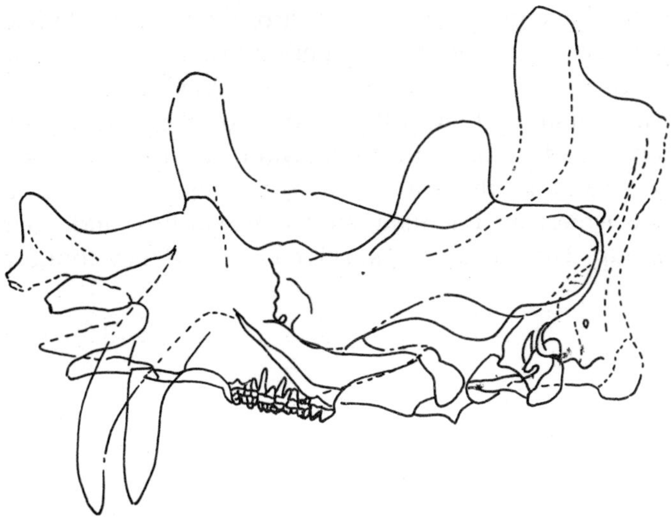

Fig. 6.12: Skulls of Uintatheres, showing that while the rest of the skull has increased markedly in size, the length of tooth row remains approximately the same. Solid line, *Uintatherium mirabile*; broken line, *Tinoceras ingens*.

Fate of the Archaic Mammals

The archaic mammals as such have long since vanished from the earth, and were it not for their remains entombed in the Eocene rocks we would be unaware that they ever existed. Theirs was a brief span compared with that of the reptilian hordes and also with that of their mammalian successors; but for a while they throve mightily until competition with creatures of a better sort became too great for them to bear. That they strove to meet this competition is evident, for certain of the later creodonts, notably *Patrifelis* and the powerful *Haragolestes* and especially *Andrewsarcus* of Mongolia, increased materially in bodily size, while the hyaenodonts actually increased the bulk of the brain, which aided in making them the sole survivors of the group after the close of the Eocene. Competition was doubtless, therefore, a prime cause which led to the extinction of these forms. We have argued racial old age in *Uintatherium,* but if that be deemed insufficient in itself, we have the noteworthy fact that where evolution of an animal runs to the development of tusks and horns, possibly favoured by sexual selection, the grinding teeth are apparently neglected and are apt to show arrested development. And bulk is fatal where correlated with inadequate feeding mechanism, and with brain power not adequate to enable the females to defend and care for the young as well as to meet new conditions of life.

Thus the fate of the archaic mammals was, first, extinction, and, secondly, transmutation of a few—a very few—into higher types. There remains yet a third possibility, and that is emigration, not of the later but of the earlier sorts, across the southern land-bridge into South America, where together with a certain admixture of other stock they may have given origin to part of the remarkable South America fauna which rose and flourished during the long period of Neogæan isolation. Others passing beyond the limits of South America, may have crossed an Antarctic land-bridge into Australia, where as marsupial they still persist. If, on the contrary, the entire marsupial order is to be considered archaic, the conclusion that they may still be surviving in these remote forms and in the American opossums is tenable. There is a further possibility that the American Edentata, sloths, armadillos, and their allies, should be included in the group. They also have survived because they found asylum in isolated South America.

Chapter—7

Proboscideans

Aside from the whales and the great dinosaurs of Mesozoic, the elephants lead the animal kingdom in size and majesty, and stand unique in nobility of physical and mental characters. Add to this the fact that their evolution since the close of the Eocene can be traced with great fullness, and their claim to our interest, second only perhaps to that of horses and mankind, is complete.

Place in Nature

The placental mammals may be grouped, principally according to the character of their foot armament, into four cohorts; the clawed or unguiculates, forms, the hoofed or ungulate, the nailed or primate, and the cetaceans or whales. Of these the Carnivora, as representatives of the unguiculates, and the whales have been discussed, and we now pass to a consideration of the hoofed creatures of which the Proboscidea form, in a sense, one of the most primitive of living orders.

Proboscidea are therefore members of the class Mammalia, cohort Ungulata, which embraces also the familiar Perissodactyla and Artiodactyla, the archaic Condylarthra and Amblypoda, and the curious South American ungulates, as well as well as the Hyracoidea, and, as an appendix to the cohort, the Sirenia or sea-cows. It is with the last two orders particularly that we are concerned, for Paleontology has shown that however, far removed from the lordly elephant the humble hyraces or conies on the one hand and the whale-like sea-cows on the other may be, they are nevertheless the nearest to the Proboscidea of all mammalian orders.

The great divergences between the ultimate representatives of the Proboscidea and Sirenia, the elephant and manatee, are merely due to environmental adaptation, the offspring of swamp-dwelling ancestors coming to a parting of the ways of which one leads to firmer ground, the other to the waters. The elephants evolution, as we shall see, largely concerns the head; the body, except for the increase in bulk and the mechanical readjustment of the limbs to bear the weight, exhibiting but little change with the flight of time.

The sea cow, on the other hand, is profoundly altered in its body contour, the hind limbs have disappeared, and in their stead there has been developed a propulsive tail. The head, however, in contrast with that of the elephant, has remained practically as it was.

Fig. 7.1: Manatee, *Manatee australis*.

ELEPHANT ANATOMY

Something of the anatomical structure of the elephant is necessary to an understanding of the evolutionary changes which its ancestors have undergone. In our discussion of this structure we will speak first of the primitive or archaic characteristics, then of the elephant's specializations.

Archaic Characters

The elephant contains within its huge body a number of primitive features, for the soft parts of an animal's anatomy, with the exception of the skeletal muscles, are less subject to mechanical stresses and are therefore rather more conservative in their rate of change than are the bones and teeth. To enumerate briefly: the stomach is simple in form, the liver has but two lobes without a gall-bladder, the lungs are simple and but slightly lobated, there are two superior venæ cavæ (the ancient number) which carry blood to the heart, the placenta by which the unborn young are nourished is primitive, and finally the brain, although huge in size, is old-fashioned in form in that the cerebrum or fore-brain does not cover the cerebellum, a notable contrast with that of man.

Skeletal Structures

The feet are five-toed, although there is a tendency to ward the reduction of the lateral digits of the hind foot especially in the African elephant. The number of hoofs may be fewer than the actual digits, as the entire structure is encased within a huge cylindrical mass of flesh and skin so that no external sign of the digits other than their terminal nail like hoofs is visible.

The carpal or wrist bones are serial, that is, placed one above another in line with the metacarpals themselves. This is the type of wrist seen in the condylarth *Phenacodus* as compared with a displaced or interlocking carpus in which the bones alternate as do the stones in a well laid wall.

On the part of all cursorial ungulates such as the horse or deer the ulna tends to reduce, especially in its lower two-thirds, the upper portion, which forms the elbow joint, being of necessity retained. In the Proboscidea, on the other hand, not only is the ulna retained entire, but it has become the dominant bone of the fore arm, the radius being much more slender and crossing over the ulna from the outer to the inner side.

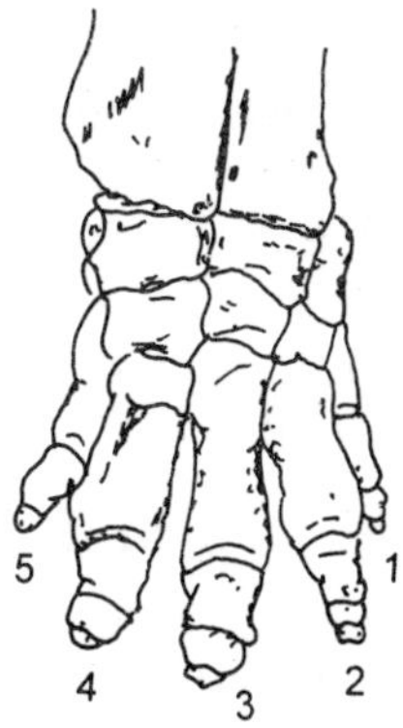

Fig. 7.2: Fore foot of elephant, *Elephas maximus*, viewed from in front, showing the seerial carpals or wrist bones.

Specializations

Size

The grandeur of the elephant is a familiar thing in these days of zoölogical gardens and circus caravans, but rarely does one see a really huge specimen. The tallest living elephants belong to the African species, as those of India are longer and lower. "Jumbo", the huge African elephant purchased by P. T. Barnum from the London Zoo, had a height of 11 feet and a weight of 6½ tons! His weight but not stature was exceeded by a huge Indian elephant in Barnum's herd some years ago. Wild African elephants are said to attain a height of 13 feet, while the largest American proboscidean was the imperial elephant of the early Pleistocene, whose stature equaled if it did not surpass that of the African form. The greatest height recorded so far is that announced in the English press for a straight-tusked elephant, *Elephas antiquus*, discovered near Chatham, England, in a Pleistocene river terrace in the ground of the Royal School of Military Engineering at Upnor. This 15-foot creature, now mounted in the British Museum, exceeds in size the great skeleton of *Elephas meridionalis* in the Paris Museum, which measures about 14 feet in height at the shoulder.

Pillar-like Limbs

While the bones of the limbs and feet are primitive in their numbers and arrangement, they are modified in one way in that they lack the angulation characteristic of ordinary ungulates and are perpendicular one above another. Thus in the horse the thigh is permanently flexed at the knee so that its long diameter is always oblique, but that of the elephant is vertical. This produces an alteration in the shape of the bone itself, for in the horse it is an elongated S with the articular faces more or less paralled with the axis; in the elephant it is I-shaped, the articulations lying at right angles with the axis of the bone. As the stress is thus transmitted through the length of the bone, the latter may be flattened without serious detriment to its strength which is impossible when the stress passes obliquely through the bone. We find this same type of limb again and again, as in the amblypod *Dinoceras* and the sauropod dinosaurs, and while these creatures have not been observed in the flesh, the inference that their limbs were graviportal as an adjustment to weight-carrying is irresistible.

Shortening of the Neck

As a rule, long limbs like those of the elephants are accompanied by a corresponding lengthening of the neck, as in the horse or more notably the giraffe, to enable the owner too reach the ground. This arrangement serves well enough where the head is comparatively small and there is no proboscis, but in the elephant the huge head could hardly be borne on a long neck, and besides, the proboscis obviates the necessity for this.

Probscis

The trunk of the elephant is in many ways its most distinctive feature and indeed gives the name to the order. It is the much elongated combined nose and upper lip, and the nostrils run the entire length, terminating at the tip, which is provided with one (Indian) or two (African) finger-like processes by which relatively minute objects may readily be picked up. The trunk is a great muscular mass with an enormous number of component muscles which by their coördinate movement shorten or lengthen the organ as a whole or curl it about any larger object to be lifted. It abundantly compensates for the extremely short.

Form of the Skull

Next to the proboscis, one of the most remarkable elephantine features is the peculiar proportions assumed by the skull, which has not only increased in actual size, but in height is all out of proportion to its length as compared with that of other animals. This is simply a mechanical adaptation to give leverage for the great weight of the trunk and its occasional burdens. The skull may be looked upon as a lever of which the occipital condyles form the fulcrum, the long axis the weight arm and the occipital plane, at right angles to the long axis, the power arm. Shortening the long axis reduces the weight arm, while the heightening of the skull, especially at the occiput, lengthens relatively and actually the power arm, thus increasing the leverage and at the same time giving greater surface for the attachment of the great elastic ligament (ligamentum nuchæ) which runs backward to the vertebral spines and bears the weight of the head, and for the huge muscles of the neck. Thus not only is the power greatly increased but it is much more effectively applied through this "bulldogging" of the skull, as it has been called. The alterations in shape thus described do not carry with them an increase in the size of the brain cavity, but the outer and inner "tables" of the cranial bones have separated from each other and the intervening space has become filled with air-cells or sinuses separated by thin apparently irregular, bony plates. To this cancellous bone with its air-cells the name diploë has been given. The sinuses are found in the skulls of other animals and of man where wide expansions of bone are developed, as in the skull of *Coryphodon* but nowhere is the diploë developed to the extent found in the elephant.

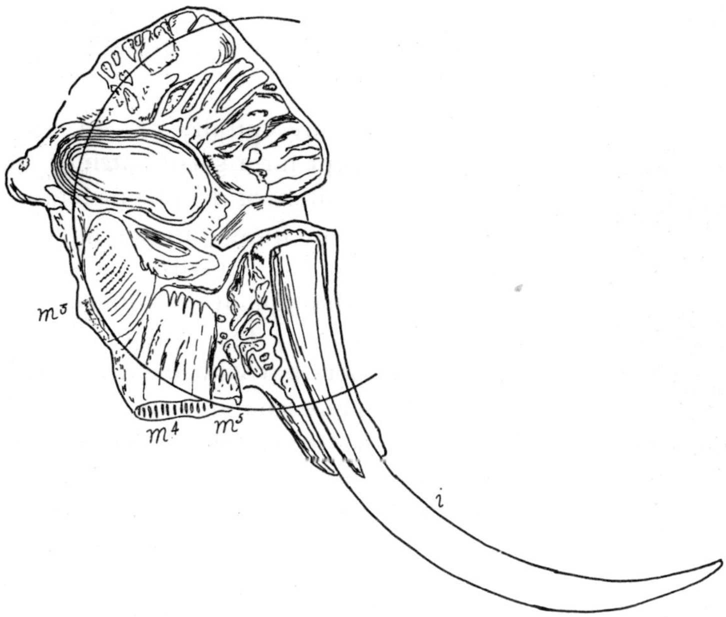

Fig 7.3: Skull of elephant, *Elephas maximus*, sectioned. *B*, brain cavity; *D*, diploë: *i*, incisor (tusk); m^{3-5}, molars 3-5.

Dentition

Another highly characteristic proboscidean feature is the dentition, remarkable in three ways: fewness of the teeth present at any one time tooth secession, and the development of the individual tooth itself.

An elephant (genus *Elephas*) never has normally more than one pair of tusks, which are the second upper incisor teeth, and one complete or two partial grinders in each half of each jaw, that is, 6 complete or 10 partially worn and partly formed teeth at any one time. Of course the total number of teeth is greater, 28 as compared with the normal 44, but instead of having a milk set, *succeeded* vertically by the permanent teeth, the teeth appear in numerical sequence. The upper milk tusks are succeeded by the permanent ones as in other mammals; the grinders, however, are formed one at a time in the rear part of the jaws and move downward and forward through the arc of a circle to replace those worn out by use. Owen tells us that the milk or deciduous tusk appears beyond the gum between the sixth and seventh month and rarely exceeds two inches in length and a third of an inch in diameter at its thickest part where it protrudes from the socket. The permanent tusk cuts the gum usually a month or two after the milk tusk is shed. The first molar tooth appears during the second week, is complete and in full use at three months, and is shed when the elephant is about two years old. The second molar has most of the plates in use at two years of age and is shed at six, the third appears at two, is at its maximum at five, and is shed at nine. These are looked upon as milk molars. The first true molar, which is the fourth grinder in succession, appears at the sixth year and is shed from twenty to twenty-five, the fifth shows its crown at twenty and is shed probably at sixty, and last molar appears at from forty to fifty years.

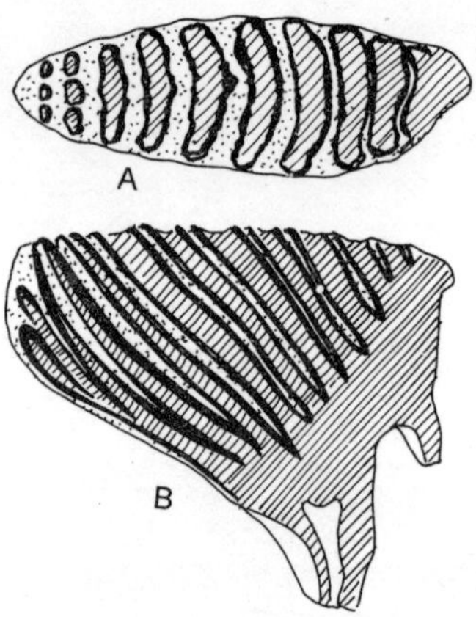

Fig. 7.4: Molar tooth of elephant, *Elephas maximus*. A, crown view; B, sectioned longitudinally. Black, enamel; oblique lines, dentine; dots, cement.

The tusks are spirally curved, elongated cones composed except for a small patch of enamel at their unworn tip, entirely of dentine or ivory of superlative fineness. They are formed from a large conical pulp at the bottom of the alveolus or socket and grow continuously throughout the animal's life. Many other creatures, such as the rodents, have continually growing incisors but with them the upper and lower teeth are antagonistic and are kept within limits by wear. With the elephant's tusks this is not true and while their use, especially that of digging, entails some wear, there is nothing to limit their monstrous growth. The tusks of the Indian or Asiatic elephant are comparatively moderate in size, the largest cited by Owen in his *Odontography* having a length of 9 feet with a basal diameter of 8 inches and a weight of 150 pounds; but, as he says, these dimension are rare in Asiatic species. The record for an African elephant, on the other hand is that of a superb pair of tusks seen in New York, of which the right one was 10 feet ¾ inches long by 23 inches in circumference and weighed 224 pounds, while the left was 10 feet 3½ inches long by 24½ inches around and weighed 239 pounds, giving a total of 463 pounds for the pair! It is said that the creature that bore these tusks was so old and the tusks so burdensome that he occasionally had to stop and rest their tips on the ground. The females of both species usually have smaller and straighter tusks than the males, although the tusks may be vestigial in the Asiatic females and in the males as well. In certain of the extinct forms, notably the imperial elephant of southern United States and Mexico, the tusks are much large those of a specimen at Yale measuring more than 13 feet on the curve, while one in the City of Mexico is said to exceed 16 feet.

The molar teeth are highly complex structures made up of a number, up to twenty-seven or more, of transverse plates or Lamellæ, each of which is composed of a flattened mass of dentine surrounded by enamel. The plates are united by a third substance known as cement so that as the crown wears away it bears a number of transverse ridges, formed of the harder enamel, separated by depressions at the bottom of which lies the softer cement or dentine as the case may be. The first teeth are relatively simple, but the number of plates and consequent ridges, two to a plate, increases with the size of the teeth until in the last molar the maximum of twenty-four the Asiatic and ten or eleven for the African elephant is reached, the latter being more primitive in its tooth structure.

Brain

As we have already seen, the brain is of an old-fashioned sort in that the fore-brain does not cover the hind-brain; on the other hand, its specializations lie in its great size, which actually twice exceeds that of man and is second only to the size of the brain of the great whales. In addition to its volume the elephant brain is noted for its convolutions but this is in part an adaptation to size, for the bulk of an object increases with the cube of its diameter, while the surface enlarges with the square, the one therefore outruns the other, and if they are to bear a definite ratio to each other the surface must be increased by infolding. Beddard speaks of the proboscidean brain as a great specialization of a low type.

The intelligence of the elephant has been exaggerated by some writers and greatly minimized by others. Elephants possess a remarkable memory of injuries, real or fancied; of misfortunes; of friends and foe; and of the time and place of the ripening of favourite fruits, as many a planter known to his cost. They also learn to perform complex labours, such as the carrying and piling of logs in the teak yards of India without direction other than the initial order; they are obedient and docile, notably those of India, and this seems the more remarkable when it is remembered that they are not domestic animals in the sense of being the product of generation of selective breeding but that practically every one is caught wild and subsequently tamed so that these qualities of which we speak are inherent in the race.

But the docility, especially of the males, is subject to rude interruption by periods of nervous excitement, apparently of a sexual nature, known as "must," during which they become very dangerous and sometime destroy the keepers in their paroxysms of rage. Ultimately all male elephants become surly and intractable; in the wild state such are known as rogues and live apart from their kind until they die. The great Asiatic elephant "Gunda" in the New York Zoölogical Park, when purchased in 1904 was so docile that children rode upon his back. In 1908 he began to show signs of surliness and the following year made a murderous attack upon his keeper. In 1912 "Gunda" was put in chains for another savage assault, and in 1913 another keeper had a narrow escape from death. Finally in 1915 the beast had become so dangerous and so unhappy that in spite of being in every other way a superb specimen he was condemned to be destroyed. His age at the time of his death was about twenty-four years. The famous "Jumbo" was sold from the London Zoölogical Gardens because he was no longer trustworthy from the same cause. He was not, however, a confirmed rogue, even when he died three and a half years later. "Jumbo" was also twenty-four years old at the time of his death.

Fig. 7.5: Embryonic elephant, *Loxodonta africana*, resting on a drinking glass.

There is a certain parallelism between the nature of human mental development and that of the elephant. One of the most potent factors in the evolution of man's mind is his ability to handle various objects and thus bring them before the eyes for examination. This is also true of the elephant, although to a less extent, and undoubtedly has aided materially in its mental development.

Elephants have been accused of timidity and cowardice, though when brought to bay rage may simulate courage, making a charging tusker a most formidable foe.

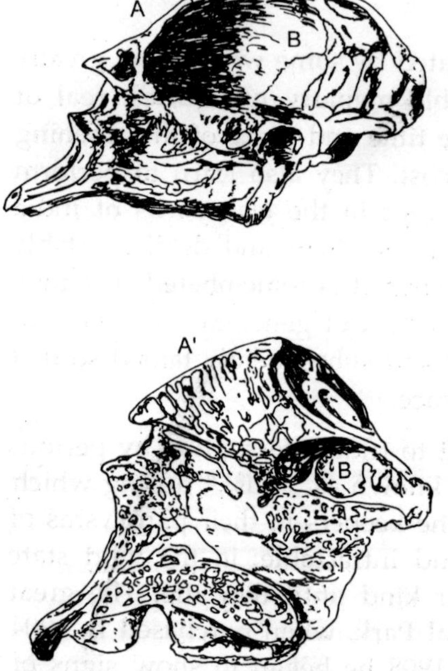

Fig.7.6: Sections of skull of (A) young and (A') adult African elephant, *Loxodonta africana*, showing development of cellular structure or diploë with age. B, brain cavity.

Senses

In common with most forest and jungle dwellers with whom opportunity for extended vision is rare, elephants are relatively dull of sight, though keen of scent and hearing, in fact, marvellously so, for Schillings, the German explorer, tells us that they either have an acuteness of some known sense far beyond our comprehension or some other sense unknown to us. The latter, however, is hardly possible and since the sentinels of the herd stand with uplifted trunk testing the breeze, it is probably in the sense of smell that elephants are thus gifted.

EVIDENCES FOR EVOLUTION

Ontogeny

But little is known of the earlier stages of elephant ontogeny owing to the great scarcity of embryonic material. The smallest and most immature embryo of which a description has been thus far published was pictured in *L'Illustration* for December 20, 1912. In this picture the creature, whose length was about 17 centimeters ($6\frac{5}{8}$ inches), is seen astride an ordinary drinking glass (tumbler), but even at this early stage is essentially elephantine, proboscis and all. About the only thing noticeable in the picture wherein the specimen departs from the normal elephant is the marked angulation of the limbs and the relatively greater length of the foot below the heel. The embryo, which is that of an African elephant from the Congo, is also long-jawed, although this is not evident from the picture.

Aside from the gradual increase in tooth complexity with age, perhaps the most notable ontogenetic change is the heightening of the skull with the development of the diploë, for the cranium of a new-born elephant is like that of other mammals, a comparatively thin-walled brain-case, the cavity of which increases but little in size with the growth of the skull as a whole, as the figure shows.

PHYLOGENY

Size

The phylogenetic changes, on the other hand, are amply recorded by the remarkably extensive series of fossil Proboscidea which have come to light. If *Maritherium* is to be considered a proboscidean in the direct line of descent, its estimated height of three feet may be taken as the one extreme in the series, that of *Elephas antiquus* of fourteen feet as the other, an increase about five diameters or 125 times in bulk.

Dentition

The dental formula of *Mœriconus* may be expressed thus: incisors, $\frac{\text{upper rt.3} - \text{1ft. 3}}{\text{lower rt. 2} - \text{1ft. 2}}$; canines, $\frac{1-1}{0-0}$; premolars, $\frac{3-3}{3-3}$; molars, $\frac{3-3}{3-3} = \frac{20}{16}$ = 36 teeth, a very slight reduction from normal 44.

In *Phiomia* the formula is: i, $\frac{1\text{-}1}{1\text{-}1}$; e, $\frac{0-0}{0-0}$; p, $\frac{3-3}{3-3}$; m, $\frac{3-3}{3-3} = \frac{14}{14}$ = 28. In *Mostodon americanus;* deciduous teeth; i, $\frac{1-1}{1\ 1}$; c, $\frac{0-0}{0-0}$; m, $\frac{3-3}{3-3} = \frac{8}{8}$ =16; permanent teeth: i $\frac{1-1}{0(1)-0(1)}$; c, $\frac{0-0}{0-0}$; p, $\frac{1-1}{1-1}$; m $\frac{3-3}{3-3} = \frac{10}{8-10}$ = 18–20. *Elephas;* deciduous and permanent; i, $\frac{2-2}{0-0}$; m, $\frac{6-6}{6-6} = \frac{16}{12}$ =28. Thus it will be seen that there is a gradual diminution in the number of the teeth during the progress of evolution; especially is this true with reference to the number present; in the jaws at any one time.

In the earliest proboscidean (*Mœritherium*) the molar teeth are small and short-crowned, with two simple transverse crests and small hinder lobes separated by open valleys. As time goes on the number of cross crests becomes greater, although in the mastodons there are never more than five or six. The mastodons have, moreover, little or no cement in the intervening valleys, although the latter may be more or less interrupted by additional cusps. In some species the worn crests are comparatively simple, in others there is a more or less complex "trefoil" pattern of the enamel produced by wear.

The transitional elephants of the genus *Stegodon* have more complicated teeth, the crests increasing in number up to ten and becoming narrower; there is also a tendency toward the filling of the valleys with cement. In *Elephas* the deep-crowned complex grinding teeth suitable for harsh herbage are perfected, reaching great intricacy in the woolly mammoth, *Mammonteus primigenius,* in which the number of crests may be twenty-five or more up to thirty. In the African elephant, *Loxodonta,* the teeth are less complex in that not only are the crests fewer, ten or twelve, but each becomes lozenge-shaped upon wear rather than having the from of a greatly compressed ellipse with parallel sides.

Tusks

The earliest form, *Mœritherium,* has three incisor teeth above on each side, the second pair of which are larger than the others and point sharply downward; the second lower incisors are in the form of procumbent tusks almost horizontal in their position. *Phiomia* the next stage, has a single pair of tusks above, with a broad enamel band, and a pair of speculate ones below at the end of the elongating lower jaw. None of the tusks are continuously growing as in later forms. From *Phiomia* on, the tusks are borne in both jaws and grow continuously throughout, the upper pair, which are curved downward, possessing an enamel band on their outer face. These are the four-tuskers or "tetrabelodonts." Subsequently most of the Proboscidea lose the lower tusks although vestiges, one or two, may be present in the male of the American mastodon. With the loss of the lower tusks the upper ones turn upward and finally lose their enamel, as in the form just mentioned and in the true elephants.

Lower Jaw

The lower jaw also undergoes a remarkable evolutionary change, elongating at the symphysis with the development of the lower tusks until in the Trilophodonts a maximum is reached. In other phyla, with the loss of these tusks, it shortens until only a spout-like vestige of the old elongation remains. In the aberrant form *Dinotherium* the lower tusks are retained, but the jaw bends downward sharply at the symphysis so that the short, pointed tusks lie at right angles to the jaw. The upper tusks are apparently lacking. Tusks seen to have had for their stimulating function that of digging—first a spade-like use of the spatulate lower tusks, the upper ones having possibly a pickaxe-like function for loosening the earth. Later when the upper ones assume the entire digging function, as we have seen, they turn upward instead of down. The African elephant to this day is a most industrious digger and one tusk, as a consequence, is almost always the shorter of the two. The use to which *Dinotherium* put its lower tusks is conjectural; there is reason to believe, however, that it may have been partially aquatic and the simplicity of its teeth points to a very succulent sort of food possibly of aquatic or swamp vegetation; if so the tusks may have been used for detaching it.

Proboscis

The presence of a proboscis is always indicated by the shortening of the adjoining bones, and the backward retreat of the nasal bones together with a strenthening for muscle attachment. There is therefore no reason to believe that *Mœritherium* bore a proboscis of any sort although it may have possessed a prehensile upper lip; but even this cannot be proved. In *Phiomia,* on the other hand, the nasals have receded and the rear of the skull has begun to heighten indicating that a proboscis had been developed probably merely for the purpose of reaching beyond the lower tusks. Thus the

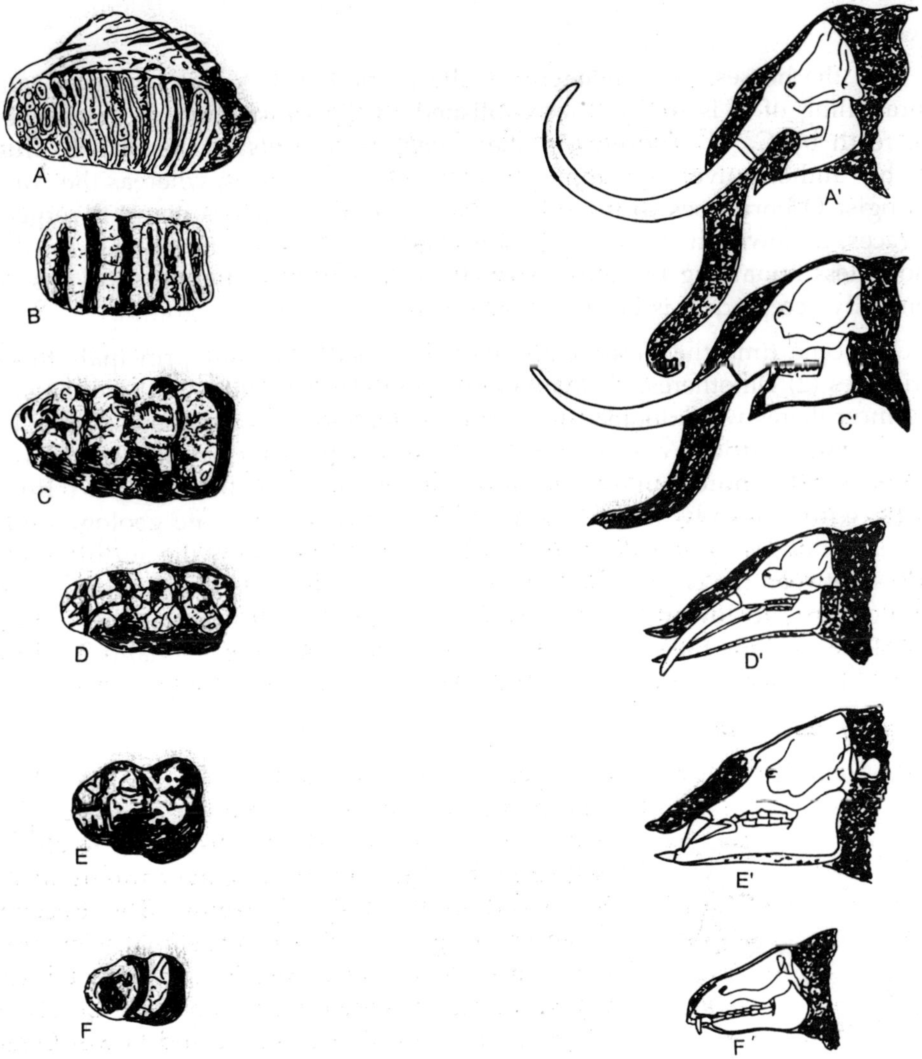

Fig. 7.7: Evolution of head and molar teeth of mastodons and elephants A, A′, *Elephas,* pleistocene; B, *Stegodon,* Phocene: C, C′, *Mastodon,* Pleistocene; D, D′, *Trilophodon,* Miocene; E, E′, *Phiomia,* Oligocene; F, F′, *Maritherium,* Eocene.

development of the latter seems to have been the prime cause of the growth of the trunk, which developed *pari passu* with the elongating lower jaw. Although the jaw was long, however, the proboscis was distinctly limited in its movement, for while it could be raised and swayed from side to side it could not be bent downward unless to one side of the jaw. The shortening of the jaw, or, as in *Dinotherium*, its downward curvature, left the proboscis as the wonderful pendent organ which the living elephant possesses.

ANCESTRY

Phylogeny

As with the horses, the phylogeny of the Proboscidea is quite complex, and the final unravelling of it is not yet accomplished to the satisfaction of paleontologists. That set forth by C. W. Andrews in the *Guide to Elephants* in the British Museum, while highly authoritative, is a simple story of successive stages, whereas the American paleontologist Osborn goes to the other extreme and recognizes many divergent and parallel races, no fewer than "twenty-eight generic and subgeneric phyla..... with other phyla doubtless remaining to be discovered." This extreme differentiation has not yet been generally accepted. It is best, perhaps to traverse a middle ground.

By Oligocene time the Proboscidea had branched into four principal stocks, the (1) Mœritheres (2) Dinotheres, (3) Mastodonts, and (4) Elephants. Of these the swamp-dwelling mœritheres were doomed to speedy extinction. The dinotheres, while longer-lived, were few, a curious aberrant side line which died out in the Pleistocene. The mastodons, vastly more numerous and widespread, are differentiated into true mastodons, with relatively simple teeth, which persisted very late geologic time and the bunomastodons whose teeth were rendered more complex by the addition of cones and trefoils between the crests. They include long medium and short jawed forms, the last having lost the lower tusks and were in general browsers. The last group (Elephantoidea) include the stegodonts or transitional browsing elephants and the true elephants which possess upper tusks only and very complex grazing teeth.

Early Tertiary Ancestors

Our knowledge of the earlier stages is due largely to the work of C. W. Andrews, who had access to the vast amount of material in the British Museum of Natural History as well as that collected by the Egyptian Government and now preserved in the Cairo Museum. The Eocene and Oligocene stages, which are entirely African, are, first, the genus *Mœritherium* (Mœris, an ancient lake near which the remains were found, and $\phi\eta\rho$, beast), which comes from rocks of late Eocene and Lower Oligocene age in what is known as the Fayûm district of the Libyan desert, some sixty miles southwest of Cairo, Egypt. The form is imperfectly known except for the

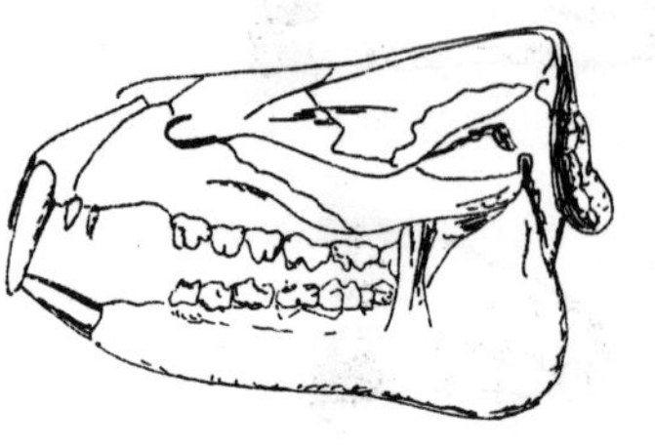

Fig. 7.8: Skull of *Mœritherium lyonsi*, Eocene, Africa (Fayûm). One-tenth natural size.

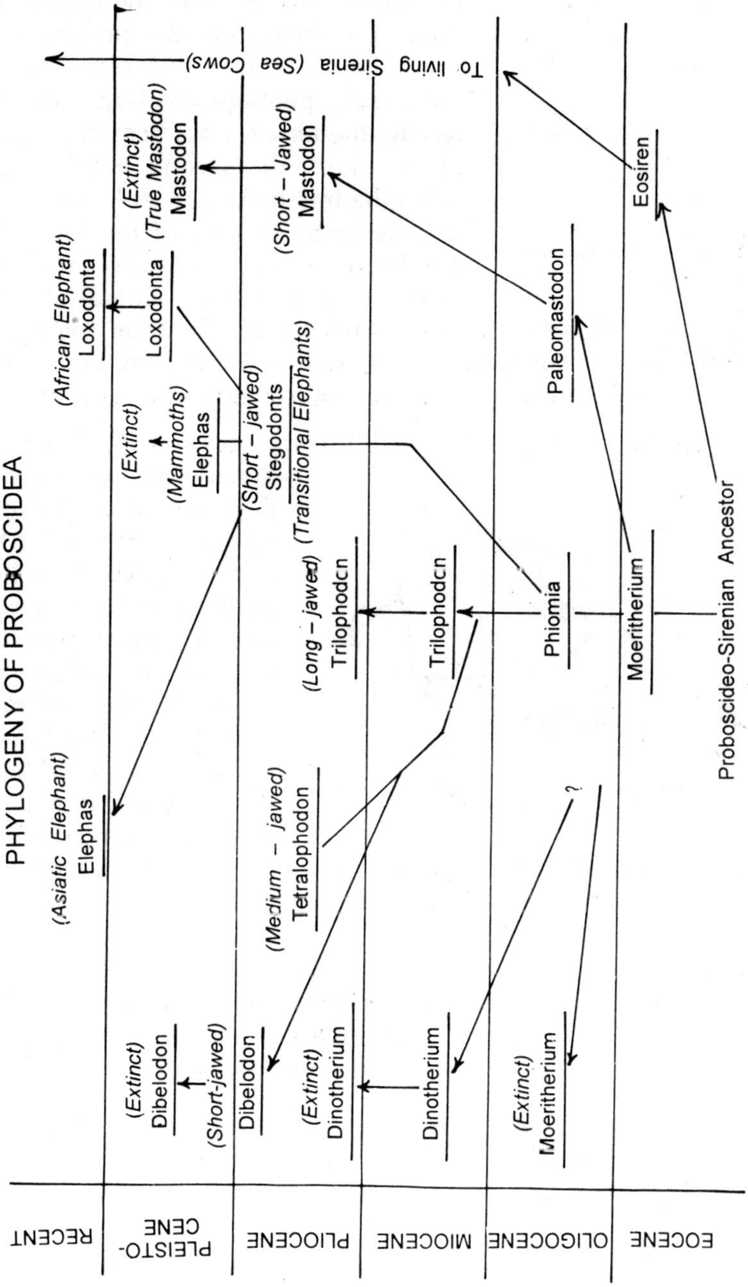

Phylogeny of Proboscidea

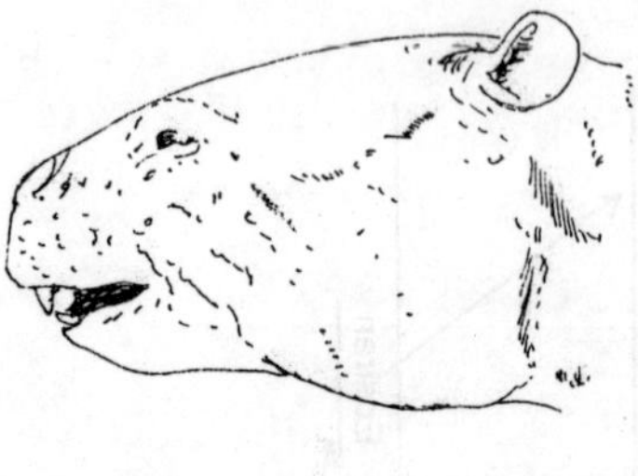

Fig. 7.9: Head of *Mœritherium*, restored.

skull, which is unlike that of any other proboscidean in that the face is short, the middle portion long, and there is no indication of a proboscis. However, it does show the beginnings of proboscidean evolution, since the nasal openings are large and are beginning to recede, the air-cells are beginning to form in the back of the skull, the second pair of upper incisors are enlarging into tusks, the molars are transversely ridged, and the anterior part of the lower jaw is elongating and becoming spout-like. So much of the skeleton as is known indicates an animal about three feet high. This creature was probably a swamp-dweller, living upon the scullion semi-aquatic herbage of the time. It is unrecorded outside of the Fayûm, but seems to have existed into the Lower Oligocene so as to become a contemporary with the next genus.

The succeeding genus in the evolutionary series is *Phiomia* which was likewise discovered in the Lower Oligocene deposits near Lake Moeris in the Egyptian Fayûm but has more recently been reported from the Gaj horizon of northern India in the Siwalik hills. It is an undoubted proboscidean of larger size than its predecessor and with limbs much like those of modern types. The skull has increased materially in height, with a considerable development of diploë, and the small nasals with their openings have receded so that they lie just in front of the orbit much as in the modern tapirs. This would imply the development of a short extensile proboscis. The upper and lower canine and incisor teeth have entirely disappeared with the exception of the second pair of incisors, which have become well developed tusks. Those of the upper jaw are large, downward curved, and have a band of enamel on their outer face. The lower jaw has elongated considerably, especially at the symphysis, and the tusks point directly forward as in *Mœritherium*. The premolar teeth have two and the molars three transverse crests composed of distinct tubercles while the hindermost tooth is tending to develop yet another crest. The neck is fairly long, although the posterior cervical vertebræ tend to shorten. The genus *Palœomastodon* is not well known,

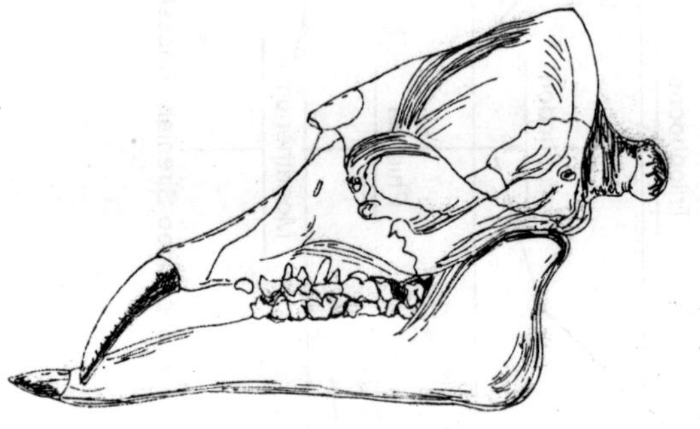

Fig. 7.10: Skull of *Phiomia*, Oligocene, Africa (Fayûm). One-twelfh natural size.

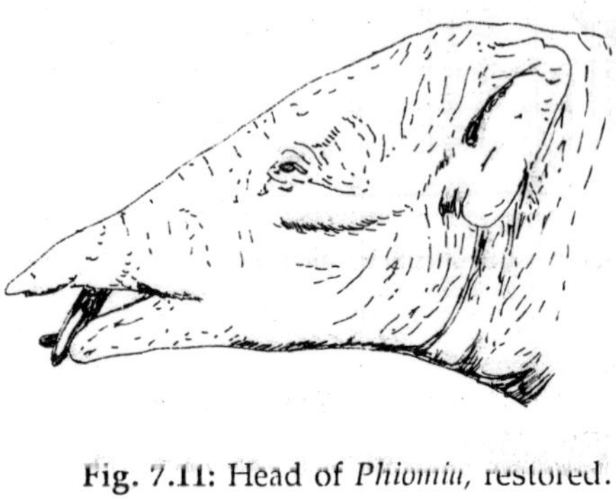

Fig. 7.11: Head of *Phiomia*, restored.

certain species being removed to the new genus *Phiomia*. The former is looked upon as the direct ancestor of the true mastodons (*i.e.*, *Mastodon americanus*) from the simple character of its teeth, although the connecting series is lacking probably because of forest-dwelling habits which do not make for ease of fossilization or subsequent discovery. *Phiomia*, on the other hand, although contemporary, has more complex molars and is believed to be the forerunner of the shovel tuskers.

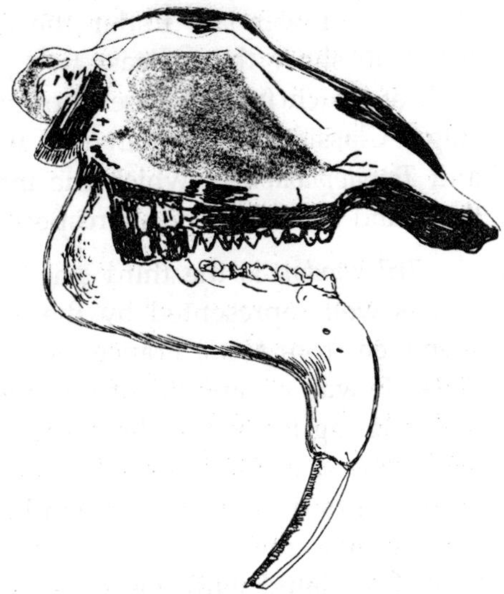

Fig. 7.12: Skull of *Dinotherium giganteum*, Lower Pliocene, Germany. One-fifteenth natural size.

Dinotherium, an extinct proboscidean whose remains have been found in the Miocene and Pliocene of Europe and India, differs remarkably from the contemporary mastodons, mainly in its dentition, in that the molar teeth, which are more numerous than in proboscideans in general and have the normal vertical succession, have but two transverse crests and a small hinder lobe. They are therefore the simplest of proboscidean molars. The upper tusks appear to be lacking, and, as we have seen, the lower jaw with its contained tusks bends abruptly downward at the symphysis. There is evidence for the presence of a well-developed trunk and the remainder of the skeleton is typical proboscidean. The occipital region of the skull slants sharply forward and contains no diploë, otherwise the skull and the body and limbs have much the proportions of the American mastodon. A gigantic skeleton from the Roumanian Pliocene, *Dinotherium gigantissimum*, is very long limbed and exceeds the largest mastodon in stature. This species could hardly have been either semi-aquatic or swamp-dwelling, since both of these habitats imply short legs like the hippopotamus. We can, in fact, conjecture little of the origin or of the habits of the dinotheres, except that their food must have been of a very succulent sort. The line must have diverged from the main proboscidean stem very early, as even *Phiomia* is too far advanced to have given rise to it. It represents an aberrant side line of fairly long duration.

Fig. 7.13: Head of *Dinotherium*.

Later Tertiary Mastodons

There is considerable confusion as to the precise relationship of the various species of later Tertiary mastodons and their exact phylogeny is not yet clearly understood, so the classification here given is tentative and subject to future revision.

It seems most natural to group together all of the four-tusked mastodons with the elongated lower jaw under the name *Tetrabeodon,* but a study of their molar teeth seems to show that a least two parallel evolutionary lines would be thus included, both of which from the trend of proboscidean evolution passed through a four-tusked stage. Classifying them in this way we recognize two principal genera, *Trilophodon* and *Tetralophodon* in which the intermediate molars (milk molar 4, molars 1, 2) have three and four cross crests respectively.

Trilophodon is the third stage in proboscidean evolution, if we omit *Dinottherium,* and is well represented by the Miocene *Trilophodon angustidens* of which a splendid specimen from Gers, France, is preserved in the museum of the Jardin des Plantes at Paris. It was an animal of considerable size, nearly as large as the Indian elephant, but differing from it in the enormously long lower jaw, which with its contained tusks had reached a mechanical limit of efficiency. The downward curved, enamel-banded upper tusks do not reach much beyond the limit of those of the lower jaw. The adult molars have attained such a size that but two can be contained in a jaw at any one time. Correlated with the great length of the jaws is a marked increase in the diploë of the skull. *Trilophodon* was a great migrant, for not only do we find its remains in Europe and Africa but it was the first proboscidean to reach North America and must have come by way of Asia early in Miocene time. Thereafter the Proboscidea formed an important element in the fauna of North America until the extinction of the American mastodon in post-Glacial time. *Trilophodon* (*Serridentinus*) *productus* is a well known Texas species from the Pliocene, while one American form, *Trilophodon lulli,*

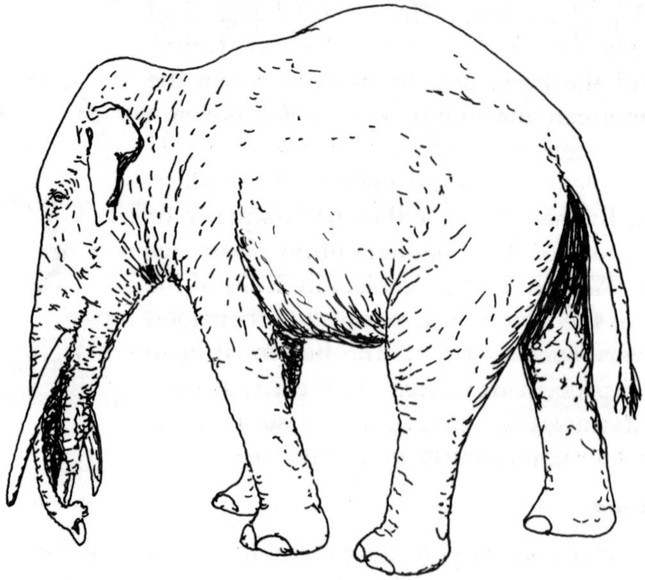

Fig. 7.14: Restoration of *Trilophodon.*

from the Nebraska Pliocene, possesses a mandible at least 6 feet 7 inches in length, and the entire animal must have been ponderous. The type specimen, that of a very old animal, has but one badly worn molar left in each jaw. These are the Longirostrines of Osborn's classification.

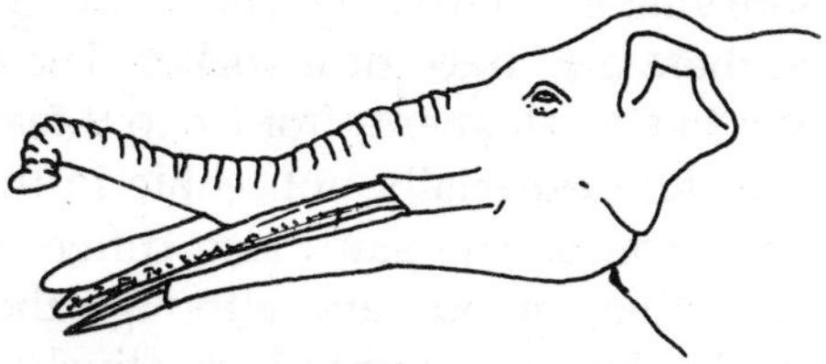

Fig. 7.15: Head of *Trilophodon lulli.* The lower jaw, one of the longest recorded in any proboscidean, measured 6 feet 7 inches as restored. About 1/40 natural size.

In the genus *Tetralophodon* the jaws are not so elongated or so effective as digging organs, but this is compensated for the greater complexity of the grinding teeth. Tetralophodonts are first known from the Miocene of Italy, but they reached India and finally North America, where they flourished and may have given rise to the peculiar South American mastodons known as *Dibelodon,* etc.

In *Dibelodon* the molar teeth are similar and, because the intervening valleys are blocked by additional cusps, form, when worn, a rather intricate enamel pattern. It differs from the tetralophodont group, however, in the loss of its lower tusks and the consequent shortening of the jaw. The enamel band of the upper tusks also tends to disappear and in its final stage we have a form not unlike *Mastodon* itself except for the greater complication of its grinders. *Dibelodon* is found widespread in the Pliocene of North America and, as far as we know, was almost the only proboscidean to reach South America, where it spread, one species along the Andean highland, another in the lower country of the east. They persisted into the Pleistocene in the southern hemisphere but in the north were replaced by the true mastodon. There are also other curious American types known as Rhynchorostrines (beak-jawed) and Brevirostrines (short-jawed), which Osborn considers separate phyla.

Fig. 7.16: Skull of *Dibelodon andium,* Pleistocene, South America.

The true mastodon, *Mastodon,* is the best known of American proboscideans. The molars lack the intervening cusps of *Tetralophodon,* so that the tooth is simpler in its appearance. The lower law is shortened in common with that of all later proboscideans, but as we have seen, vestigial and apparently functionless tusks may be present in some lower jaws, presumably those of males. The huge specimen, the so-called Warren mastodon, in the American Museum of Natural History, has one such tusk, while the

Otisville specimen, splendid young male mounted in the Yale Museum, has none, nor is there any trace of a socket. The mastodon attained a height about equal to that of the Indian elephant, from 7 to 9 feet, but was much stockier and more robust in build, a feature especially noticeable in the breadth of the pelvis and the massiveness of the limb bones. The skull also differs from that of the true elephants in its lower, more primitive contour, and although there is a large development of air-cells in the cranial walls, the brain cavity is relatively larger. The upper tusks are comparable to those of the elephants in the absence of enamel. Their length may exceed 9 feet. There are but two fully formed molars in the jaws at any one time.

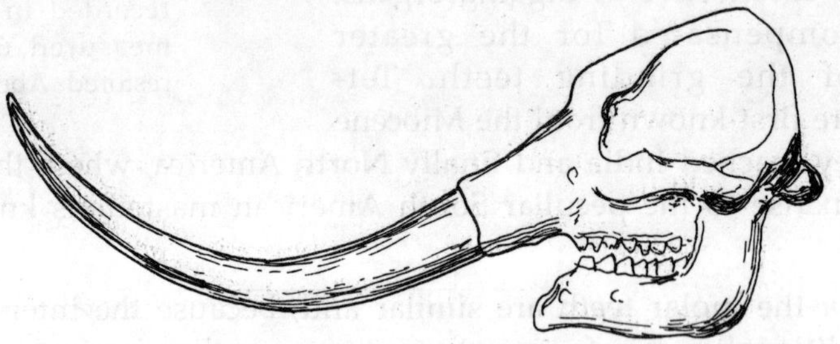

Fig. 7.17: Skull of *Mastodon americanus*, pleistocene, North America.

The true mastodons were Pliocene and Pleistocene in range, outliving the elephants in North America. In geographical distribution they ranged from Europe across Asia to Alaska and thence southward throughout the United States. They seem to have been more exclusively forest-dwelling forms than the elephants which were their contemporaries. Their remains have been found chiefly as a result of drainage excavations in the swamps and boggy lands where they were doubtless mired and thus preserved from decay. This is especially true in New York, Indiana, Ohio, Illinois, Michigan, and Iowa, where it is said that almost every bog contains a mastodon. The food consisted of twigs of hemlock, spruce and other evergreen trees, possibly other herbaceous vegetation as well, and one specimen found in Ulster Country, New York, had preserved with the bones a quantity of long, dense, shaggy hair of a dark golden brown colour.

Fig. 7.18: Restoration of the American mastodon.

True Elephants

In order to trace the evolution of the true elephants one must go back to the Upper Miocene of southern India, to the form known as *Stegomastodon laidens.* This creature gave rise to a species variously known as *Mastodon elephantoides (i.e.* elephant-like) or *Stegodon clifti,* for its transitional character is such that authorities differed as to whether it is a mastodon or an elephant.

In *Stegodon* the molar teeth, have more numerous ridges than in the mastodons and the name is given because of the roof-like character of these ridges, the summits of which are subdivided into five or six small, rounded prominences. There is a thin layer of cement over the enamel in an unworn tooth, but no great accumulation of it in the valleys, in contrast with the elephants. These teeth show how slight the transition is, however, add merely a filling of cement to bind the crests together and the elephant tooth is formed. True *Stegodon* remains have been found only in southern and southeastern Asia, which suggests that that region may have been the original home of the true elephants.

The elephants have been sufficiently defined in the anatomical section of this chapter. Aside from the living forms, many species are found in the European Pliocene and Pleistocene and two in North America, while another, the woolly mammoth, *Mammonteus primigenius,* is common to both and to northern Asia as well. The European species were *Elephas antiquus,* the straight-tusked elephant, and *Elephas meridionalis,* the former being the more primitive and showing the closest affinity with the living African species *Loxodonta.* Both these and *E. primigenius* were contemporaries of early man in Europe during the Glacial period. The American species are, first, *E. imperator,* the larger, so-called imperial or southern mammoth, as its remains are found in Mexico,

Fig. 7.19: Restoration of the imperial elephant, *Elephas imperator,* Lower Pleistocene, North America.

whence it ranged into Texas, California, and as far north as Nebraska. A single molar tooth described from French Guiana seems to pertain to this elephant; if so, it is the only other species of proboscidean, aside from the genus *Dibelodon*, recorded from the southern continent. The molars of the imperial elephant are distinctive in that the crests are relatively few and the surrounding cement very thick.

The second American species is *Elephas jeffersoni*, the Jefferson elephant, a successor of *imperator*, distinguished by its lesser stature and more numerous crests to the teeth. Each of the American species seems to have been characterized by the extreme spiral form of the tusks which in old age actually crossed at the tip so that their primal function of digging was utterly lost, nor were they efficient weapons of offence or defence. They have been cited as instances of evolutionary momentum, if such a thing exists, and certainly, so far as we can see, were detrimental to their owner rather than otherwise. *E. jeffersoni* is wide-spread throughout the United States up to the limits of the range of *M. primigenius*, which replaced it in the north. The distinction between the two species, however, is not always clear and there may have been transitional forms.

Mammonteus primigenius is the hairy or woolly mammoth, *the* mammoth of popular knowledge. It was admirably adapted to withstand the cold of the Arctic climate. This adaptation lay in the development of a coat of coarse, long, black hair with a thick coat of brown wool beneath. It was circumpolar in its range, being found in great abundance along the shores of the Arctic Ocean but extending southward into Spain and Italy in Europe and as far as North Carolina and California in the New World. The famous frozen specimens of the Siberian tundras, that of the Lena delta found in 1799, and that of Beresovka in 1901 now mounted in the Leningrad Museum, have been described. According to Matthew, the contents of the stomach show that these animals fed upon the same vegetation, grasses and sedges, birches, alders, poplars, etc., that prevails today in the far north. They must have been very numerous, as

Fig. 7.20: Restoration of the woolly mammoth, *Mammonteus primigenius*, Pleistocene, Eurasia and North America.

their tusks constituted one-half of the commercial ivory annually available and represent thus far a herd of no fewer than 40,000 individuals—not of course those living at any one time, but the accumulation of centuries.

That the mammoth was a familiar animal to prehistoric man is attested by the numerous drawings of them made by the artists of the Upper Paleolithic age on the walls of caverns. The teeth of the mammoth reached a maximum degree of complexity, doubtless an adaptation to the harsh vegetation of the north. Their tusks were of two sorts, one shorter and straighter, the other a long spiral rivalling the tusks of the Jefferson elephant. In size the mammoth, despite its name, was not great, as it rarely if ever exceeded the stature of the Indian elephant of today.

Living Elephants

There are but two well-defined species of elephants extant and these are reaching the natural limits of their racial life. They are, first, the Indian or Asiatic elephant, *Elephas maximus* which inhabits the forest regions of southwest and northwest India, Ceylon, Burma, Assam, Siam, Cochin China, Sumatra, and Borneo. During the hot season they are confined to the denser parts of the forest, generally near water, while during the rainy season they range into the open, feeding in the tender grasses.

The Indian elephant differs from the mammoths in the possession of five hind toes to four in the latter and in never having the huge spiral tusks; indeed, as we have seen, their tusks seem to be disappearing, even in the males. Their ancestry is unknown, as they could not have been derived from any discovered mammoth. The grinding teeth, however, are very similar to those of *Elephas jeffersoni.*

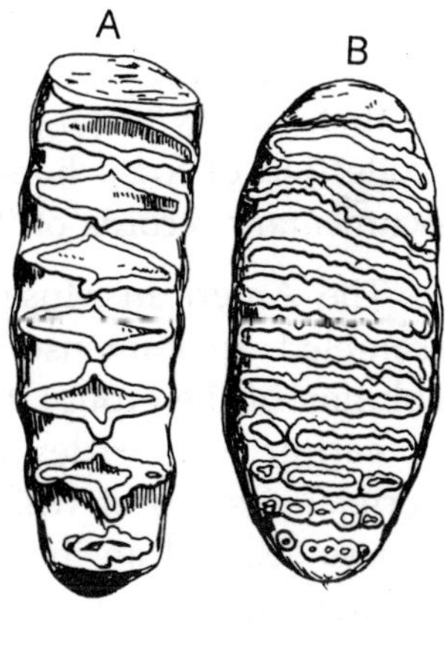

Fig. 7.21: Molar tooth patterns of (A) African elephant, *Loxodonta africana,* and (B) Asiatic elephant, *Elephas maximus.* (From British Museum Guide to Elephants.)

The African elephant, *Loxodonta africana* is distinguished by its greater size, enormous ears, lower forehead, and larger tusks, also by the character of its grinding teeth. It is confined to the wooded districts of Africa south of the Sahara and north of the Cunene and Zambesi rivers, but in many districts it is becoming extremely scarce, largely owing to the persecution of the ivory hunters, for its ivory is of a finer quality as well as being more abundant than that of the Indian species. It is probable that in the course of a few years not a single *old* individual will be left alive, and unless they are protected by law they are doomed to a speedy extinction. The African elephant is rarely tamed, although it may be fully as tractable as its relative. A number of subspecies of African elephants have been described, most of which are geographical races differing mainly in the form and proportions of their ears.

Chapter—8

Horses

The evolution of the horse has for humanity a very deep interest because of the debt of gratitude which man owes to this humble servitor and comrade and because of the fact that, largely through the unwearying efforts of Professor Marsh of Yale University, a collection of fossil horses was there assembled which was to prove one of the first documentary records of the evolution of a race.

The American Museum of Natural History under Osborn's direction has since assembled an immense amount of material which gives a very clear view of the evolution of the group; it proves, however, to be not the simple sequence of genera of the earlier authorities but includes a number of diverging lines or phyla, the interrelationships of which are not yet fully understood. Loomis of Amherst College has also secured an interesting series of forms, including several complete skeletons, and more have been added to the Yale collection since Marsh's time.

EQUINE ADAPTATION

The adaptations undergone by the horse are in their last analysis reducible to two things, the perfection of the mechanism for food-getting and of that for speed, which constitutes the principal means of defence; and the influence upon the creature of these two groups of modifications is so great that the entire body shows specialization and we cannot, as in the elephants or in the human body, point to a number of primitive characteristics with a veneer of specialization along certain narrow lines. Hence we may dismiss the consideration of archaic features in the horse and pass at once to its specializations.

Body Contour

In order to reduce air resistance the body, neck, and head are smoothly rounded, with no needless excrescences and with perfect symmetry of form, so that a running horse with head extended and ears laid back conforms to the "numerical" or stream-lines almost as perfectly as does a bird or even a fish. This same symmetry is seen in the limbs, long and slender distally, and with the powerful muscles bunched at shoulder and thigh where they blend more or less with the contour of the body, the force being transmitted to the feet by long slender tendons. This concentration of the weight high on the legs, as we have seen, quickens their rate of movement without diminishing the length of stride.

Limbs and Feet

The limbs themselves have departed widely from the ancient plantigrade gait of their primal ancestry and are unguligrade in that only the tip of the single toe, encased in its modified nail, is in contact with the ground; the wrist, the so-called fore knee, and the ankle, or hock, being raised high above the medium of support. Thus the lengthening of the distal limb segments is obtained not only by the actual elongation of the bones themselves but also by their posture. The reduction of digits is extreme, the horse being one of the few mammals which ever attained monodactyly, although the equivalent reduction in the artiodactyl or even-toed foot to the irreducible minimum of two has been reached several times.

This diminution of digits carries with it a corresponding reduction of the second bone of the lower segment of each limb, that is, the ulna of the free arm and the fibula of the lower leg. In the former case, especially, this means a restriction of motion, for it is only by the combined action of both radius and ulna that the rotary movement of the hand upon the arm is effected. Only the proximal third of the ulna, which forms the great portion of the elbow joint, is retained. All of the limb joints with the exception of the hip and shoulder are of the tongue and groove variety, their motion being thereby limited to the fore-and-aft plane. Within the limits thus imposed, however, the range of movement is very great. The shoulder girdle is reduced, as in all cursorial types, in that the clavicle or collar-bone has disappeared and there is no bony connection left between the shoulder blade and the remainder of the skeleton. This also permits great freedom of motion in the limited plane.

Lengthening of limbs implies a coördinate elongation of the neck and skull in contrast with the lack of such modification in the elephant. In general there is in a cursorial form a recognizable "speed index," as indicated by the ratio of length to diameter in the limb bones; and not only does this ratio hold for each of the several bones concerned in locomotion, but it may also be seen in the skull, vetebræ, ribs, and other skeletal elements as well. The hoof which terminates the single remaining digit is a marvel of perfection; strong, of the right rate of growth to offset a normal wear, and with the shock-absoring cushion or frog to guard the system from the great concussion produced by the impact of the foot without the ground at high speed. The perfection of the foot and limb to withstand such a shock is illustrated by the jumper "Heatherbloom," a horse which held a record of 8 feet 2 inches, in which the entire weight of the animal, coming from such a height, was repeatedly received on what is equivalent to the middle finger of the two hands. That the limit of such evolution has practically been reached, however, is evident from the fact that many a fine horse has been destroyed merely because a rutted road caused the fracture of a single bone strained beyond endurance. Bone is a wonderfully efficient material and it is utilized in what is mechanically the very best possible way to produce results, but with a very close margin of safety. It is this last fact that makes further speed adaptation for so large an animal virtually impossible.

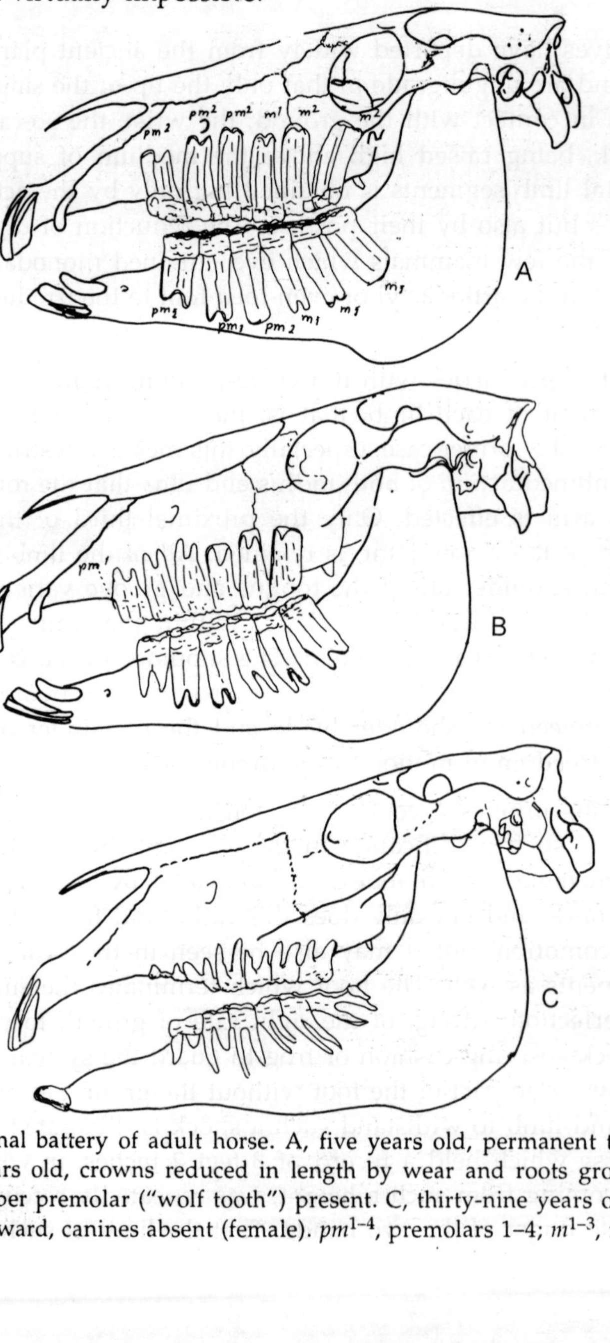

Fig. 8.1: Denal battery of adult horse. A, five years old, permanent teeth all in use. B, eight years old, crowns reduced in length by wear and roots grow longer; vestigial first upper premolar ("wolf tooth") present. C, thirty-nine years old, lower molars incline forward, canines absent (female). pm^{1-4}, premolars 1–4; m^{1-3}, molars 1–3.

Skull

The skull is characterized by a large and well-developed brain-case, orbits completely surrounded by bone, and an elongated face, the purpose of which is twofold, first the raising of the eyes as far above the ground as possible while grazing in order to increase the range of vision, so essential for safety's sake, and second to allow room for the development of the deep-crowned grinding teeth. The elongation of the jaws separates without reduction in their numbers those teeth which are concerned with the prehension of food—the incisors—from those whose function is that of mastication—the premolars and molars. Incidentally the gap or diastema thus produced forms a convenient place for the bit and thus aids in the subjugation of the creature by man, but this was hardly nature's purpose.

Teeth

There is a tendency toward tooth reduction for the first premolar, the so-called "wolf tooth," which is small and simple, is rarely present and soon shed. Then, too, the tusks or canines are rarely developed in the female, the upper ones never, although the normal male always possesses both. Sex characters are so rarely distinguishable among fossil forms that the lower jaw of a Miocene horse (*Merychippus*) in the Yale collection, in which there is absolutely no trace of canines, was at once hailed by its discoverer as that of a mare.

The incisors or cropping teeth are long-crowned and are, with the single exception of *Macrauchenia,* a peculiar ungulate of the South American Pleistocene, unique in possessing a pitlike depression or "mark" in the grinding face. This mark, which is worn away with use, is one of the best criteria of its possessor's age.

The three molars and three preceding premolars of each jaw have become deep-crowned (hypsodont) grinding teeth, having the form of slightly curved prisms strengthened by three buttresses on their outer, convex face. The teeth are composed of the three materials which characterize the elephant's tooth—dentine, enamel, and cement—elaborately interwoven when seen in cross-section. As these substances differ markedly in hardness, differential wear produces a characteristic "pattern" of the more resistant enamel upon the wearing surface. For a while the teeth continue to grow,

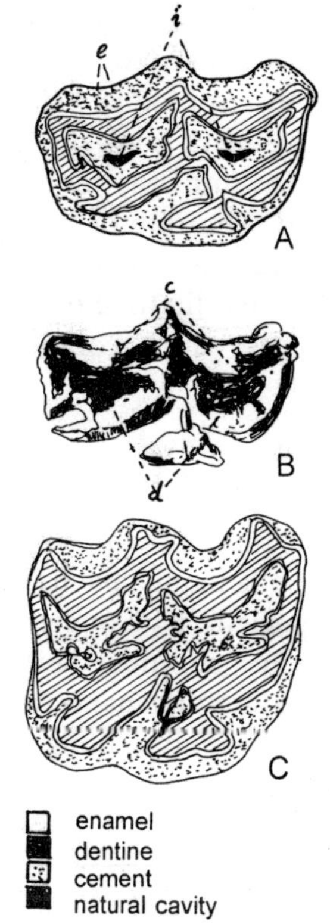

Fig. 8.2: Wearing surface of upper grinding teeth of horse, *Equus caballus*. A, worn surface of milk molar of colt about six months old: *e*, exposed enamel ridges; *i*, natural cavity in cement. B, unworn surface of milk molar of colt three months before birth: *d*, cul-de-sac to be filled later with cement; *e*, enamel. C, premolar of horse eight or nine years old. Natural size.

extending deeper and deeper into the jaw and at the same time moving slowly outward to compensate for wear. Finally, at about five to eight years of age, the dimensional limit of the jaw is practically reached, which of course makes further growth of the tooth impossible. Then the roots are formed and the tooth is completed. The outward movement, however, still continues, cancellous bony tissue filling the gradually vacated, socket until the tooth is so nearly consumed as to be of no further service, when it is shed. The rate of growth and the outward movement are in absolute accord with the normal rate of wear, and the entire tooth length is such as to last throughout the potential lifetime of the animal, about 34 years. With its final consumption malnutrition results, which, coupled with other evidences of senility, summons the horse to its final rest.

Size

Another equine characteristic is size, for aside from the elephants, rhinoceroses, and hippopotamuses, the horse compares favourably with the terrestrial animal, being equalled only by the larger bovines, the cattle, buffalo, and bison. This is of course especially true of certain domestic strains such as the Percheron horses, some of which reach a shoulder height of 19 hands or 76 inches and a weight of over 2,400 pounds. On the other hand, the Shetland ponies are reduced in size, largely due to the harsh, restricted conditions of their island home, but aided by selective breeding. The following comparative measurements are given by Chubb for two animals the skeletons of which he has most admirably mounted in the American Museum of Natural History:

	Giant Draft Horse	*Shetland Pony*
Height at shoulders	6 ft. 1 in. (18¼ hands)	2 ft. $9\frac{3}{8}$ 'in. $8\frac{1}{3}$ (hands)
Weight in life	2370 lbs.	170 lbs.
Bulk of humerus	$118\frac{3}{4}$ cu. in.	$9\frac{1}{2}$ cu. in.
Bulk of femur	188 cu. in.	$13\frac{1}{2}$ cu. in.

Brain and Mentality

The brain is not only of considerable size but is of a relatively high type compared with those of other mammals, and is richly convoluted. The intelligence of the horse is great but not equal to that of the elephant. As compared with the cattle, on the other hand, the horse is much more intelligent and is able to keep out of trouble and care for itself under trying conditions which may prove fatal to the former. The docility of the horse and its ability to learn not only from its master but also from experience are notable. On the other hand, it is emotional and its psychology is largely linked up with its normal mode of defence—flight—for the first impulse of a domestic horse upon seeing any incomprehensible thing is to run away, sometimes to its own and its owner's destruction. In the wild state this same impulse in an unarmed animal is of the greatest possible value as a means of survival.

Senses

All three of the major senses, sight, hearing, and smell, are well developed; of the three, hearing is perhaps, of the least importance to a plains-dwelling creature, just as sight is to one which is forest bred.

Evolutionary Summary

Briefly stated, the evolutionary changes which the anatomy of the horse would lead us to predict are:

Increase in size.

Lengthening of the limbs.

Reduction of ulna and fibula, with a consequent limitation of the range of movement.

Change of foot posture from semi-plantigrade to unguligrade.

Reduction and loss of digits from five to one.

Perfection of the hoof.

Perfection of the dental battery in elongation and complexity of teeth.

Premolars becoming successively molariform.

That these changes are all recorded in the paleontological record is conclusive proof of equine evolution.

PALEONTOLOGY OF THE HORSE

Place of Origin

We have spoken of the simultaneous appearance of the modernized mammals in the Old and New Worlds, and the consequent belief in their origin in some contiguous land-mass which has been called boreal Holarctica. What is true of the modernized mammals in general is true of the horses in particular, although as yet it is incapable of actual demonstration. The London Clay, however, an Eocene formation of Europe, has produced *Hyracotherium,* the Old World's most ancient known equine, while in the Wasatch rocks of western North America, of nearly equivalent age, the earliest American genus, *Eohippus,* has been found. These two genera are very much alike, but the premolar teeth of *Hyracotherium,* especially the second one of the upper jaw, are more simple than in *Eohippus,* thus stamping the Old World type as the most primitive horse-like form known.

Horses are found from time to time in Europe and Asia as one ascends the geologic column, but the sequence does not seem to be continuous as it is in North America. Hence the inference that North America was the real theater of equine evolution, while the Old World horses were merely the relics of genera which migrated thence from time to time as barriers to dispersal were temporarily removed. The earlier of these

PHYLOGENY OF THE HORSES

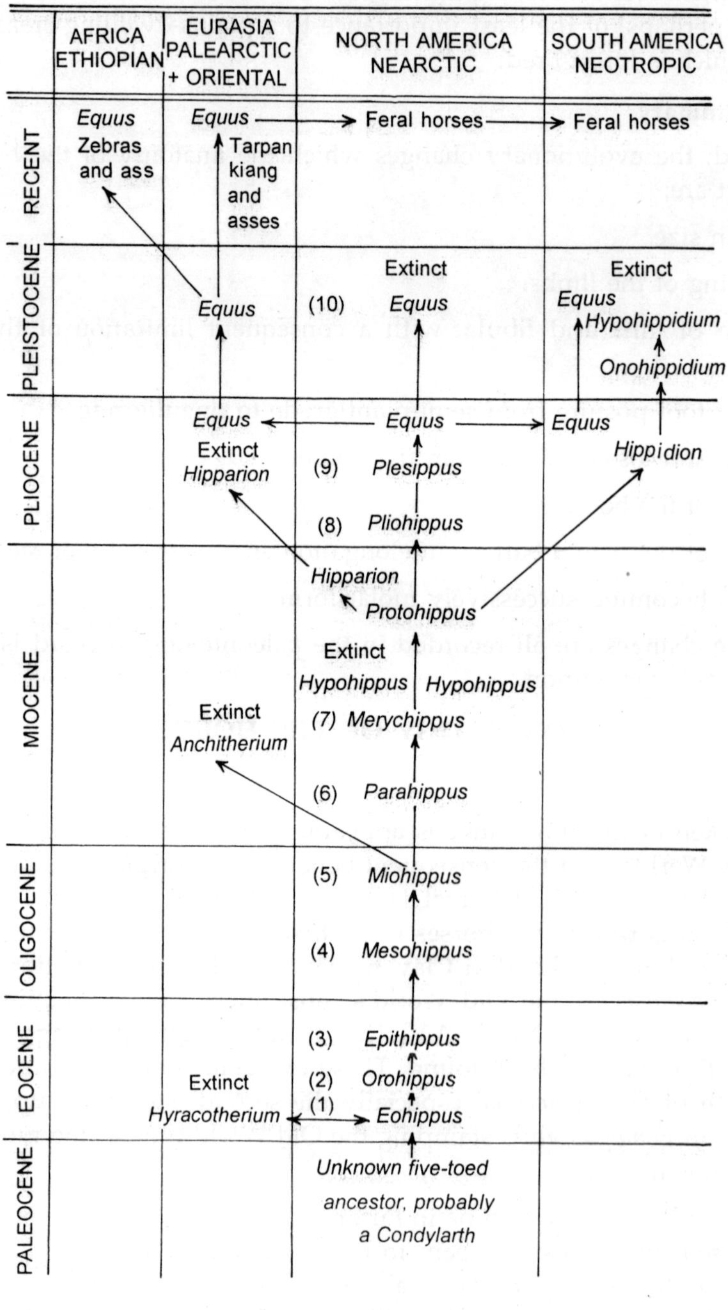

migrations, while interesting, are unimportant from the standpoint of the evolutionary continuity; had it not been, however, for the final Pliocene migration of the horses to the Asiatic and African continents within whose fastnesses they found asylum, their inexplicable extinction in the New World during the Pleistocene would have closed the book of their progress forever and we would see them only as our paleontologic vision is able to pierce the gloomy curtain of the geologic past.

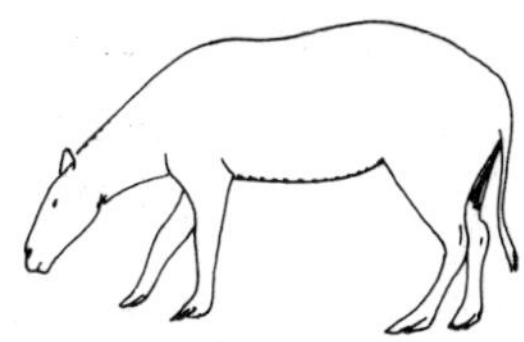

Fig. 8.3: Restoration of four-toed horse, *Eohippus*, Lower Eocene, North America.

Eocene

Several generic names have been applied to the Eocene horses, of which *Eohippus*, the dawn horse, and *Orohippus*, the mountain horse, are the best known American forms. The first comes from the Lower Eocene (Wasatch) formation, and the latter succeeds it in the Middle Eocene Bridger beds. Both are from Wyoming and New Mexico.

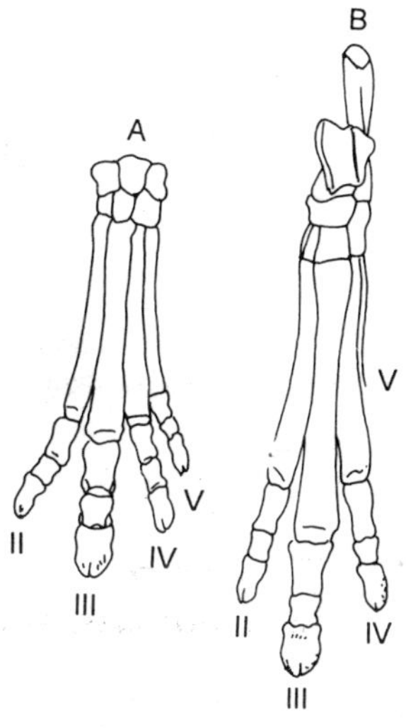

Fig. 8.4: Hand (A) and foot (B) of *Eohippus*. One-half natural size.

The Eocene was a time of warm, moist climate, during which North America was clothed with a luxuriant vegetation, forests in which grew both evergreen and deciduous trees of a distinctly modern character, and, beside the numerous streams and lakes, sedgy meadows which in turn gave rise to grassy plains. Such was the environment of the first known horses which were already somewhat advanced toward their evolutionary goal.

Eohippus, the four-toed horse, represents the first recorded stage in equine evolution. It was a small but graceful creature, averaging about 12 inches or 3 hands in height at the withers, the several species ranging in size from that of a cat to a fox terrier, with arched back, short head and neck, limbs of moderate length, and showing in the digitigrade character of the feet the beginnings of cursorial adaptation. In fact, the general proportions are much those of a dog such as the fox terrier or the whippet. The hand bore four complete toes, each terminating in a hoof-like nail, with no trace of the first digit or pollex, while the foot had but three, although vestigial remnants of the first and fifth are also seen. The advance of evolutionary progress shown by the foot over the hand is interesting, for it shows the foot to have been the main propelling organ and therefore the first to feel the influence of cursorial adaptation, and

Fig. 8.5: Upper teeth of *Eohippus*. Short-crowned, no cement, pre-molars simpler and smaller than molars. Natural size.

it also shows the reluctant relinquishment of general utility for mere propulsion on the part of the hand. Both ulna and fibula, while slender, are separate and complete. The modern tapir has feet in much the same stage of evolution as were those of *Eohippus*. The dentition is also advancing in that the molars already begin to foreshadow their future complication. The originally separate cusps are fusing into cross crests, and the fourth premolars are tending to become molariform.

In *Orohippus*, the second stage, a further advance is indicated by the loss of the splint of the fifth digit of the foot, the slight increase of size of the middle and the shortening of the outer finger of the hand, and the further perfection of the molar-like character of the third and fourth premolars. *Epihippus*, from the Upper Eocene Uinta formation, goes yet further in that the third and fourth premolars are completely molariform. The digits of the hand are still four, the outermost has diminished yet more but is still functional; the digits of the foot are three, but the middle digit of each begins to be the dominant one. There is on the part of the Eocene horses a gradual increase in size, the type skeleton of *Orohippus* mounted at Yale measuring 13½ inches in height; *Epihippus*, the third stage, was still larger, but the complete skeleton is as yet unknown. *Epihippus* may not be the direct ancestor of the known Oligocene horses to be described.

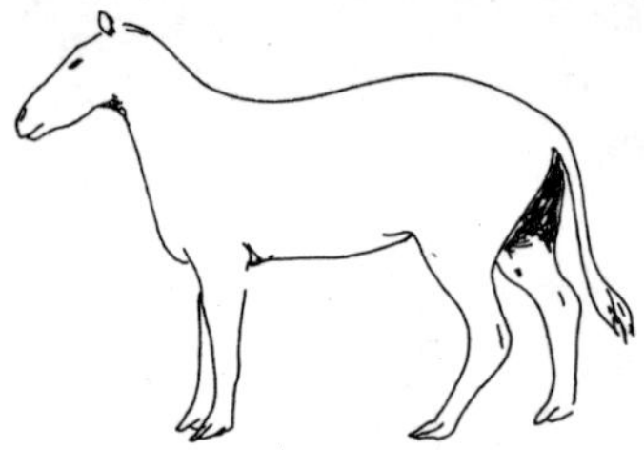

Fig. 8.6: Restoration of four-toed horse, *Orohippus*, Middle Eocene, Wyoming.

The known range of Eocene horses from Europe to New Mexico speaks for their migratory powers, always a characteristic of the equine hordes.

Oligocene

The Oligocene was a time of increased aridity due in large part to continental uplift, and while much the same conditions prevailed as in the Eocene, there was a consequent dwindling of streams and lakes which gave impetus to the development of broad meadow lands and of true prairie as well. Thus there were three conditions—woodland, meadows, and dry prairie—which seem to have given rise to several divergent lines of equine evolution, some of which terminated, being overcome in the struggle for existence, while others flourished and gave rise to the horses of the Miocene.

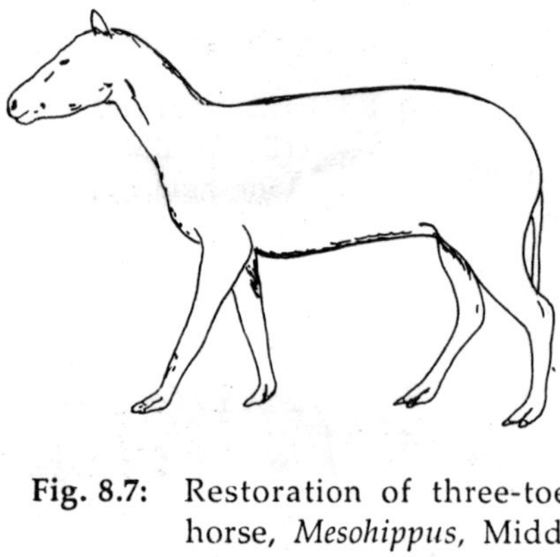

Fig. 8.7: Restoration of three-toed horse, *Mesohippus*, Middle Oligocene, North America.

But two genera of Oligocene horses are recognized, *Mesohippus* and *Miohippus*, the fourth and fifth stages, the former one Lower and Middle Oligocene, the latter confined to the Upper Oligocene. *Mesohippus,* which had attained the size of a prairie wolf, had three functional digits in both hand and foot, although a very short splint bone represented the fifth digit in the former. The middle toe in each instance was much the largest and the lateral ones in consequence bore less of the creature's weight. The shafts of both ulna and fibula are still complete but are thin and slender. *Mesohippus bairdi,* the best known form, averaged about 18 inches or 4½ hands in height and was a slender-limbed creature, very well adapted for speed. *Mesohippus intermedius* was larger, fully the size of a collie, averaging 24 inches or 6 hands in height, and was in some ways unprogressive, which, together with the conditions under which it is found, may be taken as indicative of a conservative forest-dwelling form in contrast with the progressive plains-living type. In all Oligocene horses the premolar teeth, with the exception of the small, simple, first premolar, are fully molariform. *Miohippus,* of the Upper Oligocene, is hard to distinguish from *Mesohippus,* except that the species average larger in size.

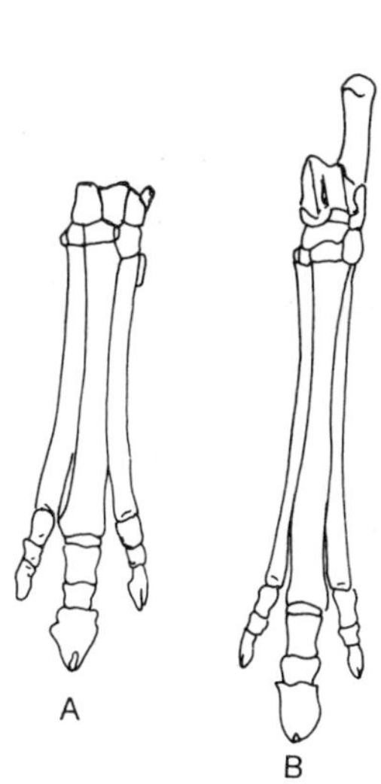

Fig. 8.8: Hand (A) and foot (B) of *Mesohippus*. Over one-fourth natural size.

Fig. 8.9: Upper teeth of *Mesohippus*. Short-crowned, no cement, second, third, and fourth premolars like molars. Natural size.

Miocene

The Mioicene was a time of great continental elevation and witnessed a wide expansion of our western prairies and a further diminution of the forestclad areas. As a consequence, many browsing animals, well fitted for survival under former conditions, could not endure the change and perished, but the grazing types, horses, camels, deer, and antelope, adapting themselves to the new conditions, throve and spread amazingly and became the dominant forms of mammalian life.

The Lower Miocene horse, *Parahippus,* represents the sixth stage, the older species being very much like *Miohippus,* while the later ones are almost indistinguishable from *Merychippus.* It is therefore fully transitional. In *Parahippus* for the first time the valleys between the crests of the teeth begin to fill up with cement formed as a deposit of "tartar" on the emerging portion of the crown. The amount of cement becomes progressively greater with the successive species. The lateral toes vary greatly; in some they are nearly as well developed as in *Miohippus,* in others much reduced.

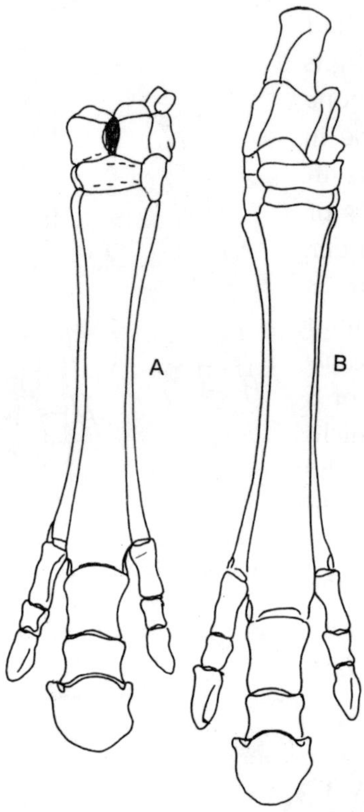

Fig. 8.10: Hand (A) and foot (B) of browsing horse, *Hypohippus equinus*. Miocene, North America. One-fourth natural size.

Hypohippus known as the browsing horse, had broad, low-crowned teeth fitted only for browsing on succulent herbage. The feet were three-toed, which was equally true of all Miocene horses, but were distinctive in their broad spreading character, with well developed lateral hoofs as though adapted, like the living caribou, to a soft yielding ground rather than hard prairie soil. In the hand, vestiges of the first and fifth digits may yet be seen as small nodules of bone at the back of the wrist. Thus in spite of its having attained the size of a pony, 40 inches or so in height, the creature was otherwise persistently primitive and did not long continue to exist. A huge form, *Hypohippus mattewi,* lately described from the Nebraska Pliocene, greatly exceeded the more typical *Hypohippus equinus* in size.

Merychippus, the seventh stage in the direct line, appears in the Middle Miocene and is of especial interest in that it marks the transition from the horse-like forms with short-crowned, uncemented teeth (hyracotheres) to the true horses whose long-crowned, fully cemented grinders are suited to the harsh vegetation of the plains. In *Merychippus* the milk teeth are short-crowned and have little or no cement and are thus reminiscent of its ancestry, while the permanent teeth are intermediate in length of crown and quite heavily cemented and are thus prophetic of the future. This is one of the most remarkable instances of the ontogenetic evidence of evolution seen among the horses.

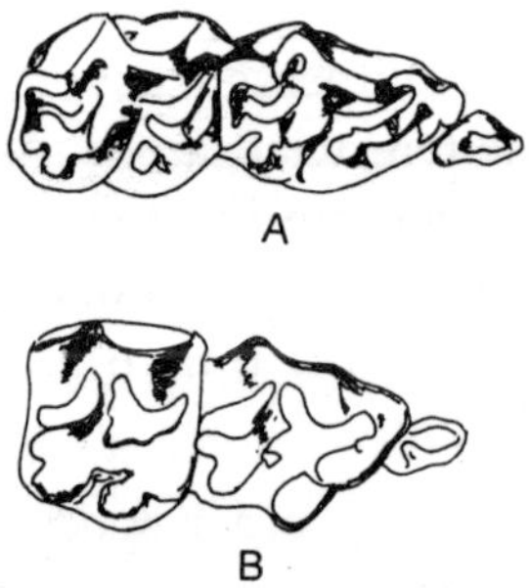

Fig. 8.11: Upper premolar teeth of *Merychippus,* upper Miocene, North America. A, uncemented milk teeth; B, cemented permanent teeth.

Merychippus is three-toed, in some instances with vestiges of the outermost digits of the hand. The lateral toes vary somewhat in the different species, though never reaching the ground, so that while structurally three-toed, the feet are functionally one-toed. The slender shaft of the ulna, which is free in the colt, is always fused with the back of the radius in the adult. The skull of this genus is the first in which the hinder border of the orbit is completed by sending downward a bony bar to join the zygomatic arch.

Protohippus and *Pliohippus* of the Upper Miocene and Pliocene are two closely related genera, in fact the distinction

Fig. 8.12: Restoration of the Miocene prairie horse, *Merychippus.*

between them is not always clear. It may suffice to say that *Protohippus* represents a from derived from *Merychippus* but differing in that the milk as well as the permanent teeth are moderately long-crowned and cemented, and in that the hand and foot still bear three toes; while in *Pliohippus,* the eighth stage, we have the first one-toed horse, although in some species the lateral toes seem to have persisted. *Pliohippus* is also characterized by having a peculiar pit or depression in front of the orbit which may have lodged a scent gland like the larmier of deer and possibly of similar function. The depressions also vary greatly in degree of development, so that they do not seem to have had an essential function. *Pliohippus* had a shoulder height of some 40 inches or 10 hands.

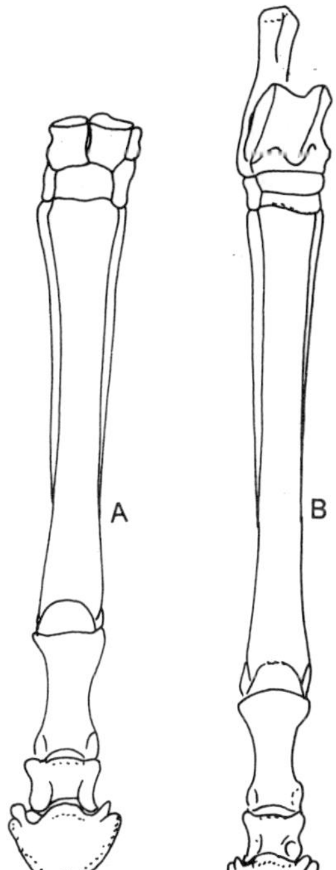

Fig. 8.13: Hand (A) and foot (B) of the first one-toed horse, *Pliohippus pernix,* Pliocene, Nebraska. One-fourth natural size.

Yet another Miocene horse was *Hipparion* closely related to the two preceding genera, from the former of which it is sometimes difficult to distinguish it. The following diagnostic characteristics are based upon a skeleton of *Hipparion whitneyi* from South Dakota, preserved in the American Museum. "This species, except for the very large head, had the graceful and slender proportions of the antelopes, but in *Protohippus* and especially in *Pliohippus* the skeleton approached more nearly the stockier proportions of the modern horses. The *Hipparion whitneyi* is regarded by Professor Osborn as fitted to live in a semi-desert country, and in contrast to the *Hypohippus,* is called the "three-toed desert horse" (Matthew). The argument for this belief is seen in the highly perfected teeth, the pattern of whose enamel is in some instances more complexly infolded than in any other horse, doubtless an adaptation to the harshest of herbage. The splendid fleetness which the skeleton implies is corroborative evidence.

Hipparion is another world migrant, as its remains are found not only in Colorado, Nebraska, and South Dakota in great abundance, but even in far-off Greece where in Lower Pliocene rocks of Pikermi near Athens they are entombed. *Hipparion whitneyi* reached a height of 40 inches or 10 hands, while *Hipparion gracilis* of Pikermi stood 44 inches at the shoulder.

Pliocene

Pliocene time was one of great unrest; conditions were becoming more and more severe, prophetic of the Glacial period, new land-bridges arose where none had existed for ages, and we find great consequent faunal interchanges recorded. It is not remarkable therefore that *Hipparion* reached the Old World just as the true elephants made their first appearance in the New.

Plesippus, the ninth stage, was formerly known from very fragmentary material, but in 1924 Messrs. Matthew and Simpson found two skeletons at Mount Blanco, Texas, that present almost every detail. The size and general proportions are about those of the Arab horse but the feet are much smaller. The teeth are an advance over those of *Pliohippus* in having a longer, less curved crown, as in *Equus*, otherwise they are more *Pliohippus*-like. There is no trace of the lateral toes, while the skull is *Equus*-like, having almost entirely lost the facial pits.

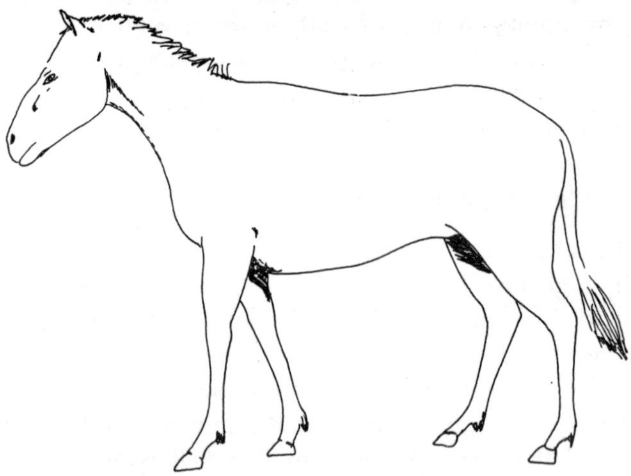

Fig. 8.14: Restoration of the first one-toed horse, *Pliohippus*.

Another notable Pliocene event was the appearance for the first time in geological history of true horses in South America, whether they went in company with the dibelodont mastodons. The South American Pliocene horse was *Hippidion*, evidently a derivative of *Protohippus* but differing in having short, stout rather than slender, one-toed feet. The teeth are like those of *Pliohippus* but the skull differs remarkably in the extremely long, slender nasal bones which, together with the great size of the head, must have given the creature a very peculiar cast of countenance. *Hippidion*, which had attained a stature of 12¼ hands, lingered into the Pleistocene, where it became *Onohippidium*, a creature but recently extinct if one may judge from the fresh-looking horny hoofs preserved in certain Patagonian caves.

The modern horse, *Equus*, the tenth and final stage, first appears in the Upper Pliocene beds of Eurasia and North America and represents the culmination of the race. The feet are one-toed, but with well-developed splints of the second and fourth digits still remaining. In some individuals these are fussed with the cannon-bone; in others they are free. The teeth are long columnar structures of intricate enamel pattern, admirably adapted to their owner's needs, and the animal has attained the maximum stature consistent with fleetness.

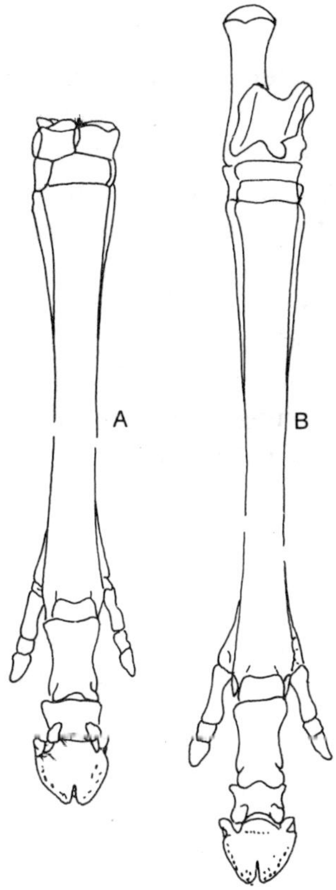

Fig. 8.15: Hand (A) and foot (B) of three-toed desert horse, *Hipparion*, Pliocene, North America and the Old World. One-fourth natural size. (After Lull).

Pleistocene

A number of extinct species of *Equus* are recorded, principally from the Pleistocene of both North and South America and the Old World. Of these the best known is Scott's horse, *Equus scotti*, from the staked plains (Llano Estacado) or Texas. This species, of which a number of perfect specimens have been found, was discovered at Rock Creek, Texas, in 1899 by an expedition from the American Museum of Natural History. Thirteen years later a party from Yale reopened the quarry and secured several more specimens, one of which is now mounted in the Yale Museum. It is of an animal about 15 hands in height, having somewhat the proportions of a western broncho, but with a very large head and with teeth greater than those of a modern day horse, although very similar in pattern. Horses of this or related species, some smaller, others larger, are extraordinarily abundant wherever the earlier Pleistocene deposits are found in North America, and they evidently survived the first glacial advance, but shortly afterward, why we cannot tell, they became extinct not only in North but in South America as well. This apparently was also true of Europe but in Asia and Africa the race found sanctuary, otherwise the horse would be included with the mastodons, ground-sloths, saber-tooth cats, and a host of other splendid creatures among the totally extinct. Glacial conditions alone seem inadequate to account for this great tragedy, for not only did the hand of death bear heavily upon the equine

Fig. 8.16: Upper tooth of *Hipparion*. *pr*, free protocone.

herds within the limits of the ice belt, but far beyond, to the uttermost confines of the western hemisphere. That no permanent change of environment occurred to render the earth unsuitable for these creatures is evident from the amazing way in which the few imported horses liberated by the Spanish Conquistadores multiplied and spread, giving rise to the great herds of wild mustangs in both North and South America. We look naturally therefore for some other cause of extinction and the one of all theories that seems most plausible is the bringing in by migrating animals of insect-transmitted disease, such as the sleeping sickness of Africa or the Surra disease which attacks domestic horses in Indian on parasitism and degeneracy. So far as we know now such an extinction cause is incapable of proof, unless it shall be found that these diseases produce a recorded change upon the bones themselves, for of course the soft anatomy of fossil horses is utterly beyond our reach for direct study.

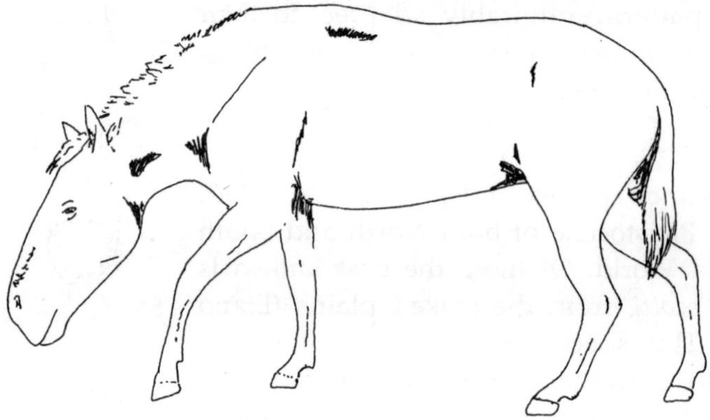

Fig. 8.17: Restoration of the one-toed Pampas horse, *Hippidion*, Pleistocene, Argentina.

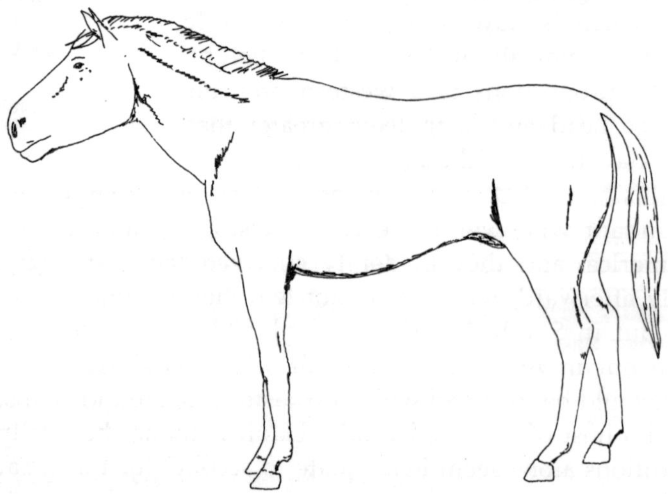

Fig. 8.18: Restoration of *Equus scotti*.

Living Horses

Several species of horse-like animals are yet alive in their wild condition in Asia and Africa, all of those of Europe and the Americas being either domesticated or feral, that is, of domestic ancestry. Of the true horses but one wild type remains the Mongolian or Przewalski horse, the tarpan of the Gobi Desert of central Asia and the neighbouring regions. It is a small animal, standing but 12 hands, of a yellow dun or "buckskin" colour, with black mane, tail, and legs, and a white muzzle. There is no forelock, the mane is short and erect, and there is a decided beard beneath the relatively large head.

At least three other primitive types of true horses are living under the fostering care of mankind, and these or an admixture of them constitute our various domestic breeds. Of them the first is the Celtic pony—pale buff, mouse gray, or even brown, with a large forelock and tuft beneath the jaw, a light coloured mane and tail, but with a certain admixture of black hairs. There is long bushy hair at the base of the tail. This horse is also characterized by a short face, broad forehead, slender legs, and small hoofs, and is found from Iceland to western Norway.

The second form is the Norse yellow dun or forest pony, related evidently to the Mongolian horse, but larger, stockier, and with fuller mane and tail. In some instances there is a dark stripe down the back and traces of barring on the legs. The face is longer and the hoofs relatively larger than in the Celtic horse. The Norse pony is the main ancestral stock for the ordinary domestic horses of northwestern Europe. The changes may be due to domestication or to the infusion of Arab blood.

The last great type is the southern horse or barb, the Arab or thoroughbred, *Equus africanus*. The colour of this creature is bay, sometimes gray, with black points and often with a white star on the forehead and one or more white legs. It has a small head and slender, graceful limbs, and possesses great docility and spirit.

Nearest the true horse comes the kiang, *Equus kiang*, of central Mongolia and Turkestan. This creature is not an ass although ass-like in many ways. It stands 12½ hands, the ears are horse-like and the hoofs broad, especially in front. The tail tuft is large and there is the rudiment of a forelock. In winter the colour is grayish, in summer chestnut, with no striping.

The zebras are exclusively African and are of course characterized by a very conspicuous striping when seen out of their natural surroundings. They are, nevertheless, generally reported to be protectively coloured when in their appropriate habitat, although this is a subject upon which Colonel Roosevelt had much to say, as he believed that the theory of protective colouration has been considerably overdrawn (*African Game Trails*, Appendix). There are two well defined species of zebra living, while a third, the quagga, was so recently exterminated that mounted skins may be seen in certain museums. The plains or Burchell's zebra is somewhat variable in the colouring but always lacks the cross-striped rump, the so-called "gridiron" of the true or mountain zebra. The former is still numerous, in fact it is said to be the second big game animal of the world in point of numbers. The mountain zebra, on the other hand, is becoming so rare that it is protected by law. It is more nearly related to the ass and has longer ears, narrower hoofs and a scantier tail tuft than the Burchell species.

The ass, *Eqqus asinus,* is domesticated the world over, in fact its subjugation by mankind long antedates that of the horse. Asses are still wild in the tropics of Africa and are gray at all seasons, with a dark back stripe. When wild the size is medium to large, ranging from 11 to 12¼ hands at the shoulder. The hoofs are small and narrow and the fore pair are no larger than the hinder ones. There are two varieties, the Nubian ass, which has a transverse shoulder stripe, and that of Somali, which lacks the shoulder stripe but has barring on the legs. The domestic variety is typical of the Nubian form.

Horses and Man

Mankind owes a profound debt of gratitude to the horse, first in savage days as an easily obtainable food, later as a partner in his labours without whose aid human progress toward a higher civilization would have been retarded immeasurably. It has been thought by some that the condition of semi-barbarism of the North American Indians, really a race of great potentiality, was in part due to the premature extinction of the American horses.

There is no record of the association of man and the extinct horses of America, but during prehistoric times in Europe, before the extinction of the mammoth, we find records of the association of the horse and man in the form of mural decorations on the walls of caverns. It is interesting to note that at least three types of horses are shown by the Paleolithic artists: one a small-headed form resembling in this regard the Arab of today, another large-headed, with the erect mane and beard typical of the living Przewalski horse, and a third which in contour closely resembles the Norse or forest pony. The presence of bridle—like markings on the head of one horse had been taken as an indication of domestication on the part of the prehistoric peoples. One finds, however, no trace of a drawing of a man on horseback or other use of the animal as a beast of burden and the idea has been advanced that possibly because of its extreme docility it may have been occasionally easier to lead home a captured horse to the slaughter than to carry home the meat.

One of the most remarkable prehistoric encampments, not in caves but in the open air, is at Solutre in Saone-et-Loire, France. Here there was a fine southern exposure sheltered on the north by a steep ridge. Encircling the south side was a kind of protective wall formed almost entirely of the bones of horses to the estimated number of 80,000 individuals! Such a wholesale slaughter of course extended over a long period of time, but might readily have been an important factor in local extermination when aided by the weakening effects of disease.

Chapter—9

Camels

The camels are another group of animals whose phylogeny has been very clearly demonstrated by the fossil evidence, the perfection of the record being second only to that of the horses. Add to this the fact that they are throughout almost their entire evolutionary career exclusively North American forms, and their title to a high place in our interest is complete.

Place in Nature

As the horses were representatives of the Perissodactyla or odd-toed ungulates, so the camels belong to the other great group of hoofed forms, the even-toed Artiodactyla. The latter, now largely eliminated from the western world, are still comparatively numerous in Africa and Eurasia, where at least 250 distinct species are known as against 34 for the Americas. That the Artiodactyla were formerly much more abundant, especially in North America, is forcibly brought home to every collector of fossil vertebrates in the West.

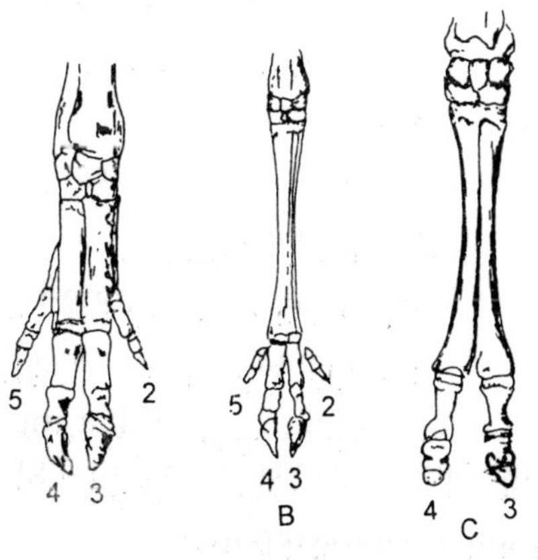

Fig. 9.1: Fore feet of artiodactyls. A, pig, *Sus scrofa;* B, red deer, *Cervus elaphus;* C, camel, *Camelus bactrianus*. To show progressive reduction of lateral digits.

The principal points of agreement of all artiodactyls are: the axis of the foot lies *between* digits three and four rather than within digit three as in the perissodactyls; hence the two digits, one on either side of the axis, are symmetrical and the number two is the irreducible minimum. While there are normally an even number of digits, the peccaries, *Tayassu,* have three remaining in the foot, and a five-toed ancestral artiodactyl is conceivable, for the oreodonts retained a well defined vestige of the first digit of the hand. Another artiodactyl characteristic lies in the astragalus, the ankle bone which articulates with the tibia or shin. In common with that of the perissodactyls, the upper or tibial facet is pulley-shaped, but in the latter the distal facet is flat, thus permitting no movement between it and the succeeding tarsal bones. In the artiodactyls, on the contrary, the distal facet is curved in such a way that a double tarsal joint is formed. This type of astragalus, a very resistant bone, is an extremely common fossil and is absolutely diagnostic of the group.

The molar teeth are invariably one of two sorts or a combination of them. The crown either bears conical cusps as in the swine, or the cusps are crescentic, the latter sort being typical of the cud-chewing forms of ruminants. Certain ancient types (anthracotheres) had buno-selenodont teeth of a transitional character. The teeth may be short-crowned or, in the grazing ruminants, deep-crowned as an adaptation to abrasive food; they never, however, reach the degree of perfection seen in the true horses.

Artiodactyls are apt to possess weapons, either tusks, which are modified canine teeth, or horns or antlers of various sorts and degree of development, or rarely both.

A classification of the artiodactyls adapted from Romer is:

Suborder Suina (swine-like)

Family Suidæ (swine)

Entelodontidæ (giant swine)

Dicotylidæ (peccaries)

Hippopotamidæ (hippopotami)

Suborder Ruminantia (cud-chewers)

Family Camelidæ (camels and llamas)

Oredontidæ (oreodonts)

Infraorder Pecora (true ruminants)

Family Cervidæ (deer)

Giraffidæ (giraffes)

Antilocapridæ (prong-bucks)

Bovidæ (antelopes, sheep, cattle, etc.)

Thus it will be seen that the camels occupy an intermediate place within the order; they are, however, an isolated group, as their connection with the others is not yet clear.

Camel Characteristics

The characteristics of living camels are as follows: The limbs are long, and two-toed; the metapodials, which are fused to form a cannon-bone, diverge distally, and the keels which serve to restrict the lateral movement of the digits in other forms are here very limited. Hence the secondarily digitigrade foot is yielding as an adaptation to desert sands. The feet are, moreover, provided with one or two cushion-like pads, hence the name Tylopoda. The stomach is three-chambered; there is a primitive type of placenta; and the red blood corpuscles are elliptical instead of circular in outline, which makes the group absolutely unique among living mammals.

LIVING GENERA

Camelus

There are but two living genera, each of which includes a like number of species—four altogether extant—one, *Camelus*, being confined to the Old World, while the other, *Lama* or *Auchenia*, is characteristic of the New.

The two species of camel are *Camelus dromedarius*, the one-humped Arabian camel or dromedary, and *C. bactrianus*, the two-humped Bactrian camel of central Asia. The two species will interbreed and the consequent hybrid or mule camel possesses the one hump of the dromedary and the brown shaggy coat of the Bactrian parent. The progeny of a male Bactrian and female Arabian camel is preferred to either of the pure breeds.

The camel has rightly earned its name of "ship of the desert," for practically all of its peculiarities are but an adaptive response to the harsh conditions of that inhospitable environment. As in the horse, two directions of adaptation stand out sharply—that of speed, ever a desert requisite, and that of teeth, for the harsh and scanty herbage. The other characteristics the horse does not possess, for they are the direct outcome of desert life.

Speed

Cursorial characteristics are well shown in the length of limb, reduction of digits, of ulna and fibula, and in the limitation of the range of movement of the limb joints. The feet, however, have retrogressed in that they are no longer unguligrade but

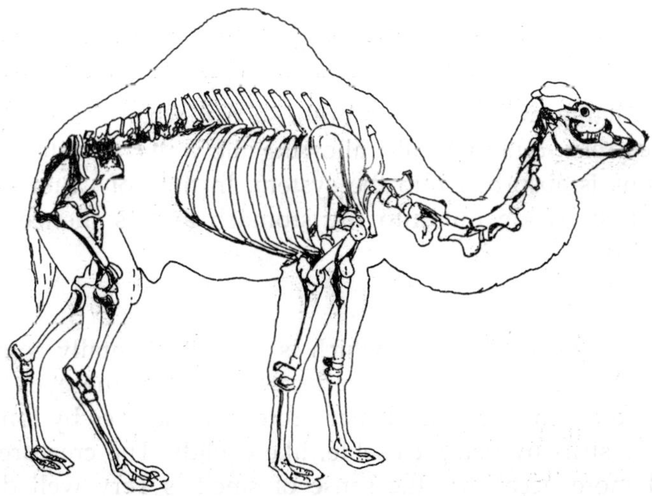

Fig. 9.2: Skeleton with bodily outline of camel, *Camelus dromedarius*.

digitigrade, for almost the entire length of the phalanges, except for the intervening pad, lies flat upon the ground. The hoofs are reduced to nail-like structures and the whole yielding foot, with its absolutely silent tread, is admirably designed to support the animal on the shifting desert sands. The foot retrogression somewhat diminishes the extreme length of limb, but this is to a certain extent compensated for by the fact that the thigh is freer from the body than in other ungulates and thus the length of stride is increased.

Teeth

The teeth of the camel have suffered a reduction in numbers, to 34, instead of the normal 44. There is but one upper incisor left on either side and it is more canine- than incisor-like. The lower incisors, on the other hand, are all present, more or less spatulate and procumbent, and the canine is somewhat similar and functions as an incisor. Behind the canine comes a short diastema and then a recurved, tusk-like premolar which has assumed the discarded form and function of the canine. This is followed by a longer toothless area, and the four cheek teeth—the fourth premolar and the three molars—form an efficient compact grinding mechanism, of long-crowned but, as compared with the horse, relatively simple teeth.

Hump

The hump which forms so very characteristic a camel feature consists of a conical mass of gelatinous fat when the animal is well fed, his nourishment stored against a time of scarcity, and can be drawn upon during the passage of the desert. Whether or no any of the extinct camels possessed such an organ we cannot tell, as it is entirely superficial and leaves no impression upon the skeleton. The hump becomes flaccid and falls over on one side in an exhausted camel.

Water Reservoirs

Another desert characteristic is the development of water reservoirs in the walls of the stomach (properly the paunch or rumen). These are small flask-shaped cavities, each with a constricting muscle at its mouth, so that when the stomach is filled with water the muscles relax automatically, allowing the water to enter the cavities, while that which remains is absorbed into the system. In time of water scarcity the stored liquid is allowed to trickle out into the stomach and is thence available for the impoverished blood.

Senses

The proud carriage of the head, which is held horizontally some nine feet from the ground, protects the eyes from the reflected heat and the eyes and nostrils from the sand. The sense organs are still further protected, the eyes by long lashes, the ears by hair, and the nostrils by being closable, like eyelids. The creature is keen of sight, but what is still more necessary, the sense of smell is very well developed so that water may be detected a long way off.

Mentality

Mentally the wild camels are sagacious, as the brain is large and well convoluted, but the domesticated ones are so stupid that their bad traits are notorious.

> Thus Palgrave observes (in Flower and Lydekker): "If docile means stupid, well and good; in such a case the camel is the very model of docility. But if the epithet is intended to designate an animal that takes an interest in its rider so far as a beast can, that in some way understands his intentions, or shares them in a subordinate fashion, that obeys from a sort of submissive or half-fellow feeling with his master, like the horse or elephant, then I say that the camel is by no means docile—very much the contrary. He takes no heed of his rider, pays no attention whether he be on his back or not, walks straight on when once set agoing, merely because he is too stupid to turn aside, and then should some tempting thorn or green branch allure him out of the path, continues to walk on in the new direction simply because he is too dull to turn back into the right road. In a word, he is from first to last an undomesticated and savage animal, rendered serviceable by stupidity alone, without much skill on his master's part, or any cooperation on his own save that of an extreme passiveness. Neither attachment nor even habit impress him; never tame, though not wide-awake enough to be exactly wild."

As beasts of burden, however, camels are entitled to respect, as one can carry 500 to 800 pounds, and their endurance is phenomenal, the Arabian breed known as the "Heirie" camel traveling from 125 to 150 miles a day for eight to ten days at a time. The distance from Tunis to Tripoli is 600 miles, yet a single camel has carried over that route a burden of rider and gear weighing not less than 250 pounds, in four days—an average of 150 miles a day!

Uses

Alive, camels are used as beasts of burden, for their milk, and for the shed hair which is spun and subsequently woven. Dead, the flesh is used as food, the hides for leather, the hair for fabrics, and even the bones are utilized. Their importance to mankind, especially to the nomads of the East and to traders between great cities, can hardly be estimated. Necessities of war have developed means of desert transportation in the form of tractors and other motor vehicles, as well as airplanes. To what extent such machines will be used to aid travel and commerce in time of peace remains to be seen. With nomadic peoples, however, who do not care for or possess means of mechanical transportation, the camel must still be highly important.

It is interesting to note that, while the American Museum expeditions to Mongolia have an adequate fleet of motor cars, they also have a supporting train of camels, which carries in gasoline and other supplies and brings specimens out.

There are wild camels in remote Turkestan, the desert of Lobnor, and in Spain, but some are certainly feral, *i.e.* of domestic ancestry, and all are probably so, as there are ruins of ancient cities in the Asiatic portion of their range of which the very traditions have vanished and to whose departed citizens the ancestors of these camels may well have belonged. It is highly probable that they have not existed wild for thousands of years.

The area of servitude includes Arabia, Persia, India, all of the country from North Tartary to the confines of China and the coat of the Persian Gulf, and the Canary Islands and Africa north of the Sahara. There were none in Africa, however, until the third century of our era. Attempts to naturalize them in Australia and North America have been made, but the lack of success in America has not been due to the climate or other physical conditions of their ancestral home, but rather to their unfavourable reception by the Americans, who greatly preferred the highly serviceable, cheaper, and more tractable burro.

LAMA (AUCHENIA)

Characteristics

The South American genus, *Lama* (= *Auchenia*), includes two wild species, the guanaco, *L. huanacus*, and the vicuña, *L. vicunia*, and their domestic derivatives, the llama and alpaca. These creatures are of much smaller stature than the camel and lack the characteristic hump of the latter. The feet are narrow and the toes more distinctly divided, with two pads beneath instead of but one. The hair is woolly as a protection from the cold of their mountain home, for instead of being adapted to sandy desert conditions, although their structure shows a desert ancestry, they are upland animals, which may well account for their attainment of South America along the isthmian land-bridge, to the exclusion of the true camels.

Fig. 10.3: Guanaco, *Lama huanacus*, living, South America.

Range

The *Lama* range is along the west side of South America from the equator to Cape Horn. Domesticated, they have their uses similar to those of their Asiatic cousins, as beasts of burden, and for the flesh, hides, and wool. They have the distinction of being America's only contribution to the list of mammals domesticated by mankind, but although their local importance may be great, they have by no means contributed to human progress and well-being to the extent that the camels have.

EVOLUTIONARY CHANGES

These are in a way comparable to the changes undergone by the horses, with the exception of the secondary retrogression of the feet. To summarize briefly, the changes

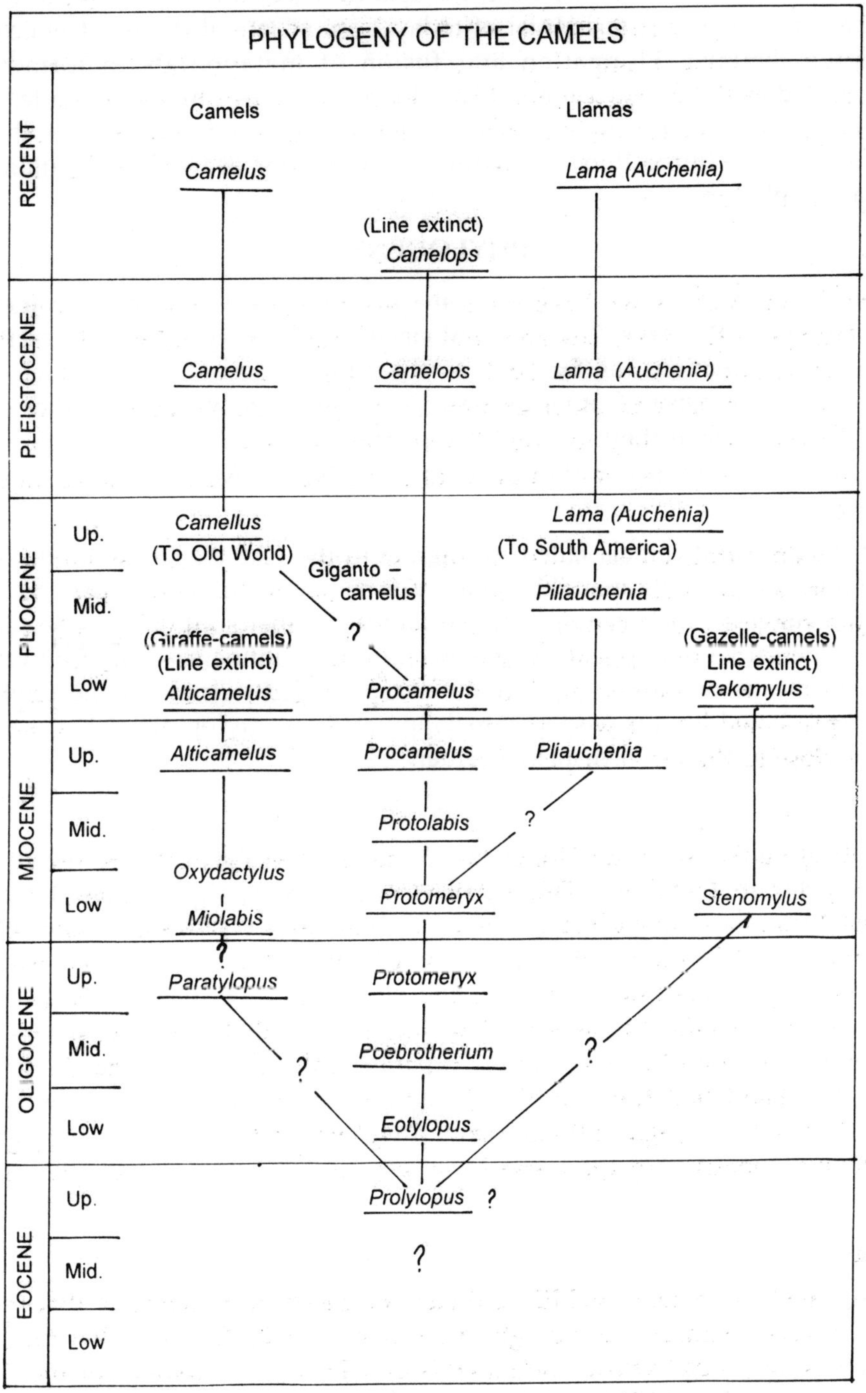

PHYLOGENY OF THE CAMELS

are as follows: Increase in stature from the size of a western jack-rabbit to one much greater than the huge Bactrian camel of today. Loss of lateral digits, of which not the least vestige remains. Elongation and fusion of metapodials to form the very characteristic, distally spread cannon-bone. Retrogression from unguligrade, deer-like feet to digiti-grade, broadening of phalanges, development of foot-pads with the coming of desert conditions in the Pliocene, reduction of number of teeth, elongation of teeth in the grazing phylum.

PHYLOGENY

North America is, as we have seen, the evolutionary home of the camel family, and, as Matthew (1915) says, "Its ancestral stages can be very fully and exactly traced in the western formations, as far back as the Upper Eocene, below which they are merged with the ancestry of other groups. They are unknown in any other continent until the Pliocene, when they invaded South America and Asia and Africa, surviving in those continents to-day, although extinct in North America since the Middle Pleistocene."

Why the ancestral camels failed to migrate to the Old World before the Pliocene, when the horses repeatedly made the journey, is somewhat obscure unless, as Matthew (1915) again suggests, their center of radiation was further south, for, as he says, "The center of dispersal would appear to have been in this continent,—how far to the north we have no means of estimating; but the exceptional directness of the phylogenetic series as represented by our western fossils indicates, in my opinion, that these fossils lived in or close to the racial dispersal center."

Eocene

Camels are unknown until Upper Eocene time when the first possible ancestor of the line appears in *Protylopus*. This small creature was no larger than a jack-rabbit, and had 44 teeth, those in each jaw forming a continuous series, the canine being only slightly enlarged. All of the molars were low crowned. The skull with its narrow face suggests that of the existing forms, but the bony orbit was incomplete behind. The fore limbs were considerably shorter than the hind so that the back sloped upward toward the rump. The ulna was entirely separate from the radius and the fibula was complete. The hand had four functional digits but the lateral toes of the foot were greatly attenuated although still complete. *Protylopus* is from the Uinta stage and is thus contemporaneous with the horse *Epihippus,* to which it is comparable in degree evolution.

Oligocene

During the Oligocene, camel-like animals increased in numbers so that they must have been a very characteristic though rare element in the fauna of that time. This is especially true of *Poëbrotherium,* the remains of which are abundant in the "Big Bad Lands" of South Dakota, whence they range across Nebraska into Colorado and west

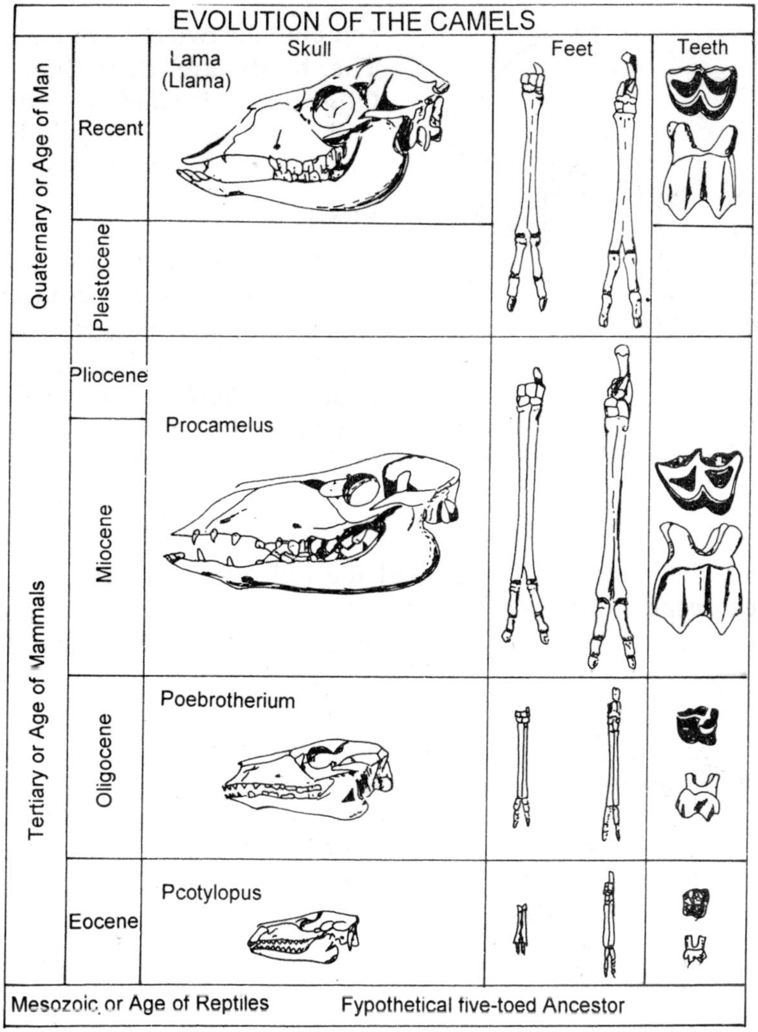

Fig. 9.4: Camel evolution as indicated by the skull, feet, and teeth.

to the John Day valley of Oregon. As *Protylopus* parallels *Epihippus,* so *Poëbrotherium* resembles the contemporary *Mesohippus* in the degree of its evolution. *Poëbrotherium,* like *Mesohippus,* attained the stature of a sheep, but the former was more lightly constructed than the sheep, with relatively longer limbs and neck and with a small tapering skull. The teeth are still forty-four as in *Protylopus* and the incisors and canines are more typical of mammals in general, not the procumbent, spatulate structures of later cameloids. The grinding teeth of the upper jaw are short-crowned, while the lower molars have begun to elongate. The jaws are very slender. The limbs show a marked digital reduction both in the hand and foot, in that small nodules only are present as the last vestiges of the lateral toes. The ulna has coalesced with the radius

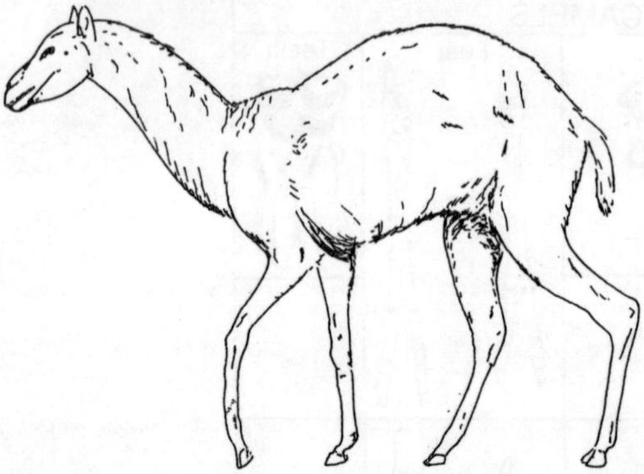

Fig. 9.5: Restoration of ancestral camel, *Poebrotherium labiatum*, Middle and Upper Oligocene, North America.

and only the two ends of the fibula remain. The hoofs are deer-like. The Upper Oligocene *Protomeryx* differs from *Poëbrotherium* mainly in the complete encircling of the orbit by bone.

Miocene

During the Oligocene there began an initial divergence into at least three phyla which became well-defined groups during the Miocene, paralleling once more the equine evolution. Of these the grazing camels, which were the main line leading to the modern representatives of the race, include the Lower Miocene *Protomeryx* and the Upper Miocene *Procamelus*.

The former, *Protomeryx*, still possessed the full quota of teeth, but the grinders show a decided deepening tendency as an adaptation to the abrasive grasses. The feet had two digits and possessed pointed hoofs like those of the deer.

Fig. 9.6: Gazelle camel, *Stenomylus*, Miocene, Nebraska.

In *Procamelus* we find the first tooth reduction, in that the first and second upper incisors are lost in the adult state. The feet have advanced, for the metapodials are beginning to fuse to form the cannon-bone. The first desert adaptation is shown in the foot bones in this genus which imply the presence of a padded foot. In size *Procamelus* must have exceeded the dimensions of the modern llama.

A very notable fossil locality in western Nebraska, of Lower Miocene age, has yielded a large number some forty or more, of a slender camel-like form known as *stenomylus* the gazelle-camel, delicate in its proportions and much smaller than any of its contemporaries. Its Oligocene ancestry has not yet been traced and except for the early Miocene *Rakomylus*, the inference is that the line soon became extinct. It has an apparent anomaly in its dentition, as there are ten incisor-like teeth in the lower jaw, six true incisors, and in addition the canines and first premolars which have assumed a similar form and function. The low-crowned molars imply a browsing habit. The head is small, the neck long and delicately built, and the limbs and feet extremely slender, with very thin-walled bones. There are but two toes on each foot and the metapodials are not fused. Apparently fleetness was *Stenomylus'* only defence, which may have accounted for its brief racial career. The known specimens, of which a group of two individuals is mounted at Yale, are almost without exception from a single quarry, where they occur in profusion, some dismembered, others in completely articulated condition as though the carcasses had drifted downstream in time of flood, to be caught in the backwater of some large cove and buried by sediment. This is in accord with the belief of the discoverer, Professor Loomis, based upon anatomical grounds, that *Stenomylus* was an upland form. The only associated remains other than those of the camel found in the quarry pertain to a large wolf-like creature known as *Amphicyon* (=*Daphænodon*) *superbus*, probably one of the forms which preyed upon the camels.

The name giraffe-camels does not imply relationship with the giraffes, which, so far as known, with one exception have been confined exclusively to the Old World, but is applied to creatures which from community of habit converged very strongly toward the existing giraffe in size and proportions. Two Miocene genera have been discovered which pertain to this family, and in the aberrant Oligocene *Paratylopus* we recognize the first recorded ancestor of the group. In the Lower Miocene the representative is *Oxydactylus*, remains of which have been collected in eastern Wyoming and western Nebraska. This form is much smaller than its successor, *Alticamelus*, and has a shorter neck and limbs. The metapodials do not fuse to form a cannon-bone and the hoofs are sharp pointed and deer-like, with little indication of the sand-adapted feet of the later camels. The teeth of *Oxydactylus* are rather short-crowned, as though fitted for browsing rather than for grazing, and they are yet 44 in number.

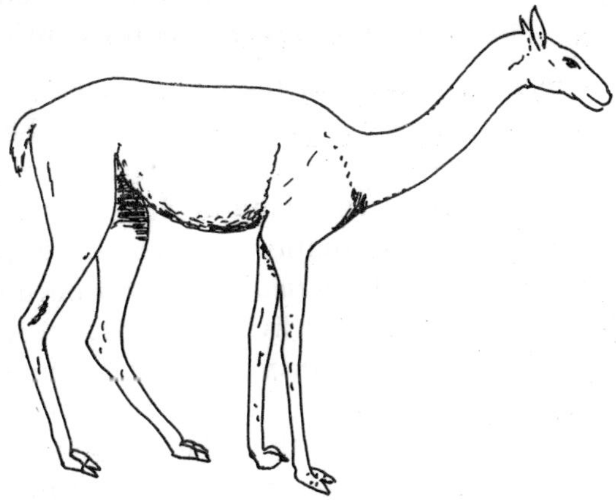

Fig. 9.7: Giraffe camel, *Oxydactylus*, Miocene, Nebraska and Wyoming.

Alticamelus of the late Miocene and early Pliocene, although clearly derivable from *Oxydactylus*, is much further advanced in more than one way, for we find that the feet show the same desert adaptation that its contemporaries of the grazing phylum do—cannon-bone, depressed phalanges, and indications of pads—a remarkable instance of convergence, the like response of unrelated phyla to a similar climatic change. *Alticamelus* was a very large animal, although the head was small and the teeth, like those of its predecessor, are low-crowned. The neck and limbs are very long as in the giraffe and were probably, as with the latter, an adaptation to permit the animal to browse upon the otherwise inaccessible foliage of high and thorny shrubs like the African mimosa, which forms the staple of giraffine diet. But while the result of this remarkable convergence was to produce the same effectiveness for such a method of feeding, the way in which it was brought about was not the same.

Pliocene and Pleistocene

Pliocene camels belong to the main grazing phylum and are the direct descendants of the Miocene *Procamelus*, from which they hardly differ at all. The Pliocene and Pleistocene camels *Camelops* and *Eschatius* are not very well known, owing largely to the fragmentary character of the material thus far collected, with the exception of fine skeletons of *Camelops hesternus* from the Rancho la Brea. The Pleistocene camels, however, were very numerous and future discoveries are sure to bring better specimens to light. Not all of the Pleistocene forms are true camels in the sense of being like those of the Old World, but are cameloid creatures, possibly more llama-like in appearance. Some of them attained an enormous size, as certain bones preserved at Yale are half again as large as equivalent elements from a Bactrian camel.

Typical camels, of the genus *Camelus*, differ from *Procamelus* in the further loss of one premolar in the upper and two in the lower jaw, and this genus is apparently the first to brave the cold of the northern route and pass to the Old World, presumably by way of the Bering land-bridge, for we have the first recorded remains of it in the famous Siwalik formation (Lower Pliocene) of India. The two Indian species of *Camels*, *C. sivalensis* and *C. antiquus*, show a peculiar tooth character found in the New World llamas but lost in the living camels, indicating that in some respects the New World types are the more primitive. Fragmentary *Camelus* fossils have been found in the Pleistocene of southern Russia and Roumania, and in Algeria not far removed from the living habitat of the race. The extinction of the North American camels is as inexplicable as that of the horses, and may well have been due to the same unknown complex of causes.

Chapter—10

South American Mammal Radiation

The South American fauna presents an extremely interesting example of what may happen in a large and varied area during long period of isolation. The continent of South America is today, owing to the existence of the Panama Canal, entirely isolated by water from the rest of the world. So it was for long periods of geologic time, due of course to natural causes. As a consequence there developed, during these periods of isolation, a most amazing local fauna, the members of which differ markedly from the more familiar creatures inhabiting the rest of the globe.

Paleogeography

Not only is South America vast, extending as it does from 10 degrees north of the equator to 55 degrees south, but it varies greatly in climate as a consequence of its latitudinal range, as well as its topography. Moreover, it is so completely surrounded by water that its isolation by the forces of nature was most readily accomplished. Of the former land connections there have been several opinions, but the authority of Charles Schuchert is taken as final, for it is backed by years of the most painstaking accumulation of data and by very judicial interpretation.

Three possible line of communication have existed in the past: that with North America, though not necessarily by way of the present Panamanian isthmus; one from the southern extremity to Antarctica and thence to Australia; and the third across what is now the South Atlantic to Africa.

The evidences of these old-time connections are in part geological but more largely biological and paleontological and are based upon present and past distribution of animals both by the land and sea. Manifestly similar and closely related terrestrial animals, vertebrate or invertebrate, normally incapable of overseas migration, point to the existence of land-bridges as migratory routes. The same is true of strictly fresh-water fishes. Marine creatures whose home is on the continental shelf can, on the other hand, only migrate where shores are continuous. Thus, where Atlantic and Pacific marine animals are similar, it points to aquatic continuity, by way of a strait or other body of water, and therefore to the absence of land connection, whereas dissimilar marine forms imply land continuity, which would of course serve as a sea barrier. It is by the accumulation of evidences such as these, not only in existing but in fossil forms, and by the careful plotting of the occurrences on successive geological maps that the old-time shore lines are indicated, for they must lie somewhere between the points of occurrence of land and marine animals of contemporary age.

The evidence may be summarized as follows:

Between the Americas there was connection during the Mesozoic, broken in Eocene time (Upper Eocene, according to Schuchert; Matthew, however, says Paleocene, otherwise the carnivores would be creodonts and not marsupials), reëstablished in early Pliocene, and practically continuous until the present day, except for the canal. Thus the period of isolation lasted during part of the Eocene, all of the Oligocene and Miocene and into the Pliocene, a duration of perhaps fifty million years.

After its reëstablishment in the Pliocene the land-bridge was evidently much broader than the present Isthmus of Panama, but soil had to form as a result of rock disintegration, and the establishment of vegetation had to precede the migration of terrestrial mammals, which in the long run depend upon the plants for food.

With Australia via Antarctica the communication, if it actually existed at all, was again an ancient one and was also severed very early in the Eocene and has never been reëstablished. The same was apparently also true of the African connection. Evidence speaks of the existence of a vast equatorial continent, extending from South, America across the South Atlantic to Africa, and again to India, the East Indies, and Australia, including part of what is now the Indian Ocean. This great area has been named Gondwanaland, and, while its former existence is denied by good authority, the evidence for it is strong and well founded. Some question has arisen as to the date of its final severance, but Schuchert's belief is that it began to break up toward the close of the Cretaceous, so that by Tertiary time all chance of migration by this route had ceased. Matthew's argument also holds here.

These old-time connections of course afforded paths of migration for terrestrial animals, both vertebrate and invertebrate, and from some one or more of them must have come the ancestors of the peculiar South American fauna, except such elements as were strictly native in origin. The edentates among mammals seem to be the only order which can be referred to this latter group, however, as they alone show no affinity with apparently ancestral forms elsewhere in the world.

Geological Succession

The following time table of South American strata is modified from Scott.

Eras	*Periods and Epochs*	*Age*	*Occurrence*
Psychozoic	Recent or Post Glacial		
Quaternary	Pleistocene (Glacial)	Pampean	Pampas of Argentina, also Equador Brazil, Chile, Bolivia Caverns of Brazil
Cenozoic (Tertiary)	Pliocene	Monte Hermoso Catamarca Parana	Extensive areas in Patagonia. Some extend into Tierrs del Fuego
	Miocene	Santa Cruz Patagonian	
	Oligocene	Deseado (Pyrotherium beds) Astraponotus beds	
	Eocene Paleocene	Casa Mayor (Notostylops beds)	
Mesozoic	Cretaceous		

Scott summarizes the successive horizons as follows:

Paleocene

There are no marine Paleocene rocks, hence the inference is that the continent was largely emergent. Connection with Africa (Gondwanaland) may have continued into this epoch, but probably was severed during the Cretaceous. Hence the difficulty of deriving the South American mammals from known African orders, such as the Hyracoidea and Proboscidea, as some have argued.

There is evidence, although questionable, of the existence of the Antarctica-Australian bridge, which might account for the remarkable similarity between Australian and South American marsupials. The South Polar region could not have had so severe a climate as now, or passage across it, even were the land connections existent, would not have been possible. Communication with the Eurasian continent was probably round about, by way of North America.

Eocene

During the Eocene, North and South America were completely severed, apparently across Costa Rica-Panama, and there was apparently no connection with any other continent. Isolation was complete. Connection with Australia seems highly improbable, otherwise there would certainly be placental mammals in Australia derived from South America. Were such animals formerly existent they have utterly died out, leaving only the monotremes and marsupials. Usually when the two groups, placental and marsupial, are in competition, it is the marsupials that perish. Our knowledge of the relative organization of the two groups makes this the only understandable result.

During Eocene time the Andes were much lower than at present. Their great height in the south was probably attained not earlier than Pliocene time.

Oligocene

Paleogeographers believe that the Greater Antilles were broadly united into a common land mass known as Antillia, which existed during most of the Eocene but, due to great submergences during the late Eocene and Oligocene, was broken up into a chain of islands much smaller than the present Greater Antilles—Cuba, Haiti and Santo Domingo, Puerto Rico, etc. At all events, the Oligocene was for South America a time of complete isolation, so that it is difficult to be sure of the relative age of the Deseado formation, for instance, as direct comparison with the Oligocene mammals of North America or Eurasia cannot be made, there being no forms in common.

Miocene

During the Pliocene, connection with North America was reëstablished, but, as we have seen, migrations of terrestrial animals followed slowly, for more than mere closing of a water barrier is necessary; hence the fauna of the Santa Cruz formation in the far south is still entirely a native evolution, with no known immigrant phase, such as appeared in the Pliocene. This, together with the marvellous abundance and perfection of the fossils many of which are those of animals apparently buried alive in wind-borne volcanic ash, makes the study of the Santa Cruz fauna one of the most interesting and profitable of all. Matthew says that the reunion is dated from the appearance of *Megalonyx* in the Lower Pliocene of North America.

The Patagonian stage is marine, as the region was submerged beneath a shallow epicontinental sea. The finding of Cetacea (whales) in the Patagonian strata is comparable to those in the Maryland. Miocene establishes not only the geologic age of the former but also the fact that no land-bridge with Africa could possibly have existed at this time or the whales could not have intermigrated, as they surely did.

After the retreat of the Patagonian sea the Santa Cruz beds, which we have mentioned, were laid down, in part river deposits but largely the air-borne volcanic ash which preserved the entombed specimens with such great perfection.

While immigrants from North America may already have entered South America, none had as yet reached far Patagonia, whence the Santa Cruz animals come. Known fossil plants of Miocene time indicate luxuriant tropical forests comparable to those alive today in Brazil and Bolivia. The Santa Cruz animals come from too far south and too high an altitude for such plants to live in, and hence there is no direct record of their botanic surroundings at all. The mild climate which is indicated points to open plains with occasional trees, as in the African veldt.

Pliocene

With the ushering in of Pliocene time the climate becomes cooler, and this increases as time goes on until the austere conditions of the Pleistocene Glacial period are reached. The latter was never so severe as in the northern hemisphere, however. The paths of intermigration were now fully open to traffic, with all that that implies, and now for the first time appears the van of the increasing tide of immigration from north to south and from south to north. Climatic barriers, ever more prohibitive of north and south movement than of east and west, must to a certain extent have prevailed, for there never is such freedom of interchange of animals as between the Old World and the New.

A large portion of Patagonia was encroached upon by the seas, which extended in places to the Andes, while the broad valley of the Rio de la Plata was a gulf. The continental deposits of the Pliocene are of Parana, Catamarca, and Monte Hermoso age, the last being the most recent.

Pleistocene

As we have seen, the ice-sheets were much less extensive than in the northern hemisphere, rather were they localized, due perhaps not so much to temperature as to general difference in elevation. The climate must have been drier, for there is evidence of great dust storms, carrying immense quantities of air-borne material, thus producing the loses of the Pampas formation. This covers vast areas of Patagonia and Argentina and entombs thousands of animals of Pleistocene age, some of which seem to be comparatively recently extinct. The relative age of the Pampas formation is equivalent to the Sheridan or Equus beds of the north, so called from the abundance of fossil horse remains, and there is, as we shall see, evidence of greater freedom of faunal interchange between the two continents than ever before. On account of the more consolidated character of the Pampas deposits they seem of greater antiquity than do the Equus beds. Their approximate age, however, is attested by the contained fossils.

Another source of Pleistocene mammals is the limestone caverns of eastern Brazil. Here, while the perfection of preservation is not so great as in the Pampas, the numbers are much greater; in one instance no fewer than 500 opossum jaws having been recovered from half a cubic foot of cavern earth.

The Pleistocene witnessed the extinction of a greater part of the ancient South American stocks which for millions of years had multiplied and passed on to leave the land to their somewhat altered descendants. The same is true of the immigrant element, such as the true horses. Why, one may conjecture, but there is no demonstrable cause or causes. As in North America, horses reintroduced by man throve mightily, showing that the physical environment other than the critical climates of the Glacial period was hardly responsible. An instance will serve to emphasize this point. Horses were first landed in Buenos Aires in 1537; by 1580, only 43 years later, their feral descendants had spread west and south until they had reached the Straits of Magellan: a remarkable development in a human generation and a half, even if they had the aid of mankind in their dispersal.

The Fauna

In studying the mammals of South America two remarkable vistas are unveiled to us of such typical, and at the same time contrasting, character, that it is well to concentrate on those, with only such mention of the other ages as may be necessary for our understanding of the emphasized two. They are the Santa Cruz Miocene, when the fauna was entirely indigenous, and the Pampas Pleistocene, when interchange with North America was fully established and the fauna was mixed, partly indigenous out of the Santa Cruz stock and partly immigrant from North America.

The origin of the native South American fauna has been alluded to, partly North American or via North America from the Old World, partly Australian (if the reverse of this last statement is not true), and partly, although doubtfully perhaps, African. Add to this the probability that certain orders, notably the edentates, were a purely local development and, while derived from some one of the primitive mammalian world stocks, nevertheless had their inception and subsequent evolution in South America, where they were first found. To be sure, there are so-called Old World edentates, aard varks and pangolins, but the feeling is that they are convergent toward the New World representatives of the so-called order and that the *Edentata* of common usage is not a natural group in that the two divisions are not blood-related.

Santa Cruz Fauna

The South American radiation had already been under way for millions of years when the Santa Cruz Miocene phase was reached. Land connection with North America had not as yet been reëstablished, hence the faunal interchange could not have been effected between such remote areas as North America and far Patagonia. The Santa Cruz animals are still entirely indigenous; but it is almost the last age of which this is true, for by Catamarca time (Pliocene) the first immigrants appear in the form of certain true Carnivora. These have been identified as dogs, raccoons, and bears, but the material is so fragmentary that precise identification is not only difficult but open to question. However, these are the first scouts of the invading army; there seems to be no doubt of that.

Scott has entirely exhausted his adjectives applicable to the strange Santa Cruz animals. They are so unlike anything alive to-day that, with very few exceptions, one would look the living world over in vain for comparisons. As he says, "It is extremely difficult to convey to the reader any adequate conception of this great assemblage of mammals, because most of them belonged to orders which have altogether vanished from the earth and are only remotely like the forms with which we are familiar in the Northern Hemisphere. To one who knows only these northern animals it seems like entering another world when he begins the study of the Santa Cruz fossils." There are gaps in the history of these forms, due in part to accidents of collecting, some of which future exploration will certainly fill, and also to the fact that Miocene Patagonia had comparatively few trees, so that typically arboreal, or even forest-dwelling forms, like the primates and others which must have existed toward the tropics, are hardly represented at all.

One rather remarkable thing about the Santa Cruz mammals, which is at once apparent from the figure, is the very moderate size of the known forms; most of them were actually small, only one or two, such as *Astrapotherium*, being actually large as compared with a pointer dog which is taken as a standard of comparison.

Marsupials

As we have seen, the true Carnivora have not yet arrived, yet every fauna has a certain balance, though not an equal one, between its plant-and flesh-eating forms, the latter being much the less numerous. This is due to the ease of obtaining plant food, whereas the carnivores not only feed upon the herbivores but each individual must necessarily destroy numbers of them in the course of a lifetime. It is only in the remarkable Rancho la Brea of Los Angles that the reverse is seemingly true of terrestrial animals, for there the carnivores, cats and wolves, are by far the more numerous. This is due to the capture of the latter by the tar death trap when they came to prey upon the unfortunate herbivores that formed the lure, and does not represent the true ratios of the time among the uncaught.

In Santa Cruz time it is the marsupials that fill the rôle of carnivores. These sparrassodonts, as they are called, seem to have a rather close affinity with the living thylacine or Tasmanian wolf. Three genera are sufficiently well known to characterize in full. There was a wolf-like form which, because of its similarity to *Thyiacynus* and its earlier age, has been called *Prothylacynus*. The representatives of this genus equalled modern wolves in size and prowess, but, if one may judge from *Thylacynus*, were far inferior to them in shrewdness and intelligence and yet could hold their own in a fauna where no true wolves existed. That they fulfilled the rôle of the wolves and were probably equally successful in their day is evident. Another interesting beast was *Borthuæna*, described as possessing a short cat-like head and looking not unlike a small puma. His rôle was probably that of the larger modern cats. The rôle of the lesser beasts of prey, like the badgers and minks, was filled by *Cladosictis* and its allies. Opossums were common, although more like the Australian phalangers perhaps than the common Virginia opossum of America. The latter must, however, be a survivor of the same general race. These forms maintained the balance of nature as the types used for comparison do to-day.

Rodents

Of the herbivorous animals the first to be mentioned are the rodents, which varied in an extraordinary manner. As genera they are all extinct, although doubtless more

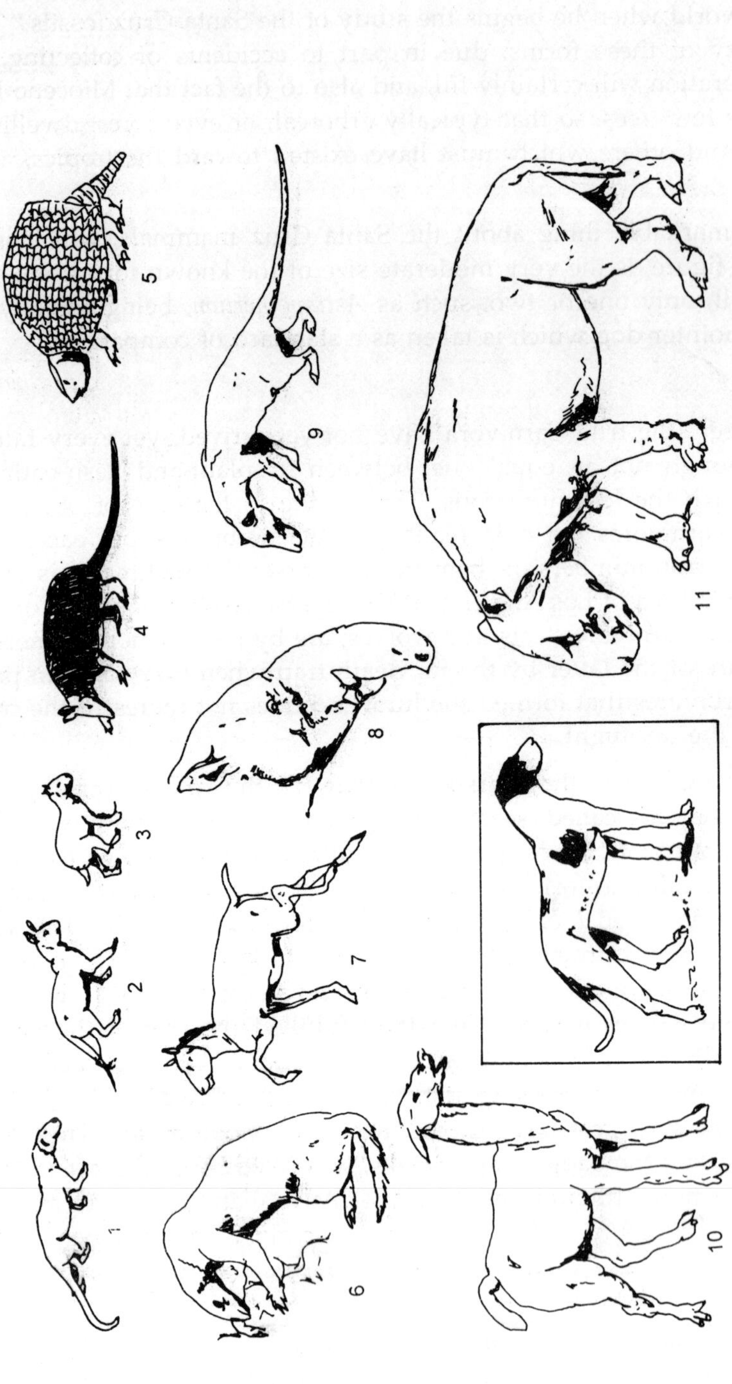

Fig. 10.1: Representatives of the South American Santa Cruz fauna. 1. *Cladosictis*; 2, *Prolypotherium*; 3, *Eocardia*; 4, *Stegotherium*; 5, *Propalæohoplophorus*; 6, *Hapalops*; 7, *Thoatherium*; 8, *Astrapotherium*; 9, *Prothylacynus*; 10, *Theosodon*; 11. *Nesodon*. Drawn to the same scale as the modern pointer dog within the rectangle.

or less akin and presumably comparable to their living descendants. Tree porcupines, chinchillas, cavies (guinea pig), coypus, and such forms, arboreal, cursorial, possibly aquatic, constitute this element of the fauna.

Edentates are perhaps the most characteristic feature of the neotropical region. They vary greatly within the group but have certain characters in common, some of which are remarkably primitive, others high specializations. Linnæus' old order Bruta (Edentata) included both Old and New World forms characterized among other things by the deficiency of their teeth, although not all are toothless, as the word edentate implies. The New World forms have remarkably complex articulations between the successive vertabræ as compared with the relatively simple ones of the Old World, so that authorities have divided the order into the Xenarthra, or New World edentates, and the Normarthra, or Old World aard varks and pangolins. As we have seen, the inclusion of two such different types in the same order may be an unnatural one. At all events there is no evidence of genetic relationship. The Xenarthra of Santa Cruz time included armadillos characterized by a bony body armour. These animals are present in the fauna of to-day, but it is doubtful whether many of the known Santa Cruz genera can be considered ancestral to those now in existence. Such a one as *Stegotherium* was probably fossorial in habits, as are the modern armadillos.

Glyptodonts were another very remarkable group of edentates, also heavily armoured, but, unlike the armadillos in which the carapace consisted of a variable number of movably articulated rings, they had a box-like shell composed of rosettes of fused polygonal plates. They also bore a head shield, as well as one protecting the tail. They were rather conservative forms, varying largely in tail armor. In Santa Cruz time the glyptodonts were comparatively small, two feet or less, although one characteristic genus, *Propolæohoplophorus,* made up for it in length of name, but that was a posthumous happening for which nature was not responsible. As Scott says, "their carapace departed less from the armadillo type than that of their gigantic descendants."

Sloths (Tardigrada). We know nothing of the Miocene tree sloths, but this is evidently due in large part to the character of the Santa Cruz environment, and not to the fact that they did not exist. It is possible that exploration further to the north, in the tropical forests of Brazil, would reveal these as well as a far greater abundance of monkeys and armadillos which are the actual ancestors of the existing forms. It is probable, however, that fossils will not be found, for it is as a rule in semi-arid or desert regions that the paleontologist finds his material, never in a tropical forest, except by the luckiest chance.

Ground sloths (Gravigrada) were leaf-eating and therefore required trees or shrubs as a food source, but not necessarily forests, if they frequented them at all. Hence the Santa Cruz environment was greatly to their liking, and their numbers and variety are amazing. As compared with their Pleistocene descendants they were small, none exceeding a black bear in size, and many were no larger than foxes. The head was small, the body long with a heavy tail and thick-set limbs, especially the hinder pair. They had long hook-like claws for uprooting and pulling down the branches of trees, and this necessitated walking on the outside of the hind foot, on the knuckles in front. Apparently they were covered with long coarse hair, at least their Pleistocene descendants were in two remarkably preserved instances, and they sometimes possessed vestiges of armour in the form of scattered dermal ossicles buried in the skin. Enamelless teeth were present in the rear of the jaws, while in front they used the toothless horny skin of the mouth, gathering their food as do the ruminants with a long prehensile tongue. Among the Santa Cruz ground sloths were probably the direct ancestors of the later giant sloths. *Hapalops* is a representative Santa Cruz genus.

Ungulates

To most of the hoofed animals the group name of Notoungulata has been given to stress their aloofness from the ungulates of the rest of the world. For it must be remembered that relative ease of intermigration makes the hoofed animals of Arctogæa, which includes roughly the northern land masses plus Africa, more or less akin, and all decidedly familiar to us, whereas Notogæa, or South America, placed its peculiar stamp on the native ungulates within its borders.

There are four suborders of Notoungulates, all out of the same ancestral stock, whatever that may have been, condylarth or hyracoid. Of these the first to be mentioned are the *Toxodontia* (*i.e.*, bow tooth in allusion to the sharp curvature of the grinding teeth). There were the most numerous in individuals of any of the ungulates, heavily and clumsily built and having but three toes fore and aft—feet hardly adequate, apparently, to the creature's weight. They ranged up to a modern tapir in size and were occasionally horned, the horns being odd in number of course to correspond to their odd-toed feet, for that and the reverse correlation generally hold, although with some remarkable exceptions. *Nesodon* is the largest of the Santa Cruz toxodonts and seems to have been the lineal ancestor of the Pleistocene *Toxodon* itself, the last of the line. There is reason to suppose that their habitats were partially aquatic, like those of the hippopotomi, and the feeding habits may have been comparable, although in the toxodonts the teeth were longer crowned and fitted for a harsher herbage.

The *Typotheria* were anomalous creatures which, while closely related to the toxodonts, were vastly more rodent-like in general appearance. In size they were rarely, if ever, larger than a fox. Two most characteristic genera are *Pachyrukhos,* not larger than the rabbit which it must have resembled, at least superficially. It had long-crowned teeth. *Hegetotherium,* another form, was considerably larger and was a long-enduring genus, its racial life extending into the Pliocene.

Entelonychia

Every now and then there has appeared an anomalous group which, while clearly of ungulate affinities, departed widely from the conventional in the character of the feet. Such were the Ancylopoda or chalicotheres, first known from the Miocene of

Europe and Asia and later from North American. The American representative is *Moropus* (*i.e.* sloth-foot, clawed like those of the sloths), known by many complete skeletons from the famous Agate Spring Quarry of western Nebraska, one of which is mounted at Yale. This creature, while suggestive of the rhinoceroses in many details of teeth and structure and of the horses in its general proportions, departed from both in having laterally compressed and deeply cleft unguals which must have been sheathed in rather powerful claws instead of hoofs. The chalicotheres are looked upon as aberrant perissodactyls.

Nature has repeated the same experiment in the Entelonychia, taking this time the toxodont stock and developing grotesque creatures, long of limb and small of head, fully as remarkable and powerful as the chalicotheres. The latter were clumsy, especially in the fore limbs, which lacked the mobility of those of the ground sloths, for instance. Hence they could hardly have been used for pulling down the branches of trees to obtain the tender twigs and leaves which formed the staple food of the sloths. Dr. Matthew has suggested that they may have been used by *Moropus* in digging for water in an arid land. This may well be but would hardly suffice as a *reason* for the evolution of such curious structures, for many African ungulates—zebras and antelope—dig for water with perfectly normal hoofs. It may be that in the several clawed ungulates the claws were used for digging, but rather to loosen the earth around the roots of the trees, which could then be borne down by the straddling animal until the foliage could be reached. At all events, the Entelonychia were convergent toward the chalicotheres and may well have had a comparable manner of life, whatever it may have been, which influenced the trend of evolution in the one group as it did in the other. As in *Moropus*, the Entelonychia were large animals, in each instance exceeding greatly the bulk of their associates. They equalled a large ox or rhinoceros in size, although *Moropus* at least was much taller.

The *Astrapotheria* were ill known, as the fossils are generally very imperfect. There is a remarkable variation of size within the group, for they ranged from one extreme to the other as among Santa Cruz animals. The high-crowned head and large tusks, together with evidence of a well-developed proboscis, make these animals the most elephant-like of Santa Cruz forms. They differ, however, in that the tusks are *canine* teeth, whereas in the true Proboscidea they are incisors. Furthermore, there are no other Santa Cruz ungulates which have canine tusks, so that in this the astrapotheres are unique. Scott says of them further, "Their limbs were long and not very massive, the feet short, five-toed and somewhat elephantine. They form an isolated group, traceable to the most ancient mammal-bearing beds of South America. Their relationships are obscure and more complete material is needed to solve the problem of their origin." Both the Entelonchia and Astrapotheria became extinct shortly after Santa Cruz time, whereas many of the notoungulates carried on until the Pleistocene.

Yet another order was the *Litopterna*, which were in many ways among the most interesting of all, chiefly, perhaps, because of their remarkable convergence toward the true horses on the one hand and the camels (llamas) on the other. Hence the members of this group would probably look less strange to us than any of the indigenous South American ungulates. The Litopterna also trace their separate lineage back to the earliest Tertiary stages. Thus they were a long-lived race. By Miocene time they had differentiated into two very distinct families, the Macrauchenidæ or llama-like forms, and the Proterotheriidæ or pseudo-horses which converged in a most remarkable manner toward the true equines.

These animals agreed with the perissodactyls in having odd-toed feet, the digits numbering either three or one in both fore and hind feet. The tarsus, on the other hand, was more suggestive of the artiodactyls, and the wrist is distinctive in its serially arranged bones, comparable in arrangement to the Condylarthra, such as *Phenacodus* a curious admixture of the characters of different orders of mammals.

The Macrauchenidæ included the strange *Theosodon*, apparently the direct ancestor of *Macrauchenia* of the Pleistocene. It looked not unlike a modern llama, except for its three-toed feet. Its teeth also differed in being curiously pointed in the front of the mouth, which must have given the animal a strange reptile-like look when the mouth was open or the lips drawn back.

The proterotheres, or pseudo-horses, Scott says, were almost the only Santa Cruz animals which would not appear grotesque. As restored they are graceful little creatures, somewhat deer-like of body but again of course with odd-toed feet, three-toed in *Proterotherium* and *Diadiaphorus* and but one-toed in *Thoatherium*. Thus *diadiaphorus* converged toward *Merychippus* or *Hipparion* among true horses and *Thoatherium* toward *Equus*. In *Thoatherium*, however, the digital reduction had gone even farther than in the modern horse, for the vestiges of digits two and four were reduced to tiny nodules, whereas in *Equus* they are long splint reaching more than half the length of the cannon-bone.

The proterotheres, unlike the macrauchenids, did not survive the Pliocene, for they were termini of a long evolutionary line, no trace of which has been found in the younger rocks.

Pleistocene (Pampas) Fauna

Another great vista of South American life is seen in the Pleistocene, where in the widespread Pampas formation there has been found a wonderfully rich fauna, although again composed of creatures of the open country and of temperate climate. Some, the indigenous element, are the direct descendants and therefore the culmination of several of the evolutionary lines which were represented in the Santa Cruz. Others were immigrants from the northern hemisphere which, as is generally the case with an invading army, were living on the country and giving the natives a severe competition, which may have been one reason for the almost total extinction of the latter. As we shall see, very few and insignificant have been their survivals after millions of years of existence.

Of the indigenous creatures of the Pampas the most conspicuous were the few surviving notoungulates and the edentates. It is interesting to note that, while some of the latter followed up the line of migration (although in the reverse direction from the invading hordes) and actually established themselves in North America, the ungulates did not and may never have been found outside the limits of Neogæa.

Of the *notoungulates* the typotheres are represented by *Typotherium,* itself the last of its race, a creature of moderate size and with chisel-shaped front teeth. *Toxodon* likewise represents the terminal member of its phylum and, like its Santa Cruz fore bear *Nesodon,* was probably semi-aquatic in its habits, a huge creature with small three-toed feet and with continually growing, sharply curved grinding teeth.

Macrauchenia is again a last survivor, this time of the Litopterna. It was somewhat like its ancestor *Theosodon* but much larger, and the recession of the nasal bones, together with certain depressions about the bony nostrils, is looked upon as evidence for the existence of a very well-developed proboscis, similar but longer perhaps than in the modern tapirs. Both neck and legs were very long, and its three-toed feet were stout and well fitted to bear its weight. The teeth have curious pit-like cavities on their grinding surface. Altogether *Macrauchenia* must have been a ludicrous animal, an efficient browser but one which could not survive the competition of the new arrivals from the north, nor perhaps the environmental changes of the Glacial period, although the present-day conditions might suit it very well.

The *edentates* reach their climax of specialization, although the species were apparently much fewer than during Santa Cruz times.

Tree sloths and anteaters, existing survivors of the order, were doubtless present further to the north, but their actual remains are very scanty, possibly through accident of preserval. Armadillos, also survivors to the present, were very many, some of them largely exceeding the existing ones in size.

Ground sloths were huge, vastly greater than any of their fore-bears. There were three families: (1) Megalonychidæ, although *Megalonyx* itself, which gives its name to the family, is recorded only from North America. (2) Mylodontidæ, represented by *Mylodon,* which is both North and South American and is next in size, a ponderous brute with dermal ossicles imbedded in the skin. There are preserved in the British Museum specimens of the related genus, *Grypotherium,* found in a cavern at Last Hope Inlet, Patagonia, almost unique in the degree of preservation in that not only were the broken bones found still bearing the shrivelled remains of sinews and flesh, but large portions of the hide as well, apparently from one individual with ossicles and dense coarse tawny hair. These give evidence of having been cut from the animal by stone knives wielded by prehistoric man. Moreover, the presence of the creature's dung, consisting of entire grass and of masses of cut grass, together with human bones, stone implements, and remains of fire, seems to imply that it and possibly other individuals were actually held in capacity by man himself and finally used as food. If so, this implies one of two things: either the great antiquity of contemporaneous man

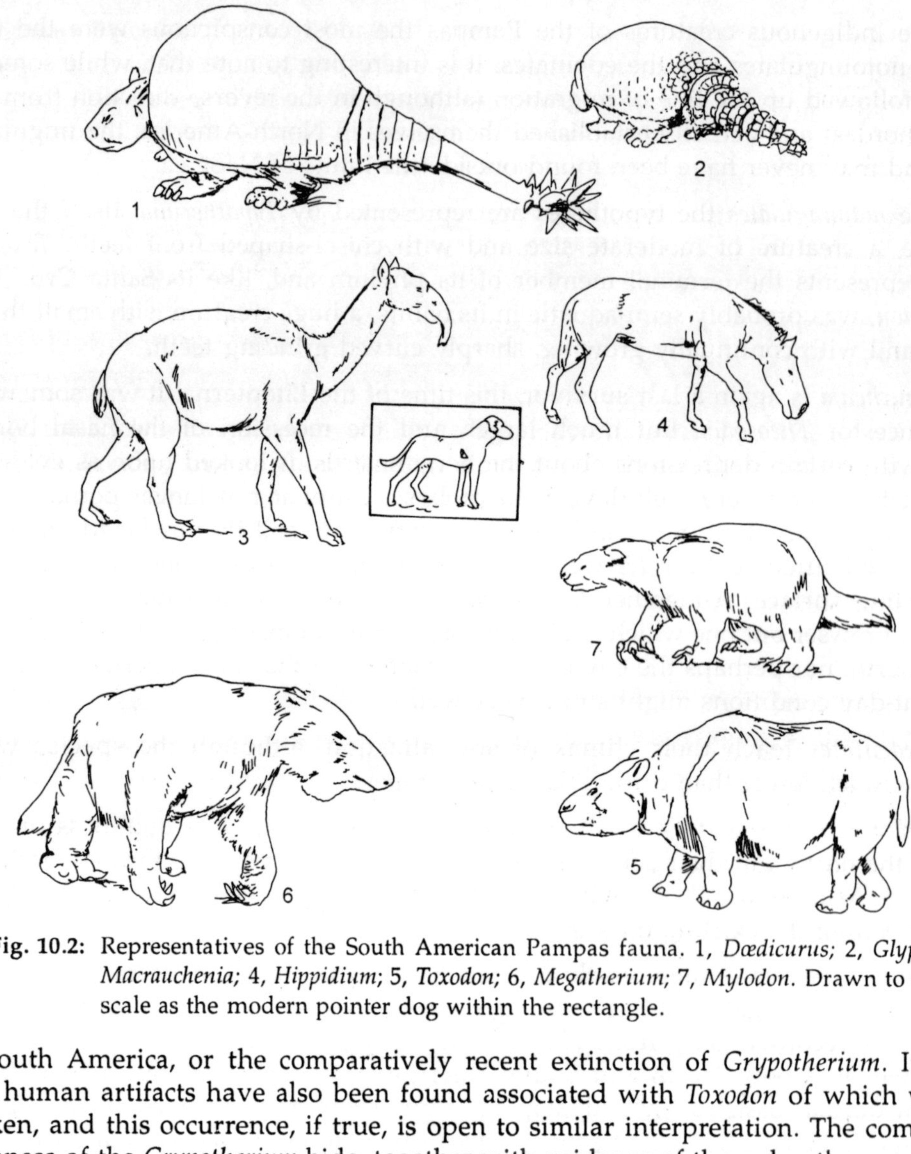

Fig. 10.2: Representatives of the South American Pampas fauna. 1, *Dœdicurus;* 2, *Glyptodon;* 3, *Macrauchenia;* 4, *Hippidium;* 5, *Toxodon;* 6, *Megatherium;* 7, *Mylodon.* Drawn to the same scale as the modern pointer dog within the rectangle.

in South America, or the comparatively recent extinction of *Grypotherium.* It is said that human artifacts have also been found associated with *Toxodon* of which we have spoken, and this occurrence, if true, is open to similar interpretation. The comparative freshness of the *Grypotherium* hide, together with evidence of the only other comparable case, that of the ground sloth *Nothrotherium,* found in New Mexico and now in the Yale Peabody Museum, points to the theory of very recent extinction as the more probable.

The third family, Megatheriidæ, which includes *Nothrotherium* just mentioned, has as its representative genus *Megatherium,* first fully described by Sir Richard Owen. This creature was as large as an elephant, although not so tall, had a long spout-like symphysis of the lower jaw, and probably a long prehensile tongue. The family possessed no dermal ossicles for defence, as the size of its members probably gave them immunity to attack by any other than the great saber-tooth cat, *Smilodon*, and he may rarely have destroyed an adult animal. *Megatherium* and the smaller *Nothrotherium* are both known from North as well as South America. These creatures were both very ponderous of hind limbs and tail and walked on the extreme outer edge of the foot, with the long claws turned inward and clear of the ground. The fore limbs were relatively much more slender and mobile, and the animal must have walked on the knuckles of the hand with the claws bent back against the palm. It is quite apparent that they could rear on the hind legs, supported by the heavy tail as on a tripod, and use the fore limbs freely to pull down the branches of trees for the leaves, on which they normally fed. In the Yale specimen the hair which is preserved over the hip is long, coarse, rather brittle, and pale yellow in colour. The living animal, however, may have been somewhat darker.

The armoured glyptodonts of the Pampas were huge, with a carapace varying up to six feet, so that the entire animal sometimes reached twice that in over-all length, ten times the size of those of the Santa Cruz. The feet and legs were short and massive, with broad hoofs unlike the claws of the other edentates. There was comparatively little variation among the several species other than in the armour of the tail, which in *Dœdicurus* was a most remarkable structure, like the battle mace of a mediæval knight. These ultimate glypodonts converged in a most amazing way toward the ancient armoured dinosaur *Ankylosaurus* of the late Cretaceous, resembling the latter in carapace-like armour, head shield, and battle-mace-like tail, and in all probability in immunity to attack except perhaps when very young. These late glyptodonts are also recently extinct, so that their shells, which once dotted the Pampas, are said to have been used by the Indians as huts or hogans in which to live. Few of the glyptodonts reached North America, as they were not particularly mobile types. They have been found, however, as far north as the Panhandle of Texas.

Primates were the typical ones found in South America to-day, such as *Cebus*, *Callithrix*, and others. Here again the fossil record does not give a complete story, as the center of their profusion was farther to the north in the tropical forests of which we have no record.

Marsupials are represented largely by opossums from the Brazilian caves, as the old predatory thylacines had gone. The latter cannot endure competition with wolves and foxes, the forerunners of which made their appearance in the Pliocene. A comparable case is found in the existing *Thylacynus* itself, which is now confined to Tasmania but whose remains are found in the superficial deposits of the Australian mainland, where it is now entirely extinct. Here the competitor was apparently the dingo dog, *Canis dingo*, the record of whose coming is lot in the obscurity of time, although it is generally believed to have been brought from Asia by the human aborigines, as it is closely related to the Indian dhole and could hardly have made the overseas migration unaided. The Australian Black fellows, whose ancestors might have brought it over, are closest of kin to the Veddas of India, which sums the evidence.

Immigrant Fauna

We have spoken of some of the South American animals which reached the northern hemisphere, edentates, rodents (Canada porcupine), and the like. The more virile North American forms, however, invaded South America in relatively greater numbers and gradually replaced many of the indigenous forms. The great extinctions of the Glacial period were ruthless for invaders and natives alike, few surviving into modern times. The Pleistocene ground sloths and glyptodonts, the notoungulates and the invading antelopes, horses, and mastodons have all been destroyed, not only in South America but, with the exception of the antelopes and horses, in the other continents as well. The immigrants did not all arrive at once but rather in successive waves of migration, the carnivores first.

Of the *Carnivores* all were of the ordinary North American types, with certain local peculiarities among those known from the Pampas. Of bears the short-faced *Arctotherium* is the only one known, but others, like those now living in the Andean highlands, may have existed. Of the cats the most interesting is the great saber-tooth, *Smilodon,* found also in North America and apparently the only enemy of sufficient prowess and armament to have attached the ground sloths and mastodons. In fact the general belief is that their evolution was in a measure a response to such prey and that they could not survive the extinction of the latter, having reached an evolutionary *cul de sac*. There were other large cats in the Pampas as well, true biting cats or Felinæ, allied to the living jaguar and puma, and smaller ones comparable to the ocelot.

The dogs were of at least three principal kinds: the bush dog, *Icticyon;* fox-like wolves, *Cerdocyon,* nearly related to those living to-day, and one, apparently *Cyon,* resembling closely the dhole of India. There were also raccoons, one very large one, and skunks and others, giving thus a comparatively well-rounded group of beasts of prey.

Ungulates

The Notoungulata, as we have seen, were relatively few and were but the last survivors of at least three of the old suborders, their place being taken by representatives of three orders of northern forms. Of the perissodactyls, there were true horses, *Equus* and the curious *Hippidion,* with two or three allied genera. There were also tapirs, which were the only perissodactyls to survive until modern times, although their range is much more restricted today. Artiodactyls include the llamas or camel-like forms, of which the wild guanaco and vicuña, with their domesticated derivatives, survive; deer and antelopes, the last being now extinct; peccaries, the only living American swine. Proboscideans of the genus *Dibelodon* are also present, mastodons with rather complex grinding teeth and no lower tusks but with well-developed upper ones, each with a spiral band of enamel. These suffered extinction along with the northern mastodons. Why the true elephants, which were so well represented in the northern continent, together with *Mastodon* itself, never penetrated South America is a mystery.

This briefly summarizes the South American faunas and their evolution. Out of more primitive stocks there evolved strange and remarkable beasts, which probably were actually stranger in appearance than our restorations seem to show, for these are perhaps unconsciously influenced by the artist's knowledge of existing creatures, without which the reconstructions would be impossible. During the long period of isolation these throve in their various ways, adapting themselves admirably to the environmental needs, both physical and biotic. Some lines died out naturally, others flourished until they came into competition with the invading hordes and then perished, yet others actually stemmed the tide of northern invades, crossed the land-bridge in the face of their numbers, and established themselves in the northern hemisphere, but never beyond the limits of the New World. The present-day survivors are very few. Among the edentates are armadillos, sloths, and anteaters; among the marsupials, opossums and *Cænolestes;* and among the rodents, tree porcupines, cavies, the chinchilla, and the capybara; and that is about all, only a remnant of a great and remarkable host.

Chapter—11

Prosimians

The classification in Table I lists most of the main genera now recognized among primates, but classification is continually changing and there is much disagreement, especially on points such as which genera should be admitted to the order Primates or to the family Hominidae. The sensitivity and disagreement among biologists and anthropologists about such questions provides an amusing insight into their humanity, though it leads to much confusion.

The number of living genera recognized in each group is given in parentheses, and also its range in geological time. The families and genera marked with a dagger are extinct.

Table 11.1: Classification of the Primates

Order Primates

Suborder I Prosimii

Infra-order I Lemuriformes

Superfamily I	Tupaioidea
Family I	Tupaiidae. Recent (6). Tree shrews *Tupaia* (tree shrew) *Ptilocercus* (pen-tailed tree shrew)
Superfamily 2	Lemuroidea
Family 1	Plesiadapidae.† Palaeocene *Plesiadapis*
Family 2	Carpolestidae.† Palaeocene *Carpolestes*
Family 3	Phenacolemuridae.† Palaeocene *Phenacolemur*

	Notharctus
Family 5	Lemuridae. Recent (6). Lemurs *Lemur* (lemur)
Family 6	Indriidae. Pleistocene- Recent (3). Indrises *Indri* (indris) *Propithecus* (sifaka)
Superfamily 3	Daubentonioidea
Family 1	Daubentoniidae. Recent (I) *Daubentonia* (aye-aye)
Infra-order 2 Lorisformes	
Family 1	Lorisidae. Pliocene-Recent (4). Lorises *Loris* (slender loris) *Nycticebus* (slow loris) *Perodicticus* (potto)
Family 2	Galagidae. Miocene-recent (3). Galagos *Galago* (bush baby)
Infra-order 3 Tarsiiformes	
Family 1	Anaptomorphidae.† Middle palaeocene-Eocene *Tetonius* *Necrolemur*
Family 2	Omomyidae.† Upper Palaeocene-Middle Eocene *Omomys* *Hemiacodon*
Family 3	Tarsiidae. Recent (1). Tarsiers *Tarsius*
Suborder 2 Anthropoidea	
Superfamily 1	Ceboidea
Family 1	Cebidae. Miocene Recent (12). New World monkeys *Homunculus*† (Miocene) *Cebus* (capuchin monkey) *Alouatta* (howler monkey) *Ateles* (spider monkey)
Family 2	Callithricidae. Recent (2). Marmosets and Tamarins *Callithrix* (marmoset)
Superfamily 2	Cercopithecoidea
Family 1	Parapithecidae.† Eocene-Oligocene *Amphipithecus* (Eocene) *Parapithecus* (Oligocene)
Family 2	Cercopithecidae. Oligocene-Recent (14). Old World monkeys *Mesopithecus*† (Pliocene) *Macaca* (macacque)

	Cercopithecus (guenon)
	Cercocebus (mangabey)
	Papio (baboon)
	Colobus (guereza)
	Presbytis (langur)
Superfamily 3	Hominoidea
Family 1	Oreopithecidae.† Oligocene-Pliocene
	Apidium (Oligocene)
	Oreopithecus (Pliocene)
Family 2	Pongidae. Oligocene-Recent (5). Apes
	Propliopithecus (Oligocene)
	Aeolopighecus† (Oligocene)
	Pliopithecus† (Miocene)
	Aegyptopithecus† (Oligocene)
	Dryopithecus† (Miocene)
	Hylobates (gibbon)
	Symphalangus (siamang)
	Pongo (orang-utan)
	Pan (chimpanzee)
	Gorilla (gorilla)
Family 3	Hominidae. Pliocene-Recent (1). Men
	Ramapithecus † *australopithecus*†
	Homo

Life in the Trees

The typical insectivores are animals that live on the ground (hedgehogs) or under it (moles). They are retiring creatures, active in the twilight or a night and therefore depending more on their noses than their eyes. Tree shrews, however, are insectivores that have left the ground and show the features necessary for tree life. They climb and jump well, with a long tail for balance. Especially interesting for us is the large size of the eyes and of the parts of the brain primarily concerned with vision. In the trees the sense of smell is of reduced value and the importance of sight is increased. In the tree shrews both senses are well developed, but as primate life evolved further, vision became increasingly important and the complexity of the nose and the parts of the brain connected with it were reduced. In the higher primates the eyes allow stereoscopic vision and have special central areas of cones, with a low sensitivity but

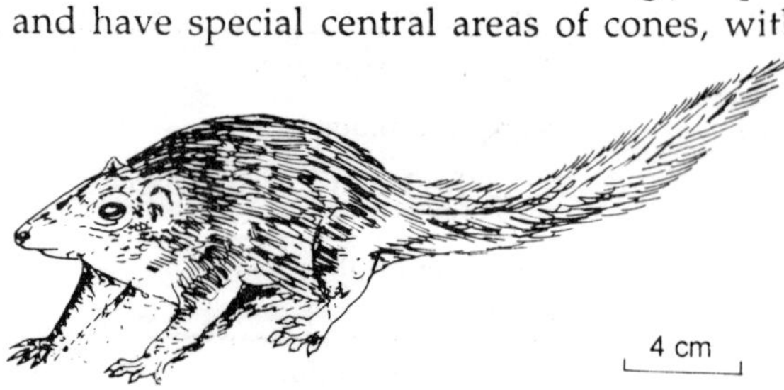

Fig. 11.1: The tree-shrew, *Tupaia glis*.

high resolution, both between points near together and of wavelengths (colour vision). This central area (Macula) has at its centre a pit (fovea) where the light reaches directly to the cones without having first to pass through nerve cells as it does elsewhere in the retina. There are about 30 000 cones in the human fovea and with these we perform all our acts of finer visual discrimination, such as reading. Thus the primate eye has developed essentially into two receptor organs, both extremely efficient. With the central area we can discriminate between points a very small distance apart. With the peripheral part of the retina, after adaptation to the dark, we can detect spots of light of almost the minimum conceivable brightness.

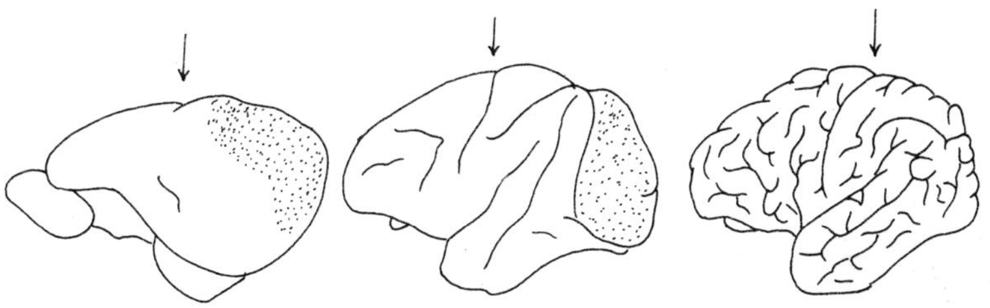

Fig 11.2: The cerebral hemispheres of (a) tree-shrew (*Tupaia*), (b) monkey (*Macaca*), and (c) man (*Homo*) drawn to the same size to show the relative sizes of the primary visual cortex (striate area), which is stippled. (In man the visual cortex extends largely on the medial surface; comparing the lateral views in this way may give a wrong impression.) The arrows point to the region of the central sulcus, separating motor from sensory cortex. (Partly after Campbell 1966).

Together with the development of the eyes there was a change in methods of communication between individuals. In most mammals this takes place largely by smell, with the development of special scent glands whose secretions allow for recognition of the sexes and appropriate stimulation. These functions are not absent in monkeys or man but they are supplemented by the display of colours and patterns of movement, especially of the face. A shrew can hardly be said to show facial expression, and indeed lacks the musculature to do so. In the later primates the facial muscles have become highly differentiated and play an important part in the interpersonal communication system which is so characteristic of primates. At the same time the whiskers (vibrissae) on the snout become smaller. As the face comes to be used for communication its sensory functions are reduced and taken over by the hand and tail.

The First Primates Prosimians

It is much debated whether the tree shrews should be classed as insectivores or as primates. The continuance of the debate shows that they lie almost exactly between the two; the non-specialist can be content to leave the argument to others. The order Primates constitutes one of the twenty or so orders into which living and extinct members of the class Mammalia may be divided. The order Insectivora is another, and others are the Chiroptera (bats), Carnivora (cats, dogs, and bears), and Cetacea (whales).

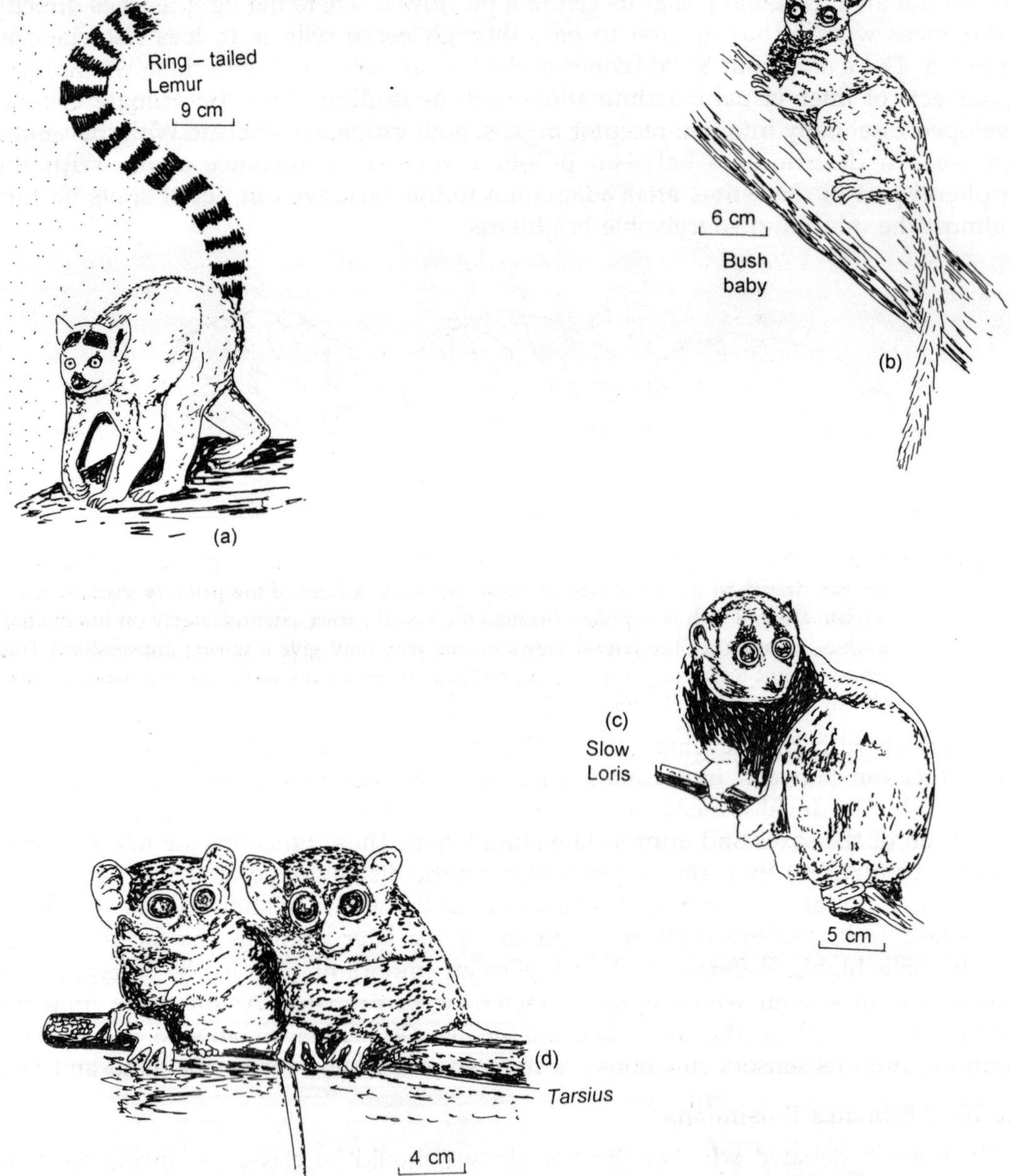

Fig. 11.3: (a) *Lemur catta,* from Madagascar, lives in woodland and although arboreal is clumsy in the trees and often adopts a terrestial habitat. The long tail is used in balancing while moving along horizontal branches. (b) *Galago senegalensis* (bush-baby). A nocturnal, jumping prosimian from the African mainland (c) *Nycticebus coucang* (slow loris) (d) *Tarsius syrichta,* the Phillipine tarsier.

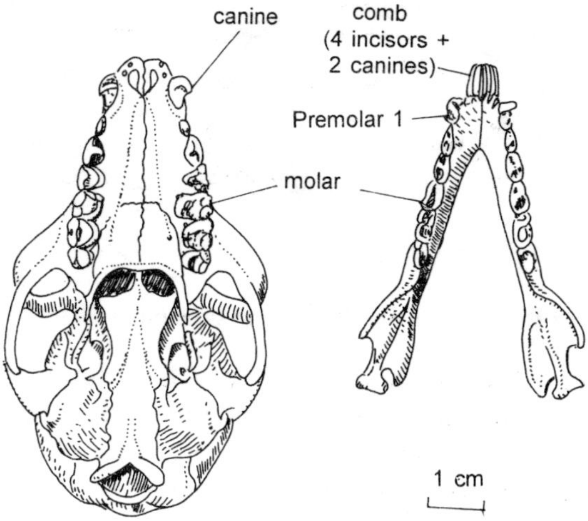

Fig. 11.4: Upper and lower jaws of *Lemur.* The incisors and canines of the lower jaw are modified to form a procumbent comb, possibly used in grooming.

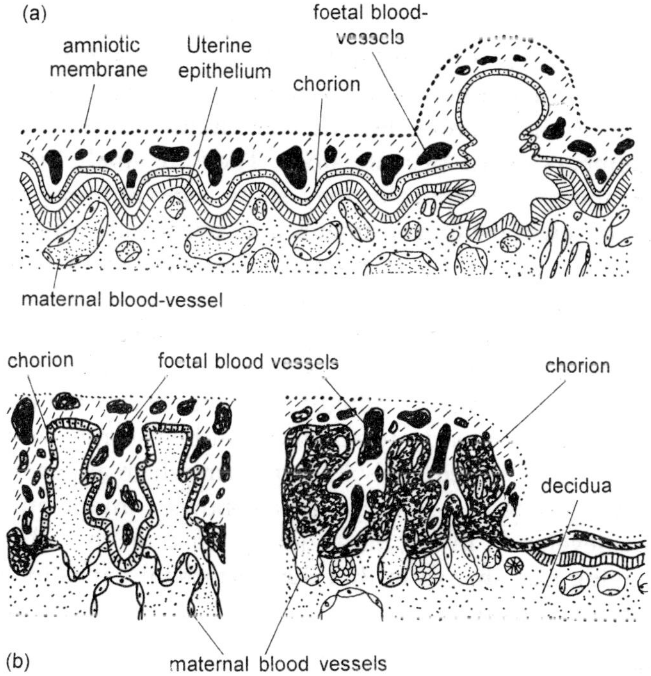

Fig. 11.5: *Placentation.* The relationship between the maternal and foetal tissues is shown in diagrammatic sections. (a) *Lemur:* epitheliochorial. The maternal tissue is bounded by an epithelial layer. (b) *Tarsius*: haemochorial. The chorionic villi of the foetus invade the uterine epithelium and come close to the maternal blood vessels. (c) Man: haemochorial. A similar situation to that in *Tarsius* which has probably evolved separately. (After O. Hill in Grassé 1955).

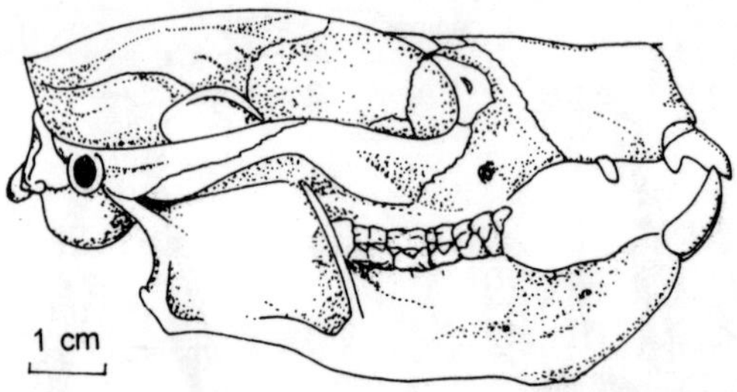

Fig. 11.6: Reconstruction of the skull of the Palaeocene prosimian *Plesiadapis*. (After Russell 1964).

The primate order is itself subdivided into two suborders, the Prosimii and Anthropoidea. The former includes the fossil remains of the earliest primates to appear and some descendants that survive with relatively little change. The Anthropoidea include all the rest, that is to say, monkeys, apes, and man.

The prosimians alive today show us what our ancestors were like some 60-70 million years ago. They include the lemurs, and tarsiers.

Tree Shrews

It is generally agreed that the living tree shrews of the Orient are forms transitional between the primitive insectivore ancestors and more typical primates. They may be reasonably placed in either order, but we here follow current tendencies in ranging them as basal members of the primate stock. They are small, arboreal animals with a squirrel-like appearance; anatomically they show a mixture of primitive features with others suggestive of relationship to primates and, particularly, the lemurs. Claws, rather than nails, tip the toes; three incisors are present in the lower jaw of living tree shrews; the cheek teeth are quite primitive in character. On the other hand, the presence of a relatively large brain in which the olfactory region is small, a postorbital bar, a middle-ear region built like that of lemurs, and considerable opposability of pollex and hallux all indicate that the tree shrews are allied to the ancestry of the more typical primates. Unfortunately, there are few early fossils that can be assigned with confidence to the tree-shrew group. The only adequately known form which appears to be closely related is *Anagale* of the Oligocene of Asia. This genus shows a minor but interesting advance in that nails had developed in the hind foot, at least; on the other hand, *Anagale* is a step below the living tree shrews in that the postorbital bar is incomplete and there is a full complement of incisors.

Eocene Lemurs

Apart from the transitional tree shrews, the living primates below the monkey level comprise a varied series of lemurs (mainly from Madagascar) and *Tarsius* of the East Indies. The lemurs are seemingly primitive forms; *Tarsius,* although specialized, is believed to represent a more progressive stage in primate evolution, transitional to the Anthropoidea.

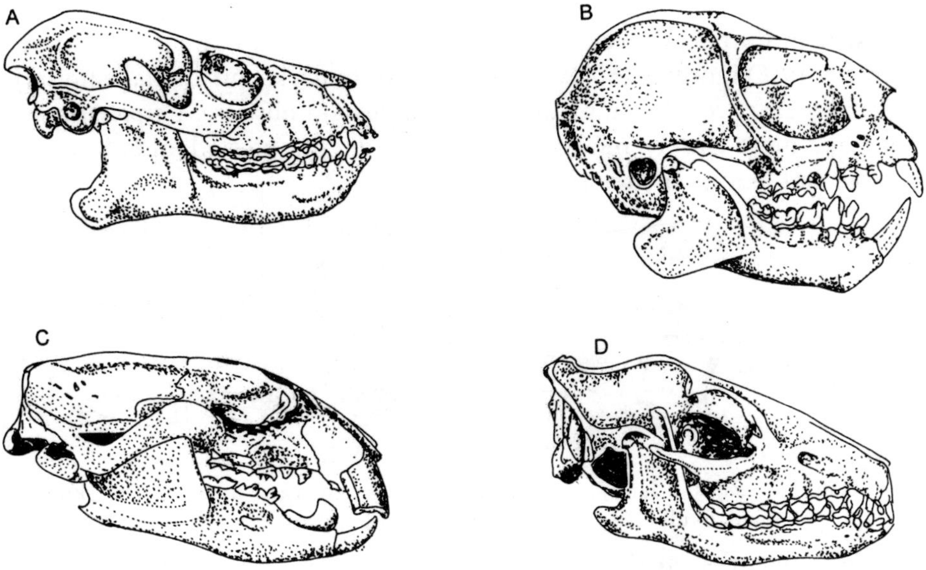

Fig. 11.7: Skulls of early primates. *A, Notharctus,* an Eocene lemur, length of skull about 3 inches; *B, Tetonius,* an Eocene tarsioid, length of skull about $1\frac{1}{8}$ inches; *C, Sinclairella,* an early Oligocene primate with enlarged incisors, length about 2¼ inches; *D, Anagale,* an Oligocene relative of the tree shrews, length about 2¼ inches.

Almost nothing is known of the antecedents of these living groups in the later Tertiary. In the Paleocene and the Eocene, however, there are, in both North America and Europe, numerous genera of small mammals which are obviously primitive primates ancestral to both modern stocks. The taxonomic assignment of many of these forms has, however, been a difficult one. Most of the remains are very fragmentary and consist of teeth and partial jaws. Primate teeth are rather primitive in nature, and dental differences between lemurs and tarsioids are not marked; it is, therefore, not surprising that many of these genera have been variously and variably assigned to the insectivores, lemurs, and tarsioids.

There was, apparently, a strong tendency among early primates for the development of long, chisel-like incisors analogous to those of rodents or of the multituberculates and diprotodont marsupials already discussed. As will be seen, such teeth occur in one of the living Madagascar lemurs, and some tendency of this sort was present in early types which are rather confidently believed to be true tarsioids. In the Paleocene and Eocene, such developments were very common. *Carpolestes, Apatemys,* and *Plesiadapis* are forms which have very highly developed incisors but differ so among themselves that each is regarded as representative of a separate family of aberrant early primates. The *Carpolestes* type, confined to the Paleocene of North America, has, in addition, developed large, shearing lower premolars, which strengthen the analogy with certain multituberculates and diprotodonts. The other two groups are present in Europe as well as in North America, and one apatemyid survived into the American Oligocene (*Sinclairella*). As our remarks on dental similarities would suggest, the exact relationships of these families are obscure. *Carpolestes* and *Apatemys* are perhaps more closely related to the tarsioids; the plesiadapids have been thought to be aberrant lemurs or tree shrews.

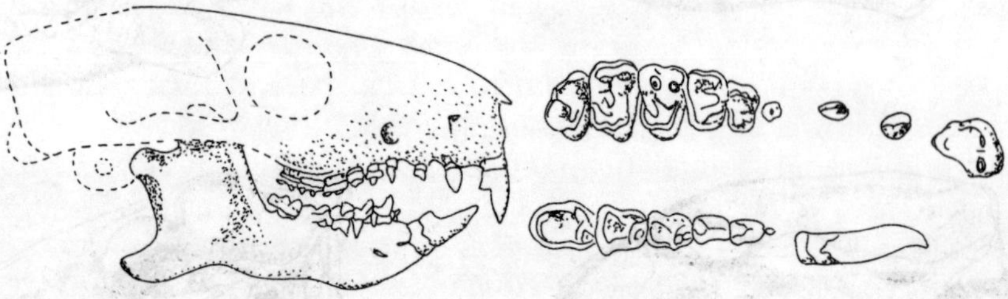

Fig. 11.8: *Plesiadapis,* a late Paleocene primate with enlarged incisors. The skull is imperfectly known. Skull about natural size, teeth ×2 approx.

As our remarks on dental similarities would suggest, the exact relationships of these families are obscure. *Carpolestes* and *Apatemys* are perhaps more closely related to the tarsioids; the plesiadapids have been thought to be aberrant lemurs or tree shrews.

Despite doubts as to the position of many genera, there are a number of Eocene forms which are generally accepted as definitely lemuroid and ancestral to the typical living lemurs. These constitute the family Adapidae, of which *Notharctus* of North America and *Adapis* and *Pronycticebus* of Europe are among the better-known types. *Notharctus* was a small mammal with a skull about 2 inches long which probably resembled the ordinary lemurs of today in general appearance. It was already quite lemur-like in its proportions and many structural features. The dentition, however, was more primitive in a number of respects. Although some members of the family had but three premolars, *Notharctus* still retained the full complement of four. Further, the upper incisors were comparatively normal; the canines were of primitive type; and the lower incisors little, if at all, procumbent. In *Notharctus* a hypocone had already developed in the upper molars, but some Eocene relatives had only slight traces of this structure. In the European forms a hypocone was developed as well; but a curious feature is that, while the two forms were closely related and had both acquired this same new cusp, *Adapis* and *Pronycticebus* had obtained it by an upgrowth of the cingulum, *Notharctus* by a budding-off from the protocone.

Madagascar Lemurs

By the close of the Eocene the lemurs had disappeared from the fossil record in the North Temperate Zone, and we know nothing of their further history until Pleistocene and Recent times. Here we find them present in Old World tropical regions, where conditions today are probably not dissimilar from those prevailing farther north in the Eocene. The main center of the group is the island of Madagascar. This region presumably was cut off from the mainland during most of the Tertiary; and there are but few carnivores—a fact which may have been responsible for the survival of these primitive primates (as well as of many insectivores).

The Madagascar lemurs may all be included in the family Lemuridae, using this term in a broad sense. They are of interest in giving us as near an approach as we can obtain today to the general appearance, habits, and structures of our Eocene primate ancestors. A typical Madagascar lemur of today is a fairly small arboreal animal, nocturnal in habits, with a bushy hairy covering which contrasts with the rather sparse coat of most higher primates. The limbs are moderately long, the ears pointed, and the eyes are directed more laterally than forward. The lemur tail, usually long, is never a grasping organ. Pollex and hallux are always widely seperably from the other fingers; the big toe is especially well developed and has a flat nail in all forms, whereas the other digits are variable in their covering. Typical lemurs have a clawlike nail on the second toe but normal nails on the other digits.

In living lemurs the upper incisors are often small or absent, while the lower ones, together with an incisor-like canine, slant forward as a fur-combing organ. The upper molars vary from the triangular to the quadrangular pattern.

The lemur face may undergo some shortening, but it is in most cases a comparatively long and foxlike muzzle. The braincase is long and low in the more primitive living types but somewhat expanded in the more advanced lemurs.

There are numerous divergences from this general type among the living Madagascar lemurs. There is, for example, a small, hopping form; and *Indris* and a few other lemurs are rather large and somewhat monkey-like in superficial features. Still other variants have been discovered in Pleistocene bog deposits on the island. *Megaladapis* was the largest-known lemur, attaining the size of a pig, with a skull a foot long, peculiarly long and decurved nasal bones, and small eyes; it seems to have resembled the Eocene lemurs closely in many features. A second extinct group was represented by *Archaeolemur,* also large for a lemur (a 5-inch skull) but with large eyes turned remarkably far forward and a large brain, there being no sagittal crest. The face was considerably shortened. This extinct form paralleled in a number of respects the higher primates and is one of several abortive "efforts" of the lemur group to reach a monkey status.

A final member of the Madagascar primate fauna is the aye-aye, *Chiromys*. It is a nocturnal form about the size of a small cat, with a short, high skull of the shape found in the higher lemurs and monkeys. All the digits except the big toe have clawlike structures rather than nails; the middle finger is very long and slim and is said to be used to extract grubs from their holes in the wood. The single remaining pair of incisors is enlarged in rodent fashion and is used in gnawing through bark in search of insect food. The aye-aye has been thought to have descended from the plesiadapids but more probably has resulted from a parallel evolution among the Madagascar lemur group.

Higher Lemurs

All the living non-Madagascar lemurs are included in a higher group, the family Lorisidae. The lorises of the Indian region are typical; in Africa the group is represented by the "bush baby," *Galago*, and the potto. The muzzle is rather short, the braincase high, the eyes turned somewhat forward, and the premolars reduced to two; the lower front teeth, however, have the typical arrangement of the Madagascar lemurs. The index finger is reduced, giving a better grasp between the thumb and the remaining fingers. The advanced skull structures of these lemurs are of the sort required to turn a lemur-like primitive primate into a more advanced type; but it is generally held that this is a development parallel to, and not on, the main evolutionary line. It is sometimes thought that the lorises, because of technical differences in construction of the ear bulla, are only remotely related to the more typical lemurs; but this belief lacks paleontological support and neglects embryological considerations.

Tarsius

A small but very interesting primate group is that which includes a single living animal, *Tarsius*, in addition to numerous Paleocene and Eocene fossil genera. These forms have often been classed as lemurs but are quite advanced in structure and may be appropriately placed in a separate suborder, the Tarsioidea, intermediate between the true lemurs and the Anthropoidea.

Tarsius is a small, nocturnal, tree-living form from the East Indies. The tail is very long, and the hind legs are adapted for a hopping gait. The braincase is large and rounded, and the foramen magnum is pushed somewhat forward onto the under side of the skull in contrast to the usual lemur condition. The visual apparatus is much more advanced than that of the lemurs. The eyes are exceptionally large and are turned completely forward from the primitive lateral position, with the orbits very close together above the nose. In ordinary lemurs there is only a superficial bar separating orbit and temporal fossa; but in *Tarsius* the alisphenoid extends outward below the frontal and jugal and closes off the greater part of the opening between the two regions, much as in the Anthropoidea.

While vision has thus advanced, the sense of smell has been proportionately reduced. The foxlike snout of the typical lemur has disappeared, and the nose is reduced to a small nubbin tucked in beneath and between the eyes. There is little projection of the muzzle (the tooth row is rather short); and we have here the beginning of the type of face found in monkeys, apes, and man. There are some specializations of the front teeth, but the cheek teeth are of primitive pattern with fairly typical tribosphenic molars.

Specializations have developed in connection with the hopping gait. Alone among primates we find in *Tarsius* a partial fusion of tibia and fibula. A third segment is introduced into the limb by means of an elongation not of the metatarsals but of the calcaneum and navicular (a similar development has occurred in a Madagascar lemur and in *Galago*). This seemingly odd fashion of developing a third joint is readily understandable when we consider the situation. The ordinary animal which introduces an extra segment for hopping or, indeed, fast locomotion of any sort (as the kangaroo, jumping mice, or any typical ungulate) has no grasping ability and is usually a ground-living type. The primate, however, is in great measure dependent for security in the trees on the powers of opposability of the great toe. If the metatarsals were to elongate, his grasp would be lost; elongation must take place proximal to the origin of the big toe, and the proximal tarsal bones give the only possibility for this development.

Apart from this peculiar limb specialization, *Tarsius* is close to the higher primates in most features. An additional suggestion that this form is a relic of a primitive group transitional to the Anthropoidea lies in the structure of the placenta. In the lemurs this structure was of a type probably present in primitive placentals. But in *Tarsius* the placental connections between mother and young are concentrated into a pancake-shaped structure which is shed at birth as in the higher primates.

Eocene Tarsioids

Prior to Recent times there is a gap in the history of the group extending to the early Tertiary. But in paleocene and Eocene deposits we find in Europe and North America numerous remains of small forms unquestionably related to *Tarsius*, as well as others of rather more doubtful position. Many of them are known merely by the dentition. This was often, as in *Tarsius*, close to the primitive placental type in molar construction, the hypocone appearing only rarely. In some genera the first lower premolar was still present as a small tooth; but normally there were, as in *Tarsius*, but three premolars, and in one case there were but two. The anterior teeth, however, were quite variable in structure. In such forms as the best-known genera, *Tetonius* of the American Lower Eocene and *Necrolemur* of the Upper Eocene of Europe, the incisors show some of the enlargement noted in various other lines of early primates.

In *Tetonius* the general structure of the skull is quite similar to that of the living *Tarsius*, with a very short face and huge, forwardly turned orbits with a developing partition behind them. Some of the others, like *Necrolemur*, appear to have been rather more lemur-like, with a somewhat less reduced snout.

These early tarsioids are customarily included in the family Anaptomorphidae, within which, however, there certainly existed a diversity of types, and some of the earliest Paleocene genera may have been close to the ancestry of the lemurs as well as tarsioids. Eocene tarsioids had specialized in a number of directions. In one or two forms skeletal material shows that the peculiar modifications of the hind legs found in the living *Tarsius* were already present. Still others may be ancestral to the higher anthropoid primates which had developed by the beginning of the Oligocene.

ANTHROPOIDEA—MONKEYS AND APES

Oligocene Primates, the Earliest Monkeys

In spite of the abundance of prosimian fossils no remains of any monkeys or other undoubted Anthropoidea have been found in the Eocene. The reasonable

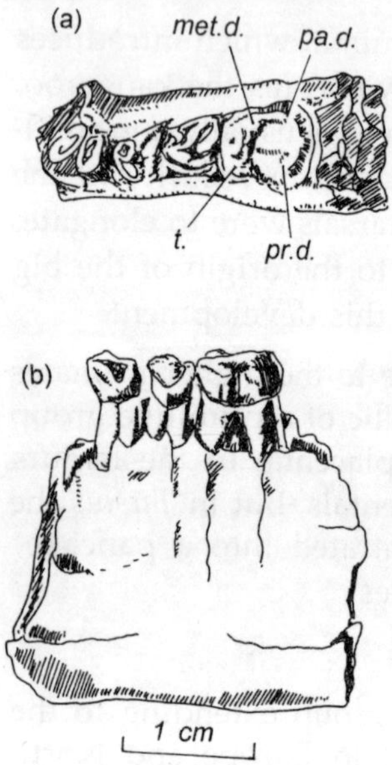

Fig. 11. 9: Drawing of late Eocene fossil *Amphipithecus*, a mandible whose molars show a pattern that may be intermediate between the tricuspid teeth of prosimians and quadrangular ones of anthropoids. Metaconid; *pa.d.*, paraconid; *pr.d.*, protoconid; *t.*, talonid.

assumption is that none existed, at least until the very end of that period. A fragment of a left mandible with parts of four teeth from the late Eocene from Burma (*Amphipithecus*) shows a cusp pattern that may be intermediate between that of the omomyid prosimians described in the last chapter and the few remains that are available of Oligocene primates. By the middle of the Oligocene, about 35 million years ago, fragments of several different anthropoidean types are found. Although we have not very much to go on it seems that the monkey and ape lines were already differentiated at that time. A form possibly somewhere near to the ancestor of the monkeys is *Parapithecus*, several of whose lower jaws have now been found in Egypt. The molar teeth no longer show the tricuspid pattern but have four main cusps arranged approximately in two rows. This is a plan that could have led to the characteristic type of tooth found in the monkeys of both Old and New Worlds, and also to the teeth of Hominoidea.

Probably from the Oligocene onwards the monkeys have evolved as a line (or lines) distinct from our own. The modern monkeys represent therefore a somewhat modified version of what our ancestors were like about 35 million years ago. This is a considerable time even by evolutionary standards and we must not expect that their structure and habits will greatly illuminate our own. In many respects wide divergence has occurred.

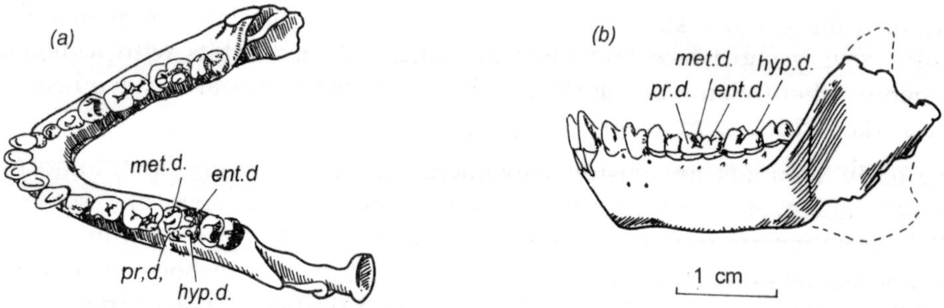

Fig. 11.10: *Parapithecus*, a mandible from the Oligocene of Egypt, with four-cusped molars that may well have belonged to a form near the ancestry of the monkeys. Of the original triangle only the protoconid and metaconid remain. The heel has a well-developed hypoconid (*hyp.d.*) and medial to it an entoconid (*ent.d.*). Other abbreviations as in Fig. 11.9.

Monkeys

The monkeys have greatly developed the arboreal capacities of the prosimians. They mostly climb and run along branches but the spider monkeys (*Ateles*) and woolly monkeys (*Lagothrix*) of South America swing from branch to branch by the arms, a mechanism superficially similar to the brachiation adopted by gibbons. This tree life is so different from our own that it is difficult to suppose that many of the features later to develop specifically in humans had appeared at the Oligocene stage. But we really know so little of Oligocene primates that it is impossible to say what they were like or whether the ceboid, cercopithecoid, and hominoid lines had yet diverged. Indeed, there is very little further fossil information about the evolution of the two types of monkey. Fossils from the Miocene of South America show that by that period the ceboid type had already evolved there. For the cercopithecoids the evidence is even more scanty. This is strange, since the Old World monkeys have generally been considered to be a relatively primitive type, diverging early, perhaps in the Oligocene. They probably did this, but they may have remained unsuccessful for a long time, at least they have left few fossils. There are reasonably abundant remains of *Mesopithecus* of the late Miocene and Pliocene, and it is clear that these were already quite like modern monkeys. The radiation of old world monkeys may therefore have occurred relatively late.

This leaves us ignorant as to how closely we are related to them, and hence how much we can learn of ourselves from their behaviour. Probably not very much, for both lines have changed greatly since they diverged.

South American Monkeys

The Platyrrhini comprise two families of monkeys now living and found as fossils only in South and Central America. Superficially, many are quite similar in appearance to Old World monkeys but seem to be more primitive in a number of respects. They are, on the average, much smaller than their Old World relatives. Almost never is there any reduction of the primitively long tail. The thumb is but little opposable to the other fingers. Three premolars are still present in all cases, whereas the oldest fossil representatives of the Old World apes had already lost the second premolar. The group name ("flat nose") refers to the fact that the nostrils are usually far apart and face outward.

Marmosets

Of the two families, the smallest both in number of forms and in the size of the individuals is that of the marmosets, the Hapalidae. *Hapale*, the common marmoset, is a small creature, squirrel-like in size and general appearance, with thick fur and a bushy tail. Squirrel-like, too, is the fact that the toes (except the hallux) have the nails compressed and clawlike, so that the progression is mainly by digging the claws into the bark; the hallux has lost much of its opposability. A peculiar feature is the fact that the last molar has been lost, the only case of its complete reduction among primates (although man is approaching this condition). The incisors, and the lower canine with them, are rather procumbent and lemur-like. A Miocene marmoset has been recently reported, but otherwise the early history of the family is unknown. Some have believed that these creatures are the most primitive of anthropoids, but such features as the molar loss suggest that they are specialized rather than primitive.

Cebids

A more important and larger group is that of the Cebidae, the typical South American monkeys. Here the size is somewhat larger, the hairy covering somewhat thinner (a tendency carried much further in Old World types), and the teeth are little specialized and with the full anthropoid complement of three premolars and three molars. In some cebids, alone among primates, a prehensile tail has developed.

Most primitive of living cebids is the group represented by *Aotus* [*Nyctipithecus*], the owl monkey or dourocouli, a small form with a presumably primitive thick woolly covering and a small brain. It is a nocturnal form, while most monkeys are active in the daytime. The owl monkey is quite similar to the marmosets and may be close to the base of the platyrrhines. This belief is strengthened by the fact that the only cebid of any antiquity, *Homunculus* of the Miocene, belongs to this group.

The owl monkeys and several other fairly primitive forms lack a prehensile tail. This "fifth limb" is developed in the remaining cebids. Of these, the commonest group is that best known through the capuchin monkey, *Cebus*, usually to be found accompanying the organ-grinder. Closely related are the spider monkeys, *Ateles*. Here the limbs are very long and slender, the thumb reduced or absent, the nails rather clawlike. These monkeys are clever acrobats, swinging from the branches by any of their five useful appendages.

A final group is that of the howling monkeys *Alouatta*. These are the largest of the New World forms, some as large as a good-sized dog, and are leafeaters with heavy molar teeth. The larynx is expanded into a huge ossified shell, with a resonating chamber within, by means of which night is made mournful in the South American forests.

Little is known, unfortunately, of the fossil history of the South American monkeys beyond the fact that both families were present in the Miocene. South America, as we have noted, seems to have been an isolated area from early Eocene times. Perhaps the ancestral platyrrhines entered South America at the dawn of the Tertiary; forest-dwelling forms of this sort might well have been present, and even abundant, in the Eocene and Oligocene and yet have failed to be preserved in the plains deposits of Patagonia. On the other hand, it is possible that, like the hystricomorph rodents, they managed to gain access to the continent at a later date. Unless some such late entry were made "overseas" from Africa, the platyrrhines must have evolved independently of catarrhines from a tarsioid stock, for no members of the Anthropoidea developed in North America, the only area with which South America was early connected.

Catarrhines

The catarrhines of the Old World form a well-defined group, including the living monkeys of Asia and Africa, the anthropoid apes and man, and their fossil representatives. The name indicates one of the many features in which they are more advanced than the South American monkeys; the nostrils are closer together and open forward and downward, with a smaller bony opening. There is a rather general tendency toward an increase in size which culminates in the chimpanzee, man, and the gorilla. Although originally arboreal, some of the higher or more specialized types, such as man and the baboons, tend toward a terrestrial life. The tail is often shortened or absent. Primitively four-footed, some have tended to an arm-swinging type of locomotion in which the body is held erect; man has attained an erect posture. Although progression on the ground is usually on all fours, the long arms bring the back into a slanting position, while sitting is the common catarrhine pose when not in motion. The thumb is usually well developed and opposable, and the big toe universally so except where secondarily reduced in man. The hairy covering is always thin and the face naked.

The brain is large; but since the forms are, on the average, larger, the ratio of brain size to body is reduced, and sagittal and occipital crests are more common than in South American monkeys. The face, primitively short, tends to elongate in many instances in correlation with a tendency toward a vegetable diet and a consequent increase in tooth size. The second premolar is always absent, the tooth formula being invariably $\frac{2.1.2.3}{2.1.2.3}$. The molars are quadrangular and essentially four-cusped.

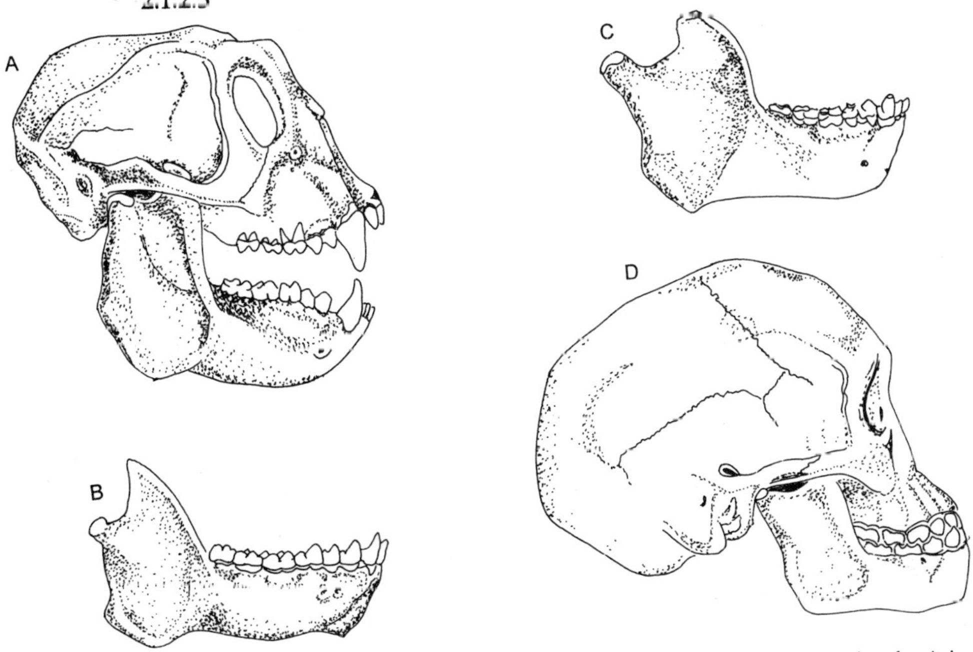

Fig. 11.11: *A*, The skull of *Mesopithecus*, a Pliocene Old world langur, length of original about 3 inches; *B*, the jaw of *Parapithecus*, oldest-known Old World monkey from the Lower Oligocene of Egypt, original about 1½ inches long; C, the jaw of *Propliopithecus*, oldest anthropoid ape, from the same deposits, original about 2¾ inches long; *D*, restored skull of *Australopithecus*, a young individual of an extinct anthropoid-ape type from South Africa.

Old World Monkeys

The most primitive known member of the group was *Parapithecus*, from the Lower Oligocene of Egypt. Although it is known only from one jaw, there is yet enough information to be derived from this to tell us that we are dealing with a small, primitive catarrhine which may have been close to the ancestry of the later Old World monkeys and apes. The canine was small, and the first of the premolars was still simple, in contrast with its usual bicuspid condition in catarrhines. There were four main cusps on the molars, but these were not connected by cross-crests, as in the living monkeys; and a fifth cusp, the hypoconulid, was present on all three molars, whereas it appears only on the last in the existing Old World monkeys. The two halves of the jaw met at a sharp angle, a primitive feature contrasting with the more rounded contour of the "chin" in all living Old World forms. The jaw proportions suggest that the skull, if found, would reveal the presence of large eyes and a large brain. As far as our knowledge goes, this small monkey may well have been the common ancestor of all the later catarrhines. In appearance it might have been similar to the more generalized African monkeys of the present day.

The living members of the Old World monkeys constituting the family Cercopithecidae include a large number of forms with a wide range of structure. They are, in terrestrial locomotion, four-footed forms, walking flat on palm and sole. The more primitive forms are arboreal, but the baboons and their relatives show a progressive tendency toward a ground-dwelling life. Toughened skin areas—the ischial callosities—are present on the buttocks. As noted above, the molars are four-cusped, the hypoconulid being present only on the last lower molar; and, characteristically, two cross-crests are developed.

Two great groups may be distinguished, the larger being that of the subfamily Cercopithecinae. This includes many primitive African and Asiatic arboreal monkeys and a series of forms leading structurally to the baboons. The primitive types are shot faced, but the baboons have a long muzzle associated with the vegetable diet and the correlated long tooth row. There are well-developed cheek pouches for the stowing-away of food, while the stomach (in contrast to the other subfamily) is simple.

The genus *Cercopithecus* includes most of the African monkeys commonly found in zoological gardens (as the green monkey, vervet monkey, and Diana monkey). Typical arboreal forms with a short face, they probably stand near the base of the monkey stock. The macaques (*Macacus*) are monkeys of fairly large size and mostly Asiatic in distribution but with a representative—the Barbary "ape" in northern Africa and Gibraltar; there are fossil forms from the Pliocene of Eurasia. The tooth row is considerably elongated, and the snout consequently projecting. These forms are only partially arboreal, some species living in great measure on the ground. From this, several intermediate types lead to the baboons (*Papio*). These forms have become completely terrestrial, with a four-footed plantigrade gait. The tooth row of these exclusively herbivorous types is much elongated, and a doglike muzzle is developed.

The tail is short and in some of the related types may be practically absent. Baboons are found in the Pleistocene of northern Africa and Asia.

The second subfamily of the modern Old World monkeys is the Semnopithecinae, a comparatively small group including *Semnopithecus*, the langurs, and their relatives (mostly Asiatic, with one African genus). There is never any facial elongation, the body is slim, the hind legs especially long, the tail usually well developed. Instead of cheek pouches, there is a complicated stomach for food storage. Cross-crests on the molars are highly developed. All the members of the group are good arboreal types. Remains of ancestral forms are widespread in the Pliocene of the Old World: *Mesopithecus* is known from the complete skeleton.

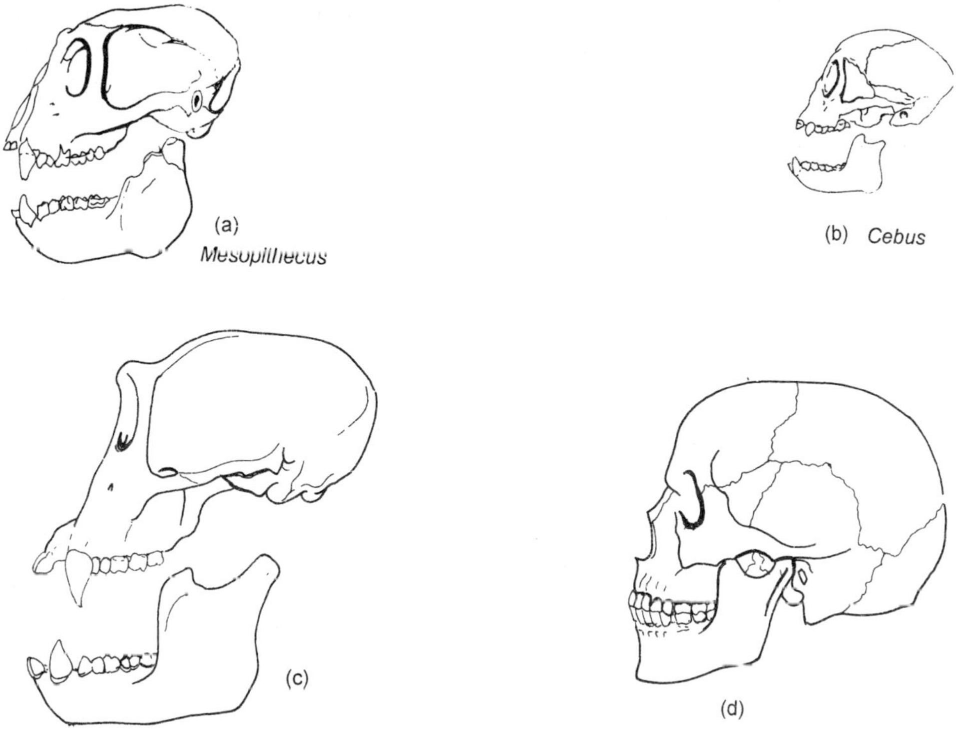

Fig. 11.12: Skull of (a) *Mesopithecus* from the Miocene, compared with those of the modern (b) *Cebus* (New World monkey), (c) *Pan* (chimpanzee), and (d) *Homo*.

The Diet of Primates

The diet of monkeys is mainly leaves and fruit, although most primates will eat meat and some do so regularly. The teeth of both groups of monkeys are rather specialized for a herbivorous diet. The incisors are flattened, perhaps for cutting fruits. The canines are pointed and larger in the male, presumably a character of sexual and social significance, and perhaps defensive. The molars are of characteristic quadritubercular ('bilophodont') shape, especially in Old World monkeys. In the upper jaw this is achieved by adding a hypocone to the trigon. In the lower jaw the protoconid and paraconid form the anterior part of the square and the talonid has moved alongside them to make the posterior part carrying two cusps, hypoconid and entoconid. This forms a grinding arrangement, characteristic of monkey teeth, which distinguishes them from the low-cusped teeth of apes and men. Fruits and seeds provide especially rich sources of nourishment, not requiring the mastication of large quantities of siliceous grasses, upon which the grazing herbivore must spend so much time. The prime need is sharp eyes to find the fruit, hands and skill to pluck it, and taste to sample it. This form of diet may well have had much to do with the development of the quick-witted manipulating primate and its differences from the 'slower-brained' ungulate types (though such differences are hard to quantify).

The supply of fruits and seeds often changes and the primate has to resort to bark, shoots, and perhaps small animals, ants, termites, frogs, and lizards. Even the terrestrial baboons who live mainly on grass seeds and roots will eat meat. Most primates are thus catholic in their choice of diet and can digest meat. This is consonant with our own omnivorous diet. Knowledge about other animals does not really help

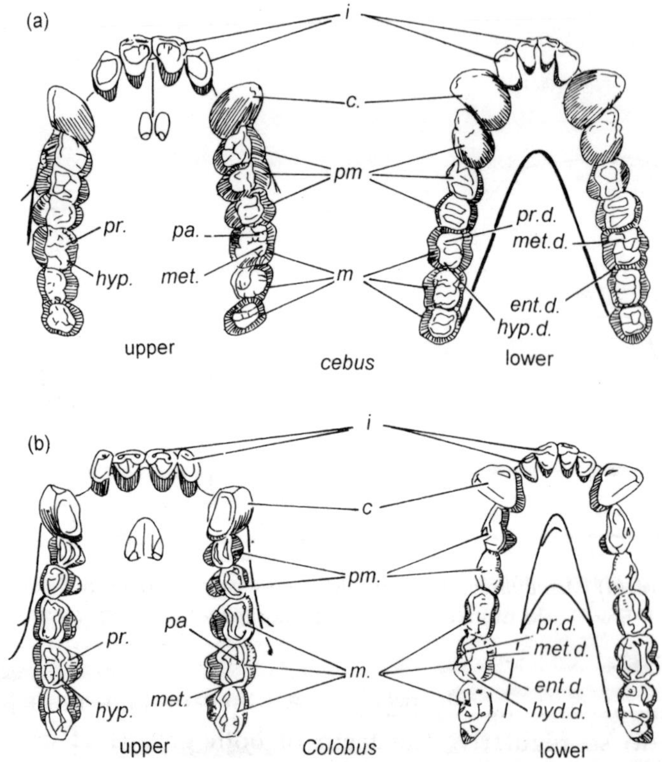

Fig. 11.13: Upper and lower teeth of (a) New World monkey (*Cebus*) and (b) an Old World monkey (*Colobus*). (After Buettner-Janusch 1966.) *c.*, canine; *ent. d.*, entoconid; *hyp.*, hypocone; *hyp. d.*, hypoconid; *i.*, incisor; *m.*, molar; *met.*, metacone; *met. d.*, metaconid; *pa.*, paracone; *pm.*, premolar; *pr.*, protocone; *pr. d.*, protoconid.

us in deciding what is a 'good' diet, let alone a 'best' one. Indeed, man himself reflects the variety of diets. He makes do with what he can get. If there is nothing but rice he will live on that, with little or no animal protein (though with, of course, the risk of malnutrition). Under conditions nearer the poles, where there is little vegetation, he becomes, as the Eskimos, almost wholly carnivorous. Others, such as Australian aborigines, bushmen, and the Hadza, eat a wide range of plants and animals. Of course, these diets are not all equally adequate either in calories, particular amino-acids, or vitamins. It is curious that man seems to have relatively little capacity to seek out such foods as shall supplement the deficiencies. Perhaps the mechanisms for doing so have been developed by herbivores to search for a relatively narrow range of requirements. The primate method has been to develop an omnivorous habit and rely upon getting something of everything. There is little exact knowledge about these special capacities and tastes in different animals.

Manlike Apes

The Old world monkeys just considered constitute two specialized lines, neither of which can be considered as ancestral to the anthropoid apes or man. To follow the history of these higher types we must return again to the Lower Oligocene beds of Egypt, in which was found *Propliopithecus,* the oldest definitely identifiable member of this group of highly developed primates. Here again we have only a jaw to deal with, but, again, one that is quite characteristic. It is not more than an inch and a half in length, indicating a form much smaller than any living manlike ape. The lower molars are definitely five-cusped, in agreement with the great apes and in contrast with the living monkeys, and the jaw has the relatively great depth of the higher forms.

The family Simiidae, of which this is the earliest representative, includes four living types: the gibbon, the orangutan, the chimpanzee, and the gorilla. They are of large

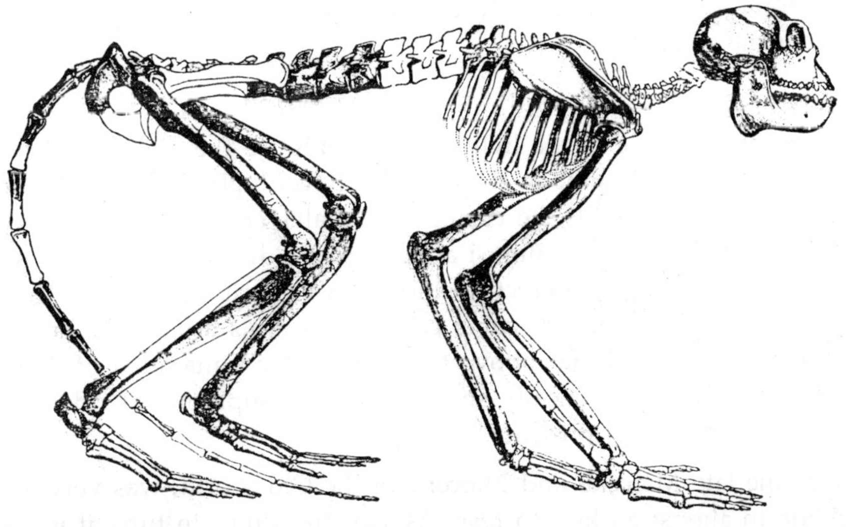

Fig. 11.14: *Mesopithecus,* a Pliocene relative of the living langurs. About 1/3 natural size.

size, ranging from the comparatively small gibbon to gorillas several times as heavy as a man. In the quadrangular molars fifth cusp (hypoconulid) is well developed below, while in the upper teeth the hypocone is comparatively small, the old trigon still being apparent. The molars tend to be rather long, and there is some lengthening of the face. The skeleton is rather close to the human type. The chest is broad in contrast to that of the monkeys; the ilia tend to broaden, as in man. The hands are rather similar to those of man, but the fingers are very long as compared with the thumb, and the arms much lengthened. The legs, on the contrary, are short; and the hallux, as well as the remaining toes, is much better developed and more finger-like than in man. These skeletal characteristics have seemingly developed in relation to the anthropoid-ape type of locomotion. With increasing size, normal four-footed progression in the trees has become increasingly impossible; and the members of the group normally progress by swinging from bough to bough with the body suspended from the hands. The feet are used more than ever as grasping organs and have so much lost their primitive mammalian character that most great apes cannot walk flat on their soles but must support their weight on the outer side of the foot.

It will be noted that in swinging by the arms the body is necessarily erect. Further, the front legs of all forms being much longer than the hind legs, the body is necessarily tilted up considerably in front in quadrupedal progression. Bipedal locomotion has thus begun among the arboreal types.

The brain is large, especially in the gorilla. But in these large forms its growth, as might be expected, has hardly kept up with the increasing body weight. Since the braincase is comparatively small, sagittal and occipital crests may develop, and the upper margins of the orbits project in most forms as supraorbital ridges above the front edge of the braincase. There is, as in most mammals, no projecting chin, the jaw sloping away under the symphysis.

Gibbons

Smallest and most primitive of the living anthropoids are the gibbons (*Hylobates*) of the Malay region. The average gibbon stands about 3 feet high in the erect position. Alone among the living anthropoids, he customarily walks erect with the arms used as balancers. The gibbon, however, is almost entirely a tree dweller and with his long arms, which can reach the ground in erect position, is a very clever acrobat. The bulk of the gibbon brain averages about 90 cubic centimeters. This seems a small figure compared with that of man but is far above the average for a mammal of that size. The braincase of the gibbon is practically smooth, and the supraorbital ridges small. Vestigial ischial callosities are still present.

Pliopithecus, of the late Miocene and Pliocene of the Old World, was very similar to the living gibbon in almost all known respects; and from this, in turn, it is a short step back to *Propliopithecus*, from which it is generally agreed that the gibbon is derived.

Orangutan

The orangutan, *Simia* (or *Pongo*), the "man of the woods" of Sumatra and Borneo, stands next above the gibbon in the scale of living primates. Considerably larger in size (adult males reaching nearly 5 feet), the orange is still a true arboreal type and is but a clumsy four-footed performer on the ground; the arms are somewhat shorter, reaching only to the ankles. The brain here is vastly larger than that of the gibbon, reaching 550 cubic centimeters in capacity. The braincase is expanded, crests rarely develop, and the supraorbital ridges are negligible. The tooth row, however, is long and the snout projecting, with a characteristic concave lateral profile to the face. Orang-like teeth are known from the Pliocene of India.

Higher Hominoids. Apes

Table 11.2: List of genera of hominoids

Age in years ago × 10^6

Superfamily Hominoidea	
Family	Oreopithecidae
	Apidium (30)
	Oreopithecus (13-7)
Family Pongidae	
	Pliopithecus (26-7) Europe and Africa
	including *Limnopithecus*
	Hylobates (gibbon). Recent
	Symphalangus (siamang). Recent
	Aegyptopithecus (?30)
	Dryopithecus (25-10) Europe, Asia, Africa
	including *Sivapithecus, Proconsul, Bramapithecus*
	Pongo (orang-utan). Recent
	Pan (chimpanzee). Recent
	Gorilla (gorilla). Recent
Family Hominidae	
	Ramapithecus (14-7)
	including *Kenyapithecus*
	Australopithecus (5-1)
	including *Paranthropus, Plesianthropus*
	Homo (man). Recent

The ape-man stock probably diverged from that of monkeys during the Oligocene period 35-25 million years ago. We have little information about the earliest members of the stock and cannot therefore say very much about the adaptive features that may have been responsible. Of course at first the divergences were slight and our ancestors in the Oligocene period were probably quite like those of the monkeys. Teeth and skull of *Aegyptopithecus* recently found in Egypt could well have been near to the ancestry of the Miocene apes. The molars still show a distinct and rather compact trigon, recalling that of earlier primates. But the lingual (internal) part of the tooth is expanded to form a broad ridge. Such an expansion is known as a cingulum and it occurs similarly on the lingual side of some teeth of *Dryopithecus*.

Of other Oligocene fossils *Aeolopithecus* is quite like a gibbon, and *Propliopithecus* could be fairly close to the lineage of the other modern apes and man.

The factors that had led to the initial separation of an ape-like line evidently became increasingly powerful, and remains of apes are relatively common in Miocene deposits from many parts of the world. We can tentatively correlate this appearance of the apes with climatic and other ecological features. During Miocene times the climate became cooler and drier and large areas of forest were replaced by open grassy plains and savannah. These were the conditions that stimulated the evolution of the horses and probably also of other running animals. Their relevance to the evolution of apes is less clear, for these are mostly partly arboreal animals and include some of the most agile performers in the treetops, the brachiating gibbons (*Hylobates*) and siamangs (*Symphalangus*). However, the other apes are all rather large animals, inhabiting mostly the lower regions of the trees rather than the canopy, though the orang goes high. Most apes spend some time on the ground. The earlier members of the group were probably not powerful brachiators. Thus *Pliopithecus* was definitely a Miocene gibbon, but without especially long arms.

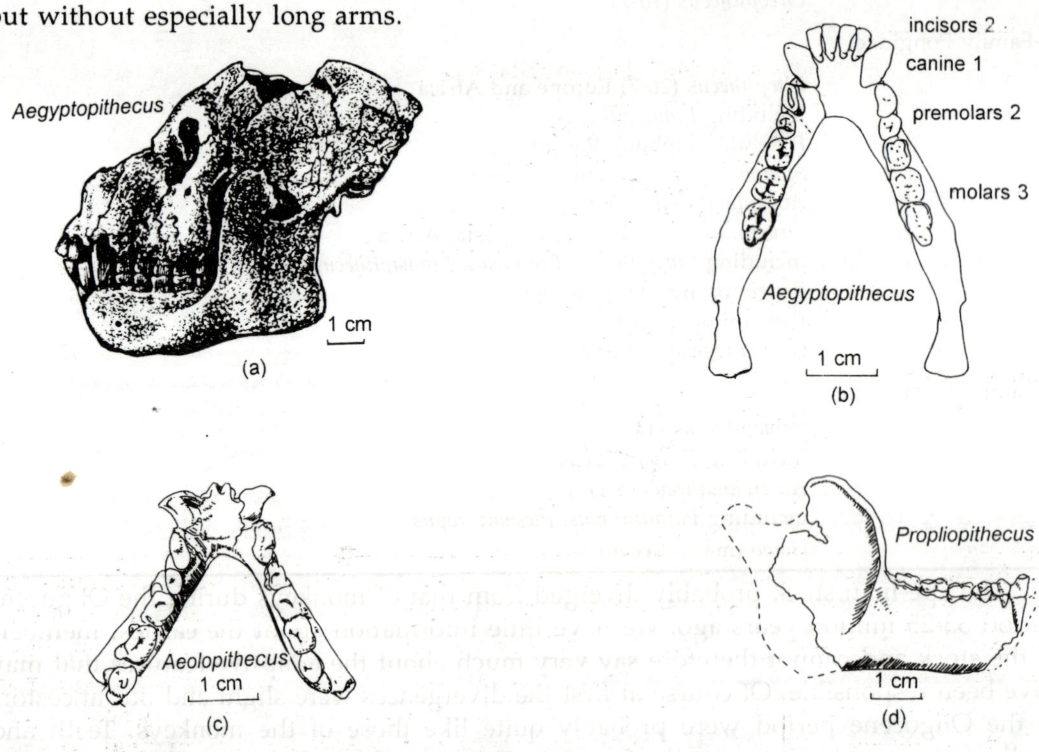

Fig. 11.15: Oligocene apes. (a) and (b) Skull and reconstructed lower jaw of *Aegyptopithecus*. (c) Lower jaw of *Aeolopithecus*. (d) Lower jaw of *Propliopithecus*.

Fossils of various Miocene apes have been known since 1856 but their affinities are still much debated. The remains are mostly teeth or jaws. More than 500 specimens have been found, from various parts of Europe, Asia, and Africa (no undoubted ones from America). They have been given a great many names but we shall here assume that all those sufficiently well known to be relevant for us can be included as follows, though this involves more lumping together of genera than some anthropologists would wish.

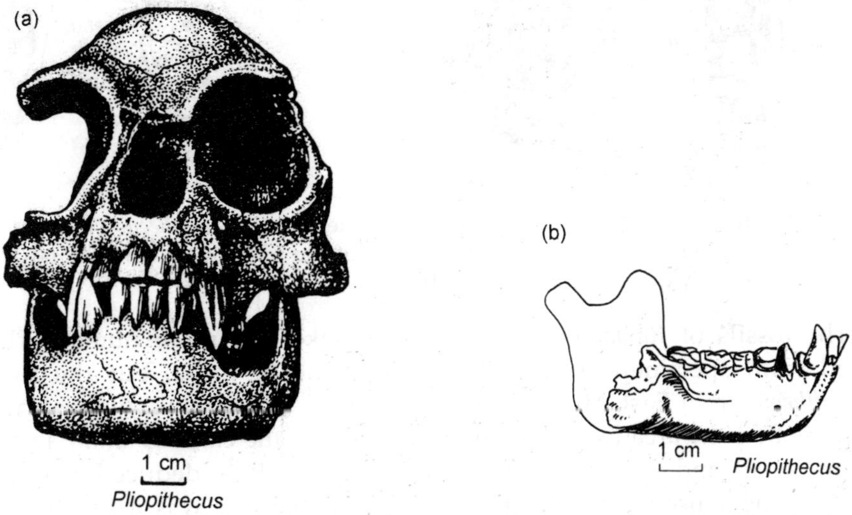

Fig. 11.16: (a) Skull of *Pliopithecus*. (b) Mandible of *Pliopithecus*.

Pliopithecus = *Limnopithecus*

Dryopithecus = *Proconsul* = *Sivapithecus* = *Bramapithecus*

Ramapithecus = *Kenyapithecus*

Members of the first group were near to the ancestors of gibbons, those of the second to the other apes, and those of the third to man.

Fossils now referred to *Dryopithecus* are the most numerous and extend from early Miocene to Pliocene (25-10 million years ago). Many were large creatures, some as big as gorillas (but '*Proconsul nyanzae*' was a small form). Their characteristic is the large development of the incisors and canines. The molars are large and rather square and increase in size backwards to the third. They have a crenellated surface pattern, a characteristic that is sometimes found in later primates and occasionally in man.

The post-cranial skeleton is reasonably well known in the form of Dryopithecine known as *Proconsul*. There were certainly no specializations for extreme brachiation or terrestrial life. In its limbs and its teeth *Dryopithecus* was close to the ancestors of the modern chimpanzee and gorilla.

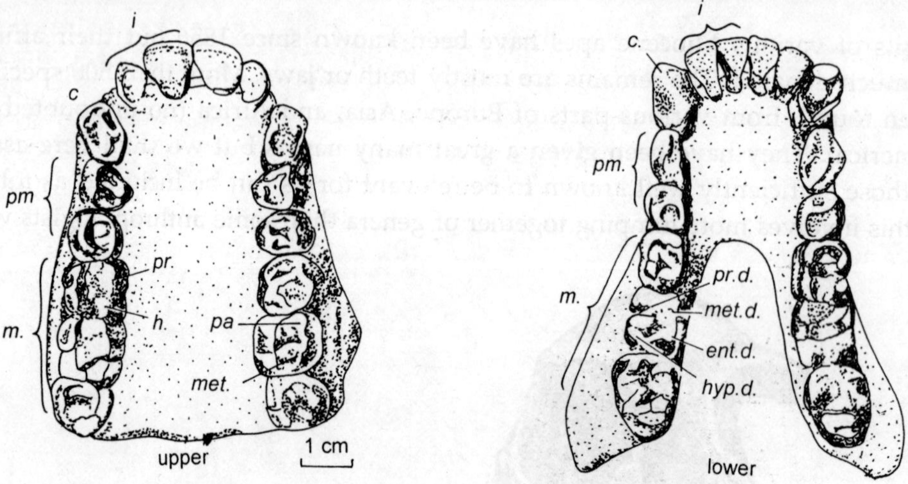

Fig. 11.17: Reconstruction of upper and lower teeth of *Dryopithecus* (=*Sivapithecus*) from the Miocene. Note the rectangular plan of the tooth rows, and the large molars with a mainly quadrangular cusp pattern.

The fossils of *Ramapithecus* are less ape-like and more human. They were first named from specimens from the late Miocene of India, but the form known as *Kenyapithecus* from East Africa (14 million years K–Ar date) is similar. The most striking feature is that the arcade of the jaw was rounded, the canines small and, probably, the incisors small and spatulate. From these features it is deduced that the front teeth were no longer used for tearing the food and that this was a function of the hands, freed by bipedalism for the task. These are obviously rather speculative conclusions, but *Rampapithecus* certainly provides a possible link between the definitely ape-like *Dryopithecus* and the later Pliocene and Pleistocene hominids. The molar teeth of

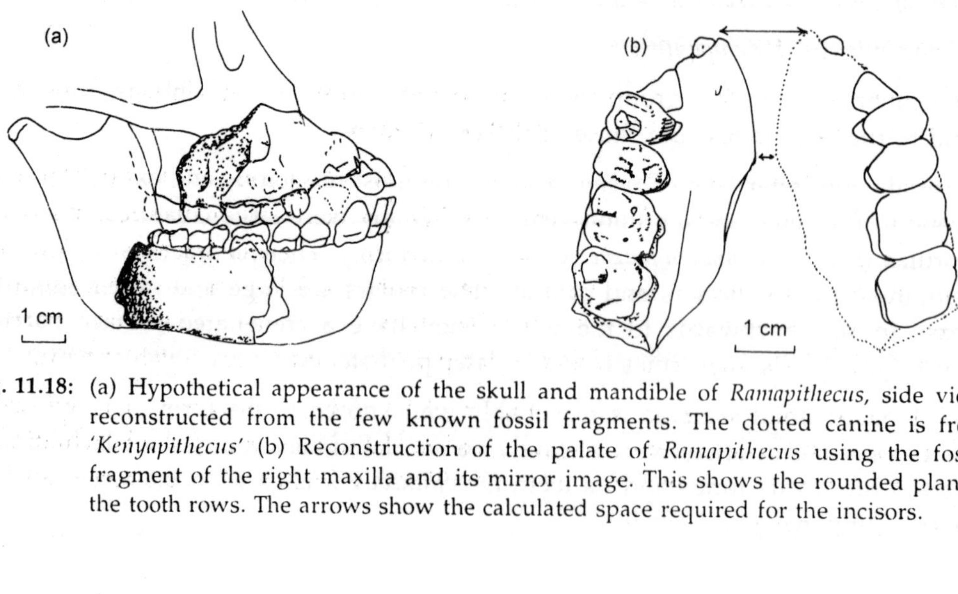

Fig. 11.18: (a) Hypothetical appearance of the skull and mandible of *Ramapithecus,* side view, reconstructed from the few known fossil fragments. The dotted canine is from *'Kenyapithecus'* (b) Reconstruction of the palate of *Ramapithecus* using the fossil fragment of the right maxilla and its mirror image. This shows the rounded plan of the tooth rows. The arrows show the calculated space required for the incisors.

Ramapithecus are, of course, relatively much larger than those of *Homo*, but are smaller than those of *Dryopithecus* and the third molar is smaller than the second. The whole animal was the size of a gibbon. *Ramapithecus* thus 'occurs in the proper time and place to represent a forerunner of Pleistocene Hominidae' (Simons 1961). As Simons remarks, it is illogical to treat *Ramapithecus* as a parallel development rather than close to our ancestral line.

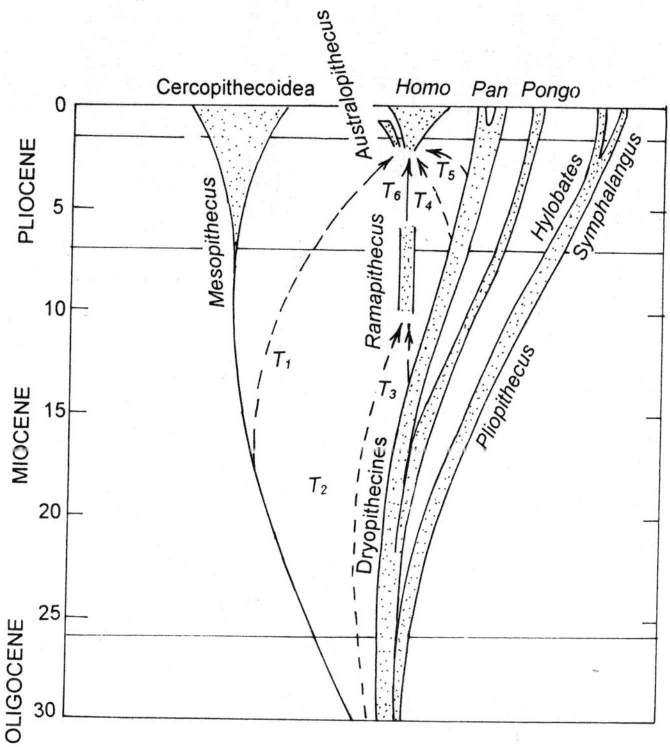

Fig. 11.19: Diagram of possible courses of evolution of the old-world Anthropoidea. T_1-T_6 show possible origins of the Hominidae, in approximately increasing order of probability, according to the evidence.

The study of the Miocene and Pliocene apes thus serves to show us rather precisely our relationship to modern apes. We and they probably possessed a common ancestor in the Oligocene, such as *Propliopithecus*, some 30 million years ago. By 20 million years ago there were divergent lines. The gibbons became specialized first. The common stock of the other pongids and hominids were probably some pre-*Dryopithecus* of the early Miocene. By late Miocene, 10 million years ago, the hominid stock had probably diverged from the pongid as a *Ramapithecus*-like form.

The question of when the ape and human lines diverged is, however, still debated. Some of the possible alternatives are shown in Fig. 11.13. It has been traditional to suppose that the lines have been distinct since Miocene times, but some evidence suggests that the separation may have been much more recent. Studies of protein

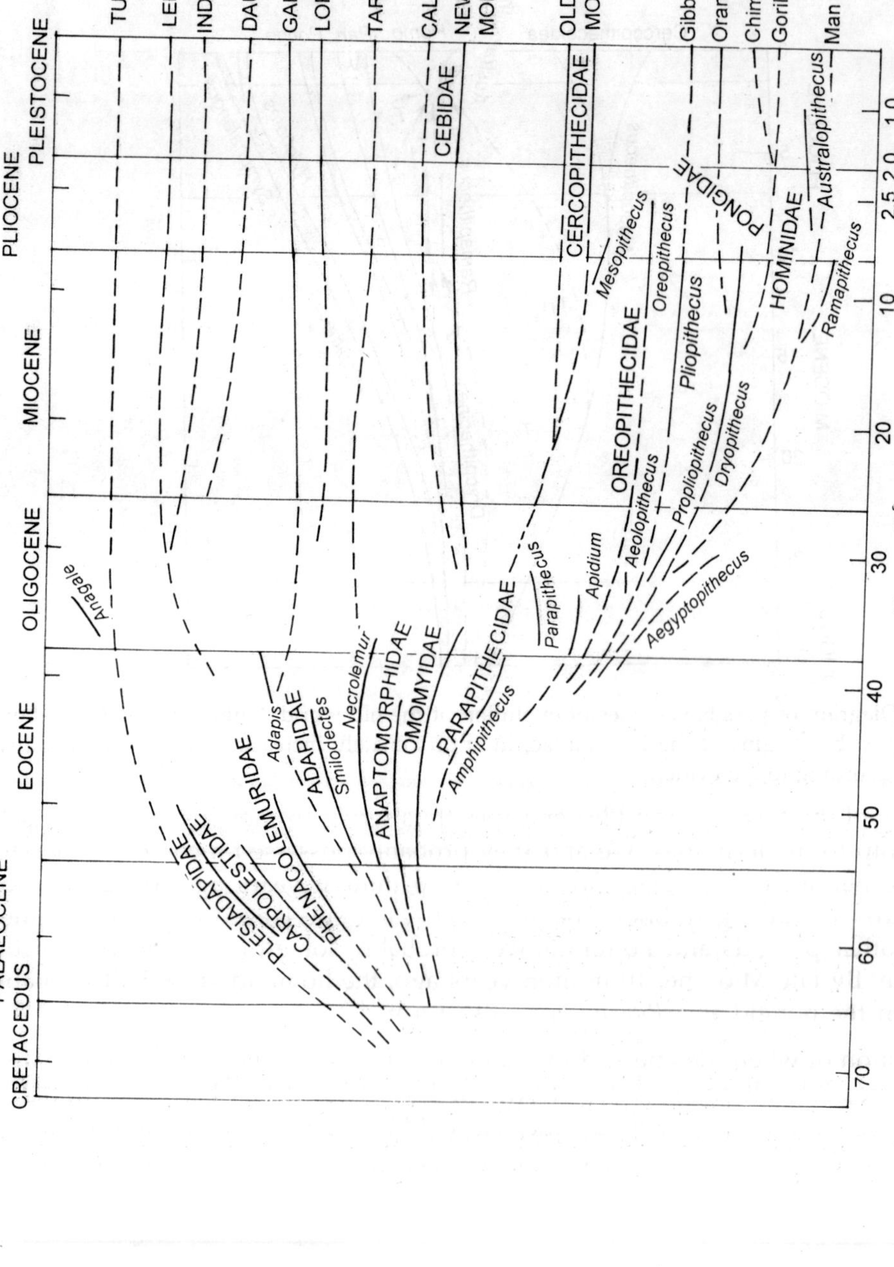

Fig. 11.20: A summary of the evolution and relationships of the main primate groups. The time scale changes at 2.5 10^6 years.

structure have emphasized the close similarity between man and the gorilla and chimpanzee. This has been shown both by immunological studies of serum proteins and of haemoglobins. For example, there is only one established difference between the sequence of the 141 amino-acids in the α-chains of human and gorilla haemoglobins and one out of 146 in the β-chains. There are seventeen differences between human and horse haemoglobin, showing that quite large changes can be tolerated without loss of functional characteristics. From such data, and others such as the similarities in chromosome structure, it is concluded by some people that the divergence of apes and man has been relatively recent. The fossil evidence does not decisively contradict this. The brain certainly remained ape-like for a long time. Further, 'In the lower Pleistocene the ischium is long and ape-like; the ilium and hand do not yet have the full characteristics of *Homo*. Most of the characteristics of *Homo* seem to have evolved well within the Pleistocene, and there is no need to postulate an early separation of man and ape'. Perhaps we have been too ready to emphasize the distance between ourselves and our cousin apes, by supposing that our common ancestors lived long ago. It is just possible that the common population existed not much more than two million years B. P., but evidence such as that of *Ramapithecus* is usually held by most anthropologists to put the date of divergence not later than the end of the Miocene, ten million years ago. The difference is considerable and emphasizes the extent of our ignorance about this period in our ancestry. However, with discoveries such as those of *Australopithecus* and *Ramapithecus* we have much more information today than was

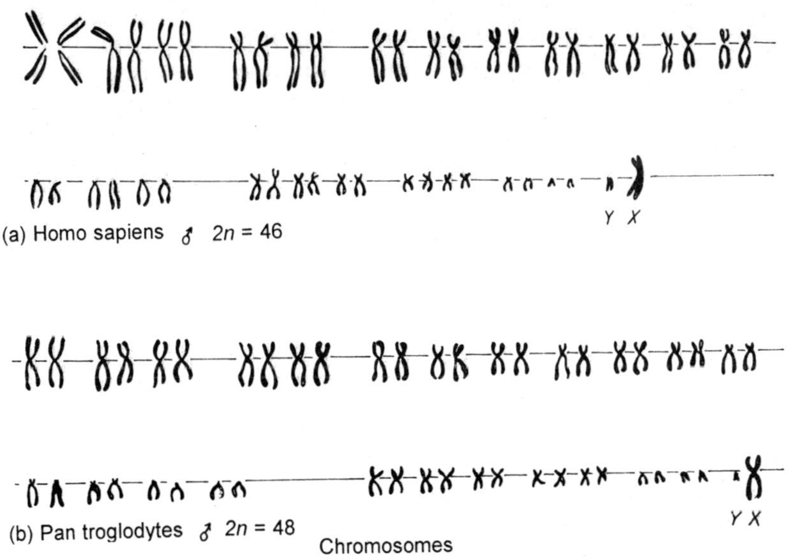

Fig. 11.21: A comparison of the chromosomes (karyotypes) of males of (a) *Homo sapiens* and (b) *Pantroglodytes*. The chromosomes are classified by arranging them in groups by the decreasing order of size. The first twelve pairs can usually be recognized individually and are very similar in man and chimpanzee. The remaining autosomes are placed in four groups, and here there may be some differences.

available twenty years ago. We may feel confident that future discoveries will narrow the area of uncertainty still further.

It is still not possible to say at what stage such human features as the upright posture appeared, let alone even more important human characteristics. *Ramapithecus* may have walked upright, as *Australopithecus* almost certainly did—but perhaps ten million yeas later. Uncertainty about these times of origin remains one of the weakest features of our knowledge of man's ancestry.

FELLOW HOMINIDS

Ramapithecus may be close to the human line of evolution but if some of them were found alive should we regard them as human? Of course we know nothing of the post-cranial skeleton and cannot say anything about their brains nor even whether they walked on two legs. The problem of assessing both the intellectual status and the gait is somewhat less difficult for the next group of remains that are available, those of the *Australopithecines* of Africa, because we can estimate the size of their brains and associated with the fossils are found the earliest evidences of fabricated stone tools. Yet although these creatures made tools, almost certainly did stand upright, and showed many human features in their skeleton and teeth, the brain was no bigger than that of an ape. What were they like? It is obviously worth looking carefully at the evidence about them.

There is some difficulty about describing concisely either their characteristics or the dates when they lived. This is because discovery of the fossils is relatively recent and experts are still not agreed as to how to group and name the various types. Description is difficult without agreed names. Moreover, the effect of these discoveries is to show us that hominid creatures existed a good deal earlier than had previously been supposed. Some of the earliest remains from the Olduvai Gorge in Tanzania have some remarkably human features and are referred to by the discoverer, Leakey, and his co-workers as '*Homo habilis*'. It may be that ultimately some or all the forms now called *Australopithecus* will be referred to as various species of the genus *Homo*. For the present we prefer to keep a somewhat more conservative (multigeneric) classification.

These hominid remains are associated with fossils marking the fauna known as Villafranchian. This includes in Europe members of the modern genus *Elephas*, zebras (*Equus stenonis*), and cattle (*Leptobs*). The Olduvai fauna includes *Elephas* and also the extinct *Mammuthus*, *Bos* (modern cattle), zebra-like horses, *Lycyaena* and other hyaenas, and sabre-toothed cats (*Machairodus*). The fact that the faunas are similar does not mean that they were contemporary in Europe and Africa. The change in fauna from Pliocence to Pleistocene in Europe is supposed to have been associated with the onset of glacial conditions. The corresponding change in Africa was perhaps to a rainy climate (pluvial). The fact is that we still do not know precisely when (or why) either of these changes occurred, and our capacity to write a coherent history remains therefore weak—and this should be openly admitted.

Characters of the Australopithecines

There is much controversy over classification, but we shall refer all the early Pleistocene fossils so far found in Africa to one genus *Australopithecus*. They seem to fall into two species as shown in Table 12.3.

Table: 11.3 African Australopithecines

Species	*Dates B.P.*	*Some synonyms*	*Localities*
A. robustus	?4 000 000	*Homo transvaalensis*	Kromdraai, South Africa
	to	*Paranthropus*	Swartkrans, South Africa
	? 1 000 000	*Zinjanthropus boisei*	Olduvai, Tanzania
		A. boisei	
A. africanus	? 5 000 000	*Homo transvaalensis*	Taungs, South Africa
	to	*Homo habilis*	Makapansgat, South Africa
	? 1 000 000		Garusi, Tanzania
			Olduvai, Tanzania
		Plesianthropus	Sterkfontein, South Africa.

Among the earliest remains are those found in the lowest Bed I at Olduvai Gorge. This is now given a radiometric date of up to 2.0×10^6 years; an earlier estimate was 1.7×10^6 years. On either estimate this is much earlier for the beginning of the Pleistocene than had previously been supposed. These specimens have been called *Homo habilis* by Leakey and his colleagues (1964), who gave for this purpose a new definition of the genus *Homo*. As we shall see, these fossils show many human characters but have only a small brain. The difficulty of classifying them shows *again that there are no sharp divisions*. It provides evidence that the characteristics of man emerged gradually, and not all at the same time. Further specimens of *Australopithecus* have recently been found in Ethiopia and may be as much as 3.5×10^6 years old, or even more.

We shall assume that the finds in Olduvai Bed I known as *Homo habilis* can reasonably be included in the species *Australopithecus africanus,* the name that was chosen by Dart in 1925 for the first Australopithecine skull found. This was discovered at Taungs in South Africa in a Lower Pleistocene deposit probably over 1 million years old. Other skulls found at Sterkfontein and Makapansgat, which are also in South Africa, and recently at Garusi in Tanzania can be placed in this species too. *A africanus* thus seems to have lived from at least 2 to 1 million years ago, though the dating of the South African remains is very difficult.

There was another type of Australopithecine living in Africa for much of this time. In 1938 Broom found a heavily built skull and some other bones at Kromdraai, South Africa, and named them *Paranthropus robustus*. This genus is now often included in *Australopithecus*. It has been strongly argued, however, that its members stand so far apart that the genus *paranthropus* should be retained. The creatures were probably a line of herbivores, existing parallel to the carnivorous or omnivorous *africanus* type. It is urged by Robinson that since the latter presumably went on to become *Homo*

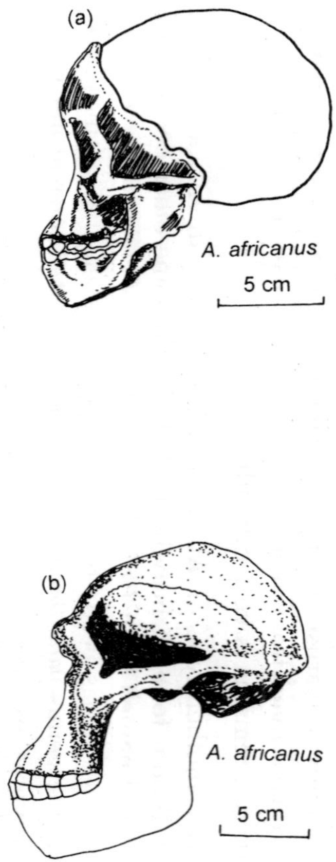

Fig. 11.22: (a) drawing of lateral view of the skull found at Taungs (South Africa), the first Australopithecine to be found (*A. africanus*). This is of a child of perhaps 15. Date over 1 10^6 years B.P. The back part of the skull is missing, but there is a remarkable cast of the brain. Drawn from a cast in the British Museum (Nat. Hist.). (b) Adult *A. africanus*, skull found at Sterkfontein (South Africa). Possibly 3 10^6 years old. From casts in the British Museum (Nat. Hist.)

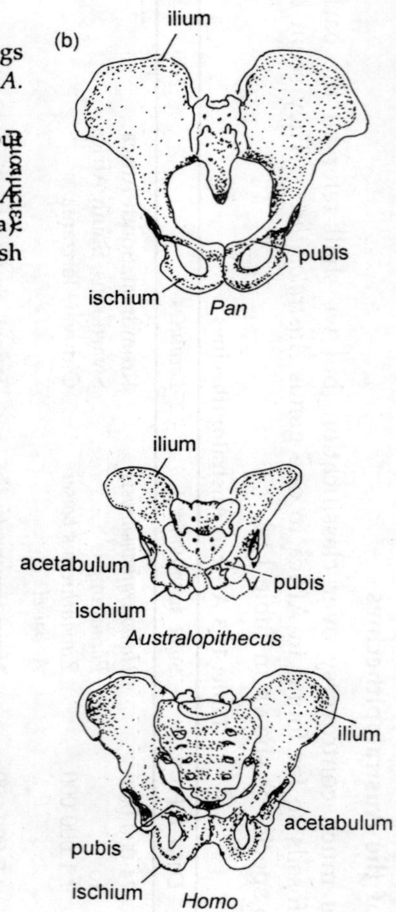

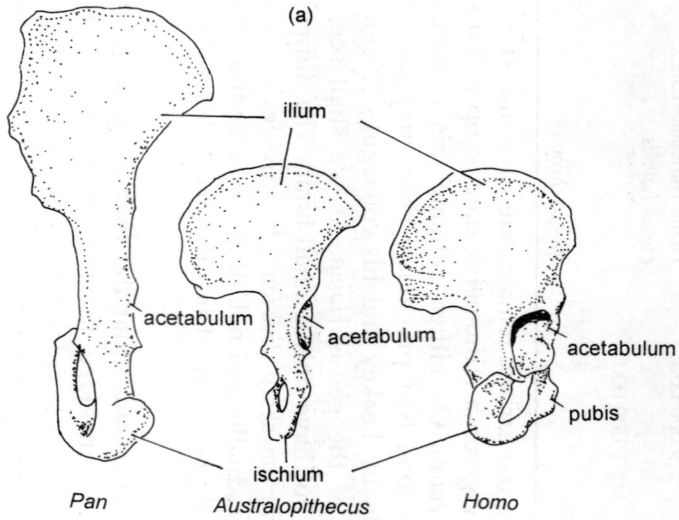

Fig. 11.23: Drawings of the pelvic bones of *Pan*, *Australopithecus*, and *Homo* to show the changes in sacro-iliac articulation, acetabulum, and the length of the ischium. (After Broom and Robinson in Grassé 1955). The ilium of *Australopithecus* resembles that of man, but the ischium is much longer.

they should be included in that genus and that the genus *Australopithecus* should be dropped. This may well happen, but we shall follow the usual current practice of keeping *Australopithecus* and its two species. They certainly differ in many ways but both have very much smaller brains than the creatures usually placed in *Homo*. It must be admitted, however, that describing the characters of 'Australopithecines' in general is a rather superficial procedure. In *A. robustus* can be included also a more complete skull found at Swartkrans. This like the Kromdraai remains, is dated rather later than Taungs, perhaps 1 × 10^6 years B.P. We shall also include in *A. robustus* some much earlier remains from Olduvai, though this is more controversial. These fossils, found by Mrs. Leakey in 1959 in Olduvai Gorge, Bed I, at a level 2 ft above that of the '*H. habilis*' remains, were called *Zinjanthropus boisei* ('nutcracker man'). Few people would probably still wish to keep the genus distinct. Some would like to keep the species as *A. boisei*, but we shall follow Simons in putting these remains with *A. robustus*, and date them provisionally at 1.9 × 10^6 years B.P.

The characteristics of the genus *Australopithecus* as a whole may be seen from the original skull found at Taungs in South Africa, though it is of a juvenile. A similar skull was found at Sterkfontein in the Transvaal. The general appearance of the face is certainly not very human, with rather protuding jaws ('prognathous'). The skulls referred to the other species, *A robustus*, are more heavily built, often with massive brow-ridges, and sometimes even a sagittal crest. The foramen magnum of the Australopithecines is usually said to point forwards, which is consistent with an upright gait. This, however, was disputed by Zuckerman and his colleagues at Birmingham, as were many other assertions of human affinities for these fossils. However, the evidence of some form of bipedalism is cumulatively rather strong, though this does not mean that the gait was identical with that of modern man.

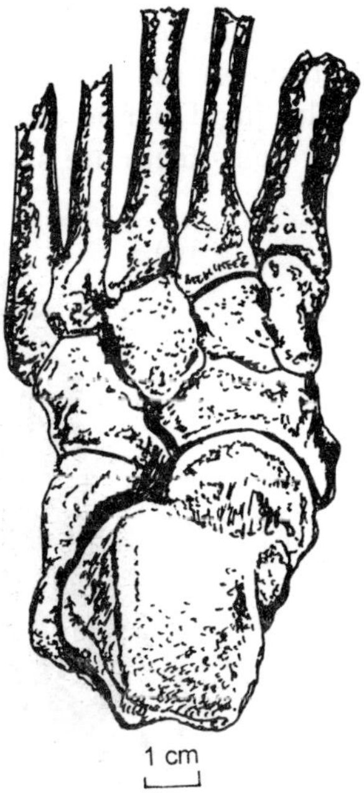

Fig. 11.24: Foot of *A. africanus* from Olduvai Bed I, 2 × 10^6 years old.

The mastoid process is large (though smaller than in man). The ilium is known from Sterkfontein, Swartkrans, and Makapansgat and is short and broad, as in man, but with a small sacro-iliac articulation and an acetabulum different from that of man. The ischium also differs from that of man quite markedly in being rather long, thus carrying the attachment of the hamstring muscles a long way from the acetabulum. This is a character of quadrupeds, giving a large mechanical advantage to the hamstring muscles for extension of the hip, drawing the leg backwards. In modern man, where these muscles have less part in propulsion, the ischium is short. It has been suggested that the Australopithecine gait was more a run than a walking stride.

A foot is known from the remains in Olduvi Bed I referred to *H. habilis*. Its general arrangement is very human. The first metatarsal is larger than the rest, but not as markedly as in *H. sapiens*. There is clear evidence that the foot was arched and the great toe not opposable.

The hand is also known from the remains in the bottom of Bed I and was very human, with a saddle-shaped joint on the trapezium, indicating an opposable thumb. There were still some ape-like features, however, including curvature of the phalanges, and it is suggested that preceding the full bipedal habit, the weight may have rested on the knuckles (as in apes), while still allowing use of the hand for grasping. The opposable thumb was much shorter than in modern man, reaching little further than the base of the first phalanx of the index finger. The hand could exert a strong power grip, but not a precision grip.

The dentition of *Australopithecus* provides features that are also interestingly suggestive of man, yet not quite human. The dental arcade is rounded, the canines are small and spatulate, and there is no diastema. The premolars and molars are much larger than in man and they increase in size to the third molar. These large teeth suggest a herbivorous diet, especially in *A. robustus*. In this connection, *A. robustus* teeth have many pits on the grinding surface, a typically herbivorous feature due to the grit on freshly dug roots. In this species there are very large jaws corresponding to the large brow-ridges and in some specimens there are quite prominent sagittal crests for the attachment of the temporalis muscles, which must have been powerful. The jugal bar is also thick, indicating a large masseter. On the other hand, the *A. africanus* teeth are smooth (like those of, say, a leopard), the jaws are lighter, and the

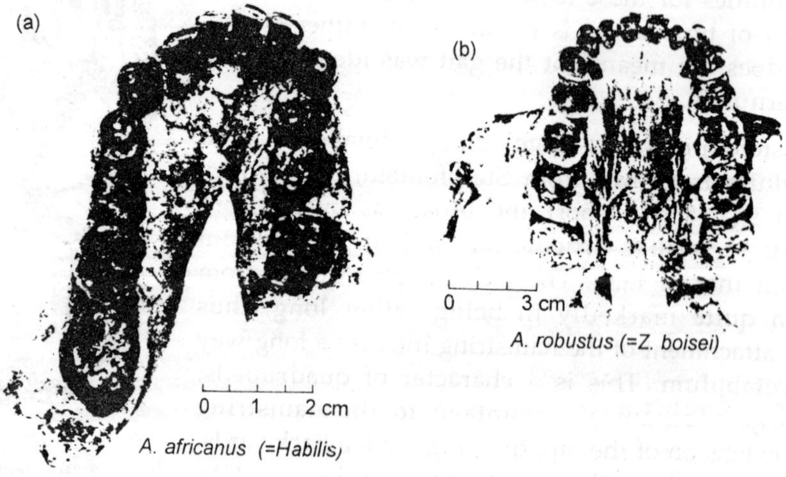

Fig. 11.25: (a) Mandible of *A. africanus* (=*habilis* Olduvai Hominid 7). The jaw has been fractured on the right, before fossilization. Age: ?2 × 10^6 years. (b) Palate and dentition of *A. robustus* (*Zinjanthropus*) Age: ?1.5 × 10^6.

skull has no crest. These facts suggest that *A. africanus* was carnivorous and that the two species occurred in different habitats occupying different ecological niches from at least 2 × 10^6 years ago.

Further evidence for the carnivorous diet of *A. africanus* is furnished by the enormous piles of fossilized bones found in several caves investigated by Dart, such as one at Makapansgat, South Africa. Although it has been asserted that these bone collections were the work of hyaenas, it seems reasonably certain that they were made by *A. africanus*.

The question now arises, was *A. africanus* a scavenger, eating carrion and collecting bones for the marrow or did he *kill* animals? Two main lines of evidence indicate that he did kill them, probably with weapons. The bone pile contained the fossilized bones and teeth of a wide range of animals from guineafowl and hares to baboons, hippopotami, and giraffes, and including *A. africanus* himself, a total of 4560 identifiable bones, portions of bone, and teeth. First, there is very strong statistical evidence of systematic selection of bones *in the field* and their subsequent transport to the cave. The bones collected tended to be, with the exception of some whole heads, those that would either serve as knives (the mandible of a small antelope) or clubs (the distal portion of the humerus of a medium-sized antelope). The emphasis seemed to be on *functional* rather than *food* value—an example of foresight? The second line of evidence lies in the fact that of the crania collected the majority (including some belonging to *A. africanus*) appeared to have been struck with a heavy implement. It is not proved

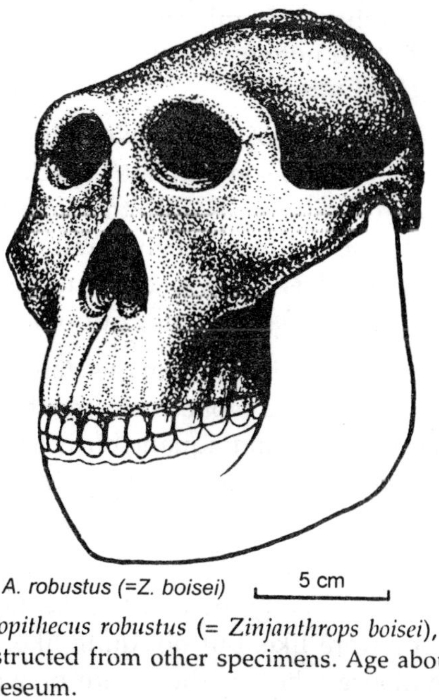

Fig. 11.26: Skull of *Australopithecus robustus* (= *Zinjanthrops boisei*), from Olduvai Bed. I. The jaw has been reconstructed from other specimens. Age about 1.5 × 10^6 years. From a cast in the British Meseum.

that this was the cause of death, but it is certain that the blow was not struck either after fossilization or well before death (i.e. there was no healing). Some outstanding examples were found: the skull of a baboon with an antelope humerus bone (distal end) jammed through the open mouth and lodged in the base of the skull; a skull of *A. africanus* with a rock still embedded in it; another *africanus* skull with two small holes in it, close together, entering at different angles. Of forty-two examinable baboon skulls, every one showed damage to the muzzle or cranium. Moreover, on many specimens there was a double depression at the impact point that exactly fitted the protuberances on the distal end of an antelope humerus bone, which was the commonest bone in the Makapansgat pile. Likewise it is interesting that most of these baboons received their (presumably fatal) injuries from the front. However it has also been suggested that the depressions on the skull were made by leopards.

In summary, it has been argued (especially by Dart and Ardrey) that *A. africanus* was a carnivore, killed his prey, and killed his own kind. It is also agreed that the Australopithecines are somewhere near to the line leading to *Homo*. It does not follow that men have a 'need' to fight.

The size of the brain of all Australopithecines was probably in the ape range of 400-600 ml. Higher estimates (up to 700 ml) have been given (i.e. for '*H. habilis*'), but are based on reconstruction of fragments and must be regarded as uncertain at present. Attempts are made to stretch the range of brain size of modern man in order to show that the Australopithecines 'nearly get there'. An immense range of modern human material is available, extending from 1000 to 2000 ml, but a useful mean is 1400 ml, with a standard deviation of perhaps 150 ml. The Australopithecines may have reached the *H. erectus* level of brain but do not come within the range of *H. sapiens*, and this is one of their most interesting features.

It is clear that the interpretation of the '*habilis*' remains as 'human' rather than Australopithecine is a question open to some doubt, especially as some of the specimens may be of juveniles. The controversy serves only to suggest that we are nearer than is often realized to having complete evidence of the gradual transition from an ape-like form to man. Since some decision has to be made as to how to name these fossils we shall include them with *Australopithecus africanus*, but others would keep *A. boisei*.

A further special interest of the *habilis* remains (and those of '*Zinjanthropus*') is that they were associated with chipped stone instruments. In addition there were the bones of numerous tortoises, catfishes and aquatic birds and the jaw of a sabre-toothed tiger. One of the bones had probably been fashioned into a tool. There is thus some further evidence of a carnivorous diet and it seems that these are the remains of a population that used tools for hunting.

To debate whether they were 'really men' is to miss the point. The whole interest of these creatures is that they were like, but not quite like, modern man. What would be interesting would be evidence whether they were or were not on the line leading

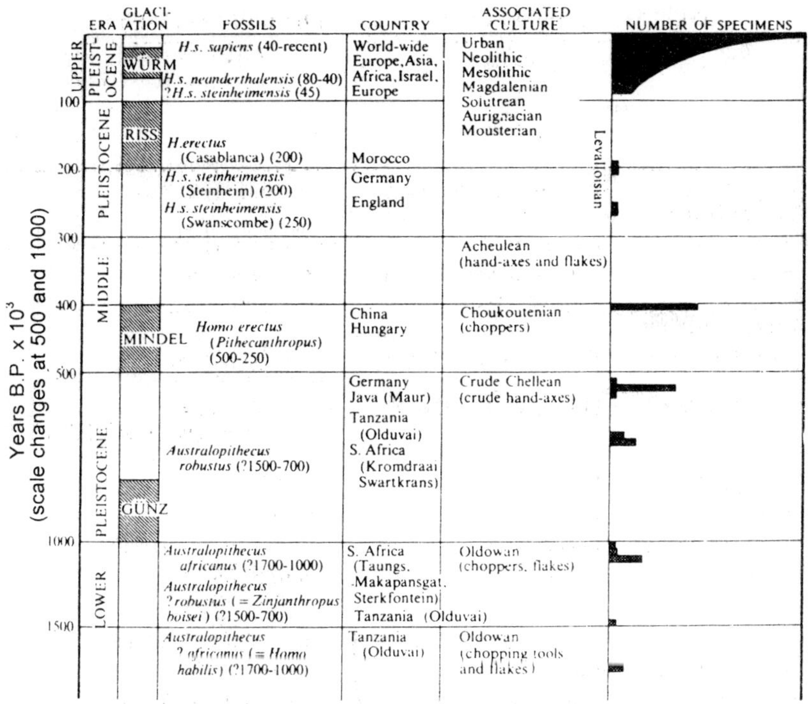

Fig. 11.27: Scheme showing relations of some Australopithecine finds to each other and to more recent hominid fossils. The last column gives the approximate number of individuals represented by the various finds; the smallest squares indicate one specimen. Notice the small number of specimens available and that they are widely scattered in time. Dating is still very tentative, especially for the Australopithecines of South Africa, which were fossilized in a limestone breccia that does not allow radiometric dating.

directly to ourselves, but this we cannot decide. If they did indeed make the primitive Oldowan flints then this shows that tool-making does not depend upon a large brain—but we know already that children make 'tools'.

Some important conclusions from these most interesting discoveries seem to be:

1. That human evolution has been 'mosaic'; hand and foot reached the human stage before brain size.
2. That creatures with ape-size brains could make and use tools.
3. That this stage was reached at least between 2.0 and 1.7 million years ago, much earlier than had been supposed.
4. That gradual changes proceeded along several related lines. It may well be that the *A. africanus* line went on to produce *Homo* while *A. robustus* remained herbivorous and became extinct.

Table 11.3: Summary of some data on hominid evolution

Approx. age (years) × 10⁶	*Site*	*Authority*	*possible name (and specimen number)*	*Estimated average cranial capacity (ml)*
>9	Baringo, Kenya (Ngorora Beds)	Bishop and Chapman	(?) *Australopithecus*	
5	Lothagam, Kenya	Patterson	(?) *Australopithecus africanus*	
4	Kanapoi, Kenya	patterson and Howells	(?) *A. africanus*	
3.5-3	Baringo, Kenya (chemeron Beds)	King; Tobias	(?) *Australopithecus*	
3	Garusi, Tanzania	Kohl-Larsen; Robinson	(?) *A. africanus (Meganthropus)*	
4-1.75	Omo, Ethiopia	Howells; R. Leakey	(?) *A. robustus* (?) *Zinjanthrpus boisei*) (L7-125) *A. africanus* (W. 23)	
?3	Sterkfontein, S. A.	Broom: Robinson	*A. africanus*	440
?2	Taungs, S. A.	Dart	*A. africanus*	440
1.9	olduvai, Tanzania (Lower Bed I)	Leakey; Tobias; Napier; Oxnard	*A. robusius (Z. boisei)* (hominid 5)	530
			A. africanus (Homo habilis) (hominids 6, 7, and 8)	>600
2-?1.5	Makapansgat, S.A.	Dart	*A. africanus*	435
?1-25	Olduvai, Tanzania (Lower Bed II)	Leakey and Leakey; Tobias; Day	*A. robustus (Z. boisei)* (hominid 20)	
			A. africanus (H.habilis) (hominids 10 and 16)	>600
1.5-?1	Swartkrans, S.A.	Broom; Robinson	*A. robustus (Paranithropus)* (?) *A. africanus (Telanthropus)*	530
1.5-1	Kromdraai, S.A.	Broom	*A. robustus (Paranthropus)*	475
1.0	Olduvai, Tanzania (Upper Bed II)	Tobias	*A. africanus (H. habilis)* (hominids 13 and 14) *A. robustus* (*Z. boisei*) (hominid 3)	590
?1.0	Peninj, Tanzania	Issac and R. Leaky; Tobias	*A. robustus*	
?1.0	Olduvai, Tanzania	Leakey	*Homo erectus* (hominid 9)	1000
1.0	Lantien, China		(?) *H. erectus*	780
1.0	Djetis, Java	von Koenigswald	*H. erectus*	750
0.7-0.5	Trinil, Java	Dubois	*H. erectus*	775-975
0.5	Pekin, China	Black; Weidenreich	*H. erectus*	1050 (850-1300)
0.4	Vertesszollos, Hungary		*H. erectus*	1400
0.2	Steinheim, Germany Swanscombe, England		*H. sapiens steinheimensis*	1200
0.07-0.04	Dusseldorf, Germany Mt. Carmel, Israel Shanidor, Iraq Broken Hill, Zambia		*H.s. neanderthalensis*	1600
0.04-0.00	Niab, Borneo Cro-Magnon, France, etc.		*H.s. sapiens*	1400

There has recently been a rapid increase in knowledge of the African Pliocene and Pleistocene deposits, necessitating caution over dates mentioned in the text. Table attempts to summarize new and old data and to present a view of current knowledge of hominid evolution.

There is new evidence from Omo of numerous teeth of hominids existing up to 4 million years ago and a mandible at Lothagam about 5 million years ago. A single molar crown at Ngorora is more than 9 million years old. There is now evidence of tools as much as 2.6 million years old.

On the other hand the date of Upper Bed II at Olduvai and of the Peninj deposits is now uncertain and may be not later than 1 million years. In that case it may be that the Australopithecines did not continue, as such, much beyond that time. However there is also evidence that some of them may have then transformed into early types of *H. erectus*. Further facts about these questions can be expected shortly.

Giant Ape-men

In recent years there have been found fragmentary jaws of giant primates from the early Pleistocene of Java and southern China. The remains show features indicating that these animals (*Meganthropus, Gigantopithecus*) were close to the ape-human boundary and quite possibly primitive hominids. It would appear that nature "experimented" with gigantism in the human stock, as she did with many other types of animals in the Pleistocene.

Human Characters

Man and his close fossil relatives are usually placed in a separate family; but, as a matter of fact, the anatomical features which distinguish men from apes are comparatively few, although, from our own point of view, important. Of great significance has been the growth of the brain. The average European male has a brain capacity of about 1,500 cubic centimeters—much more than twice that of any of the great apes. This increase is mainly in the neopallium, the "gray matter" of the cerebral hemispheres. Here new centers are developed, such as those having to do with speech; and especially prominent are areas, absent in the apes, which seem to be the seat of higher human faculties. This enlargement of the brain results in a much expanded braincase and the consequent disappearance of crests and supraorbital ridges. In lower primates the prominent canines form a marked angle in the tooth row; in man the tooth row, *D;* is shorter and rounded, with the canines much reduced. In relation to this reduction of the tooth row, the face is less projecting, although there is a tendency in modern man for the retention of a long jaw in the development of a chin. With reduction of the face in general, the nasal region tends to develop into an isolated projecting structure.

Other important differences are associated with the change in locomotor habits, Even among the higher apes we have seen a trend toward terrestrial life; man lives on the ground as a biped. Man's arms are short compared with those of the apes; his legs, long. It is not probable that we have descended from such a long-armed, short-legged form as the orang but rather that, starting from a common ancestor with arms of rather indifferent length, the apes during millions of years of tree-living have elongated their arms to varying degrees, while man has retained more of the original proportions. The human hand, except for somewhat greater breadth and flexibility, is of a type probably present in the ancestral apes; the living great apes have tended to elongate the fingers, except the thumb, for hooking over limbs. A greater difference is in the foot. The usual primate foot is similar to the hand. This, however, is an awkward structure for ground life. In man the toes have greatly shortened, the hallux has lost its primitive possibility, and the calcaneum expands to form a prop at the back (the last is true of the higher anthropoid apes as well). In further adaptation to the bipedal gait, the back of man is much more sinuously curved than in the apes, with the effect of swinging the center of gravity up above the hips and raising the head.

The features listed above are among those which distinguish modern human types from the apes. Fossil human types tend to bridge this structural gap and exhibit intermediate conditions; but our knowledge of human ancestors is still a very imperfect one, and we know almost nothing of the evolution of the postcranial skeleton.

It is probable that the human line diverged from that of the great apes at about the *Dryopithecus* stage, in late Miocene times. There are, however, no certain skeletal remains of man in the Pliocene; it is not until the early Pleistocene that we find man present and already branching out into a number of types.

Early Human Cultures

Man is a maker and user of tools, and this trait may well have appeared early in his development. Flint is a material which appears to have found favour for tool-making because it is easily fractured and worked and yet is extremely hard. Flints, thought by some to have been used by manlike creatures, have been reported from Tertiary rocks. While such finds are of dubious nature, flints were certainly being worked in early Pleistocene times, and a series of human cultural stages, based primarily on stone implements, has been established for the Pleistocene of Europe and, less surely, for other areas of the Old World. In the Lower Pleistocene, covering essentially the first of the four advances of the European glaciers and the following interglacial stage, implements were very crude ("eoliths"), and the stage maybe termed (1) the Pre-Paleolithic. No associated human remains are known. The Middle Pleistocene, covering the second and third ice advances and a lengthy intervening stage, is the time of (2) the Lower Paleolithic, the first accepted subdivision of the Old Stone Age. In much of the Old World this was the time of the use of the large hand ax or *coup de poing*, associated with Abbevillian (Chellean) and Acheulean cultures; in other parts of Europe and Asia the time was one of the use of simple chopping tools and flint flakes, such as may have been used by *Pithecanthropus* and the related Peking man. (3) the Middle Paleolithic is the time of the third interglacial stage and the early part of the final glaciation of Europe. The dominant culture is that of the Mousterian, characterized by well-shaped flint flakes and often found associated with remains of Neanderthal man. At the end of the Würm glaciation there is in Europe a characteristic series of cultures-Aurignacian, Solutrean, and Magdalenian-which constitute (4) the Upper Paleolithic (related cultures are found in other areas). In addition to expertly made flakes, bone has become an important cultural material. These cultures are the

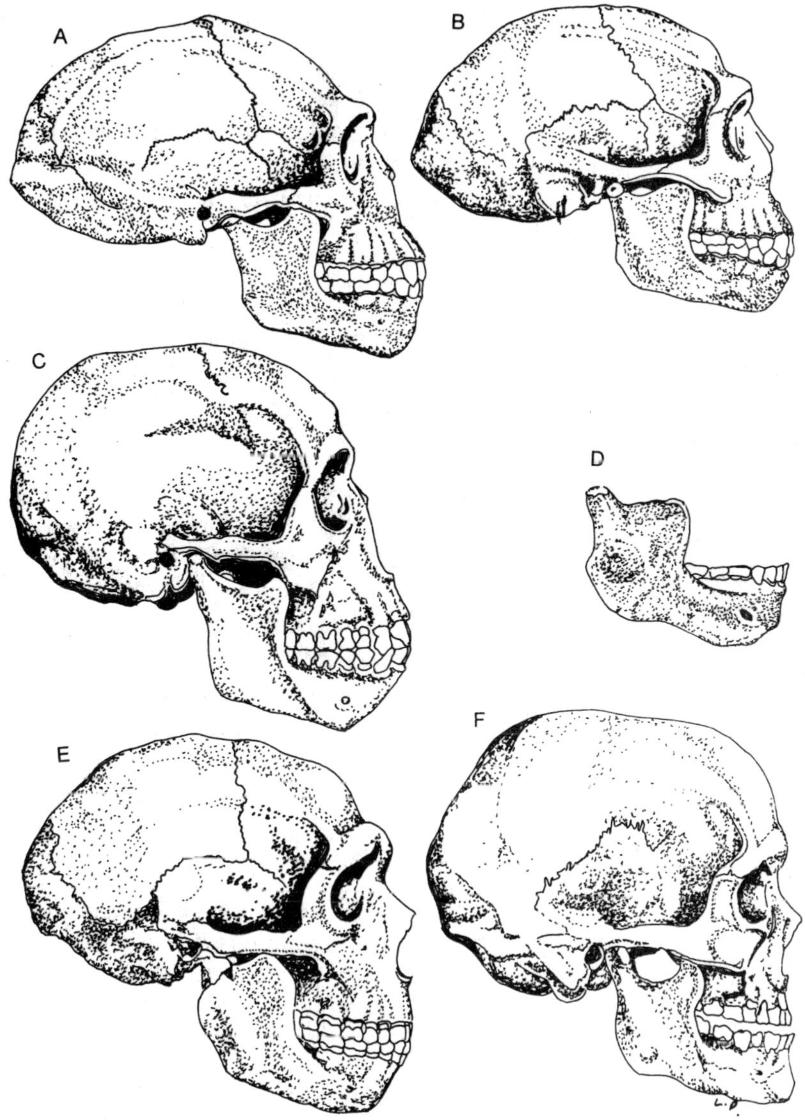

Fig. 11.28: Skulls of fossil human types. *A Pithecanthropus*, part of face restored; *B*, *Sinanhropus*, face and back part of jaw restored; *C*, *Eoanthropus*, face, some small areas of the braincase, and the tip of the jaw restored; the jaw doubtfully associated; *D*, the Heidelberg jaw; *E*, Neanderthal man; *F*, Cro-Magnon man from Grimaldi.

work of our own species, *Homo sapiens*. To Recent times rather than to the Pleistocene (and hence beyond our province) are to be assigned later human cultures—the Mesolithic; the Neolithic, in which pottery, agriculture, and domestication of animals were added to domestic economy; and the Bronze and Iron Ages of historic times.

Early Men of Eastern Asia

Most adequately known of early human types are the finds from Java and China customarily termed the "Java apeman" and "Peking man." *Pithecanthropus erectus* was first discovered half a century ago on the banks of a Javanese river near the little village of Trinial, associated with an abundant mammalian fauna of Middle Pleistocene date. The original find consisted merely of a skull cap, with which were more or less doubtfully associated a thigh bone, several teeth, and jaw material (from another locality). In recent years several other partial skulls and jaws have been found, so that the cranial anatomy is fairly well known. The braincase, including the forehead region, was extremely low, and the brow ridges over the orbits enormous. Further, the skull is rather narrow compared to its length, a feature common to most early human forms (skulls in which, as here, the breadth is 75 per cent or less of the length are termed dolichocephalic; short-headed types, with the figure over 80 per cent, are brachycephalic; intermediates are mesocephalic).

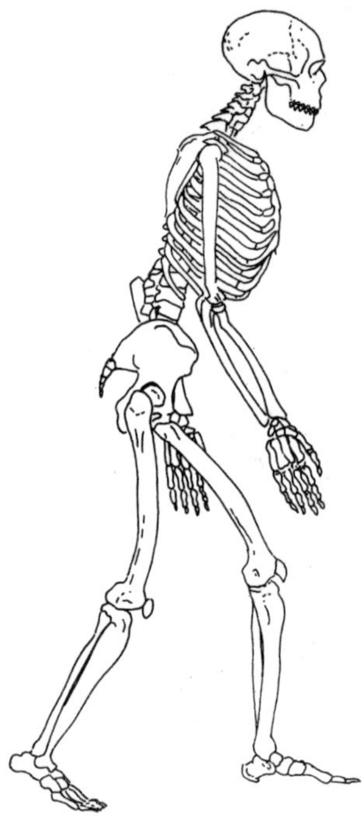

Fig. 11.29: Skeleton of Neander-thal man.

Endocranial casts suggest an essentially human type of brain, although a small one; the average figure for endocranial capacity of 890 cubic centimeters is far below that of any modern human type, although well above the highest of apes. In contrast with modern men, the face was projecting and broad, with a large (and probably flattened) nose. The teeth were in most respects of human type, but there is evidence suggesting that the males had somewhat enlarged canines. There was no chin projection. If the femur is properly associated, *P. erectus* (as his specific name would imply) had already assumed an upright posture.

During the last fifteen years a considerable number of skulls and jaws of a primitive human type have been recovered from Middle Pleistocene deposits at Choukoutien, near Peking, China. Their study has revealed another early type of man, similar to *Pithecanthropus* in many respects. The only conspicuous difference lies in the fact that the average of four endocranial cavities measured is about 1,050 cubic centimeters in the Peking man, compared with a much lower figure in the case of the Java form. This is, however, thought to be due to the fact that the Chinese specimens may be for the most part males, the East Indian skulls measured those of females; sex differences in brain size are marked in many human races. The Chinese form was given a separate generic name as *Sinanthropus*, but students of these forms are currently agreed that they are closely related, if not specifically identical, races and that the general *Pithecanthropus* stock was a widespread early human type, in Asia at least.

Early European Human Finds

Pithecanthropus is a primitive form, readily derivable from an advanced ape and, on the other hand, a good ancestor for Neanderthal man of the late Pleistocene. This suggests that early human evolution was a relatively simple and straightforward story. Finds from the early and Middle Pleistocene of Europe, however, complicate the picture.

Chapter—12

The Evolution of Man

MAN'S PLACE IN NATURE

We have come to the culmination of our course and approach the high estate of man, the loftiest pinnacle of the evolutionary fabric, but one to which all of the laws governing the survival of the other creatures have nevertheless applied. In our research we cannot consider man, that is, physical man, as a being apart from his fellows of the animal kingdom; but whatever our prejudices, we must look the facts in the face and consider him merely as one, perhaps a very special one, among the great hosts of animal life. As in the case of the other creatures whose evolutionary history we have endeavored to understand, so with man we must enquire into his place in nature, learning what his nearest relatives are and whence and why he came.

Man a Vertebrate

A brief diagnosis of a vertebrate mentions a number of characteristics such as the notochord, hollow, dorsally situated nerve-cord, perforated pharynx, and the like, all of which without exception are possessed by man the individual during some period of his career just as surely as by the horse, the dinosaur, the amphibian, or the fish. The evidence of man's vertebrate inclusion is therefore unquestionable.

Man a Primate

Of the several classes into which the vertebrates are divided, man shares with the horse, the elephant, and even the saber-tooth the several characteristics—hair, warm blood, midrif or diaphragm separating the chest and abdominal cavities, young born alive and nourished by mammary glands—which make them mammals. He cannot therefore be debarred from that group any more than the others. Within the class Mammalia he is excluded from the egg-laying monotremes and the pouch-bearing marsupials, and included with those whose unborn young are nourished by the placenta; he is therefore a placental mammal. And of the four cohorts of Placentalia he is by a process of elimination narrowed down to the nailed arboreal forms, or Primates, for he can be neither a clawed unguiculate, nor a hoofed ungulate, nor a finned cetacean.

Primates in General

This name was given to the group by Linnæus, who wished thereby to emphasize the headship of the animal kingdom, the first in the sense of the highest. The word has, however, a deeper significance than this, for the primates are also among the first of the placental mammals in their antiquity and primitiveness. Along their chosen line of specialization, the feet, the horses and camels are, it is true, vastly further advanced than are the primates, and the same thing can be said of tooth modification in elephant and *Smilodon*. It is only in the brain and such correlated modifications as mental development entails that the primates may justly lay claim to superiority, for in other respects they are as humble and generalized a group, with very few exceptions, as the mammalian class contains.

Primates may be defined as nearly all arboreal with prehensile limbs, having a more or less opposable pollex (thumb) and hallux (great toe). The five digits usually terminate in flattened nails, rarely claws; there is a clavicle, the orbit is completely surrounded by bone, the stomach simple, and the mammary glands are nearly always thoracic.

CLASSIFICATION OF PRIMATES

The classification of the primates has of course been subjected to the same vicissitudes as those of other orders, especially when fossil forms are found which link together apparently isolated living groups. An admirable study of primate interrelationships is that of W. K. Gregory (1916), from whom the following classification, somewhat abridged, is taken:

Order Primates

Suborder Lemuroidea (lemurs or "half-apes").

Suborder Anthropoidea (monkeys, apes, and man).

Series Platyrrhini (New World monkeys).

Family Hapalidæ (marmosets).

Family Cebidæ (capuchins, howler monkeys, spider monkeys, etc.).

Series Catarrhini (Old World monkeys and apes).

Family Cercopithecidæ (monkeys, baboons, macaques, etc.).

Family Simiidæ (man-like or anthropoid apes).

Family Hominidæ (men).

Lemuroidea

The lemurs are the most ancient of living primates and as such have departed least from the ordinary quadruped. They are, however, exclusively arboreal, mostly nocturnal, and of comparatively low organization, which is manifest not only in their body but also in the brain, for the fore brain is relatively small and smooth and does not completely cover the hind brain as in the higher primates. The second digit of the foot bears a claw, the rest terminate in nails.

The present home of the lemurs is, above all, Madagascar, of which they are so highly typical that they constitute perhaps one-half of the total mammalian fauna. Lemurs are also distributed through the tropical forests of Africa and the Oriental realm. They have found fossil in the Eocene rocks of North America and of Europe. Two interesting relic animals belonging to this group still survive—*Chiromys,* the aye aye, now living in Madagascar but having near allies among the long-departed fossil forms of North America; and *Tarsius,* the tarsier, now confined to Sumatra, Borneo, Celebes, Java, and the Philippines, but which also had relatives in the American Eocene.

Fig. 12.1: Lemur, the aye, aye, *Chiromys madagascoriensis,* living, Madagascar.

Anthropoidea

The anthropoids are the most highly organized of primates, with 32 to 36 teeth, a completely closed orbit, two pectoral mammæ, feet usually prehensile and generally the hands also, pollex sometimes vestigial, and cerebral hemispheres richly convoluted, covering the cerebellum. This suborder includes all primates other than the lemurs and this of course means man as well as the monkeys and apes. It is divided into two sharply marked series, the Old and New World primates, and these, so far as our evidence goes, represent parallel evolutions which, because of the long period of South American isolation, must have diverged from a common ancestry in early Eocene time.

The Platyrrhini may be distinguished by the broad nasal septum; the thumb is not opposable, and sometimes reduced; the tail may be prehensile; there are no cheek-pouches nor ischial callosities. The family *Hapalidæ,* the marmosets or squirrel-monkeys, are small monkeys with a long, hairy, non-prehensile tail. The pollex is elongated, but the hallux very small. The latter bears a flat nail, while all of the other digits are armed with curved and pointed claws. These creatures are no larger than squirrels and are active forms living among the trees in small groups. Their food consists of fruit, to which eggs and insects are added, a very common dietary.

Fig. 12.2: The tarsier, *Tarsius spectrum*, living, East Indies and Philippines.

The *Cebidæ*, the common South American monkeys, differ from the marmosets in the possession of an additional molar tooth in each jaw, making thirty-six teeth all told, in having flat nails instead of claws, and frequently a prehensile tail. These forms, none of which is as large as the larger Old World monkeys, are exclusively confined to the tropical forests, notably those of Brazil. Among the more remarkable are the slender spider-monkeys whose prehensile tail is an organ of the greatest use; the howler monkeys (*Mycetes*), whose prodigious voice arises from an especially modified vocal apparatus; and the capuchins (*Cebus*), whose pathetic figures, garbed with human habiliments, are so often seen with itinerant musicians.

The Evolution of Man

The group Catarrhini includes all of the Old World apes and man, excluding of course the lemurs. They are characterized by the possession of a narrow nasal septum with the approximated nostrils directed downward, thirty-two teeth as in man, and a non-prehensile tail, which may, however, be vestigial or entirely absent. The hallux, except in man, is fully opposable and the pollex as well although often less developed.

The *Cercopithecidæ* are the monkeys and baboons, exclusive of the man-like apes, from which they differ in the fore-and-aft elongation of the molar teeth, the presence of ischial callosities on the rump, occasional cheek-pouches, a narrow breast-bone, and in the absence of the vermiform appendix. The baboons (*Cynocephalus*) are almost the only primates with the exception of man which have forsaken the arboreal for a terrestrial mode of life; but unlike man this has not resulted in an erect posture but a typically quadrupedal one. Their head is more dog-than ape-like, hence the generic

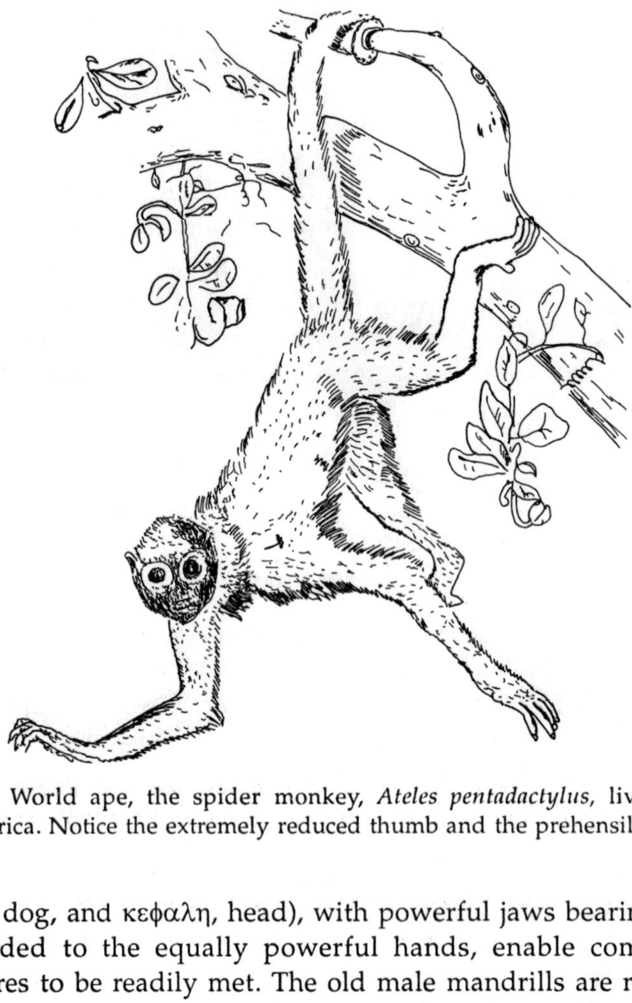

Fig. 12.3: New World ape, the spider monkey, *Ateles pentadactylus,* living, northern South America. Notice the extremely reduced thumb and the prehensile tail.

name (Gr. κυων, dog, and κεφαλη, head), with powerful jaws bearing immense canine teeth which, added to the equally powerful hands, enable competition with the terrestrial creatures to be readily met. The old male mandrills are remarkable for their ferocity. These creatures are coloured most gorgeously on the cheeks and ischial callosities, but colours which in themselves are beautiful—blue, scarlet, lilac—are in combinations which seem grievously misplaced. Thus while the fur is often beautiful and the colours lovely, the general effect is such that, as Cuvier says, "Il serait difficile de se figurer un être plus hideux que le mandrill." The mandrills, which are typical baboons, like the rest of their race, appear to be somewhat indiscriminate eaters, feeding upon fruit, roots, reptiles, insects, scorpions, etc., and inhabit open rocky ground rather than forests. Their present range includes Africa and Arabia.

Of the baboons Ditmars says: "It is fortunate for general animal life that their tendency to develop size and massive frame stopped where it did. It they had reached the size of the anthropoid apes, they would be among the most frightful creatures the earth has ever known."

Hominidæ

The family to which man belongs, the Hominidæ, bears the stamp of close relationship with the Simiidæ the differences being mainly the direct outcome of terrestrial life, the assumption of the erect posture, and the development of the brain. The erect posture has coördinated with it the alternation in the curvatures of the spine, the more complete adaptation of the hind limbs to bear the weight of the body, the loss of the power of opposition of the great toe and its more complete development in the thumb, and the greater length of the hind as compared with the fore limbs.

"The anthropoids are chiefly frugivorous and typically arboreal; when upon the ground they run poorly and (except in the case of the gibbons) use the fore limbs in progressing. Thus they are confined to forested regions. Man, on the other hand, is omnivorous, entirely terrestrial, erect, bipedal and cursorial, an inhabitant primarily of open country. The anthropoids use their powerful canine tusks and more or less procumbent incisors for tearing open the tough rinds of large fruits and for fighting. Primitive man, on the contrary, uses his small canines and more erect incisors partly for tearing off the flesh of animals, which he has killed in the chase with weapons made and thrown or wielded by human hands. These implements and weapons also usually make it unnecessary for man to use his teeth in fighting and functionally they compensate for the reduced and more or less defective development of his dentition" (Gregory).

There is but one living genus, *Homo,* included within the family Hominidæ, and all existing men of whatever race or colour are given but a single specific name, *sapiens* (*i.e.* to know). The divisions of this species into its various races, of which no fewer than 26 subspecies are recognized by Gregory, are, perhaps, unnecessary to our purpose, other than to enumerate the following:

Australian race: skull long (dolichocephalic); extremely prominent eyebrows; large teeth, especially the canines; tall, long-limbed; skin chocolate-brown; hair black, long and woolly. Habitat: Australia, Dekkan, Hindustan.

Negroid race: dolichocephalic, forehead round and childish, nasal bones flattened, teeth sloping (dental prognathism); skin and eyes brown or black, hair the same colour, short, woolly, not abundant. Habitat: Madagascar and Africa from the Sahara to the Cape of Good Hope.

Mongolian race: brachycephalic (short-headed), flat nose, small and oblique eyes; short and thick-set; golden-brown skin; sleek, coarse black hair; scanty beard. Dwell east of line drawn from Lapland to Siam; Chinese, Tartars, Japanese, Malays, Eskimos, North and South Americans.

Caucasian race: A, Mediterranean, short, slender, long-headed, hair and eyes dark brown to black; B, Alpine, of medium height, stocky, round-headed, hair and eyes dark brown to black, eyes often hazel or gray in western Europe; C, Nordic, tall, long-headed, hair flaxen, red, light brown to chestnut, eyes blue, gray, or green. Habitat:

Mainly Europe and North America; includes also Moors, Berbers, Egyptians, Kurds, Persians, Afghans, Hindus, Turks, Armenians, Africanders, and Australians (non-native, for natives see above).

ANATOMICAL AND ONTOGENETIC EVIDENCES FOR HUMAN EVOLUTION

Anatomical Evidences

Limbs

As with the elephant, man's body shows a number of primitive characteristics in addition to his specializations, the latter being developed within comparatively narrow limitations. The radius and ulna in contrast with those of the horse are both well developed and freely articulated so that the range of movement is ample, not only the hinge motion at wrist but the rotary one known as pronation as well. In the wrist the several bones are distinct, such as the scaphoid and lunar of the proximal row, which, in the Carnivora for instance, tend to coalesce into a single bone. The fibula of the lower leg is well developed and the foot has the primitive plantigrade position of the most archaic mammals.

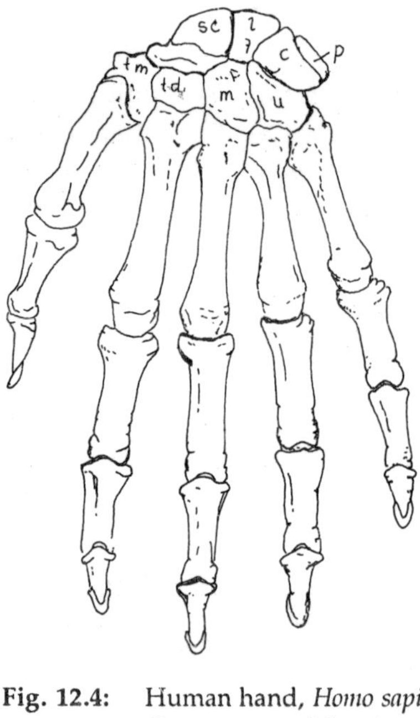

Fig. 12.4: Human hand, *Homo sapiens*. Compare with that of *Phenacodus primævus*. *c*, cuneiform; *l*, lunar; *m*, magnum, *p*, pisiform; *sc*, scaphoid; *td*, trapezoid; *tm*, trapezium; *u*, unciform.

The number of digits is unreduced from the primal five, nor are the phalanges either increased in number as in the whales or diminished, so that the primitive formula of 2, 3, 3, 3, 3, for the number of bones in both fingers and toes still prevails.

Skeleton

The shoulder girdle yet retains the clavicle, linking the scapula with the sternum in contrast with all ungulates and with the carnivores. The atlas which bears the skull, the sacrum or that portion of the vertebral column which lies between the hips, and the scapula or shoulder-blade all show certain reptilian characteristics.

Teeth

The teeth are also primitive, short-crowned, of simple structure and pattern, bearing relatively few, low cusps. The two premolars which are present are simpler than the molars, all of which is in marked contrast with the forms we have studied—the horse, elephant, and camel.

Soft Anatomy

Man's soft anatomy also bears the mark of great antiquity, especially those organs having to do with his pre-natal nourishment (placenta). And the means whereby the intestine is attached within the body cavity, the mesenteries, have still the same arrangement, seen in the quadrupeds.

Specializations

These are, first, the erect posture, which has reacted upon the skeleton in several ways, for instance, the four *curvatures of the spinal column* which are reduced in number and degree in infancy and extreme old age. Second, the *basin-shaped pelvis* in contrast with the flattened form seen in the anthropoids. The form of the human pelvis aids in

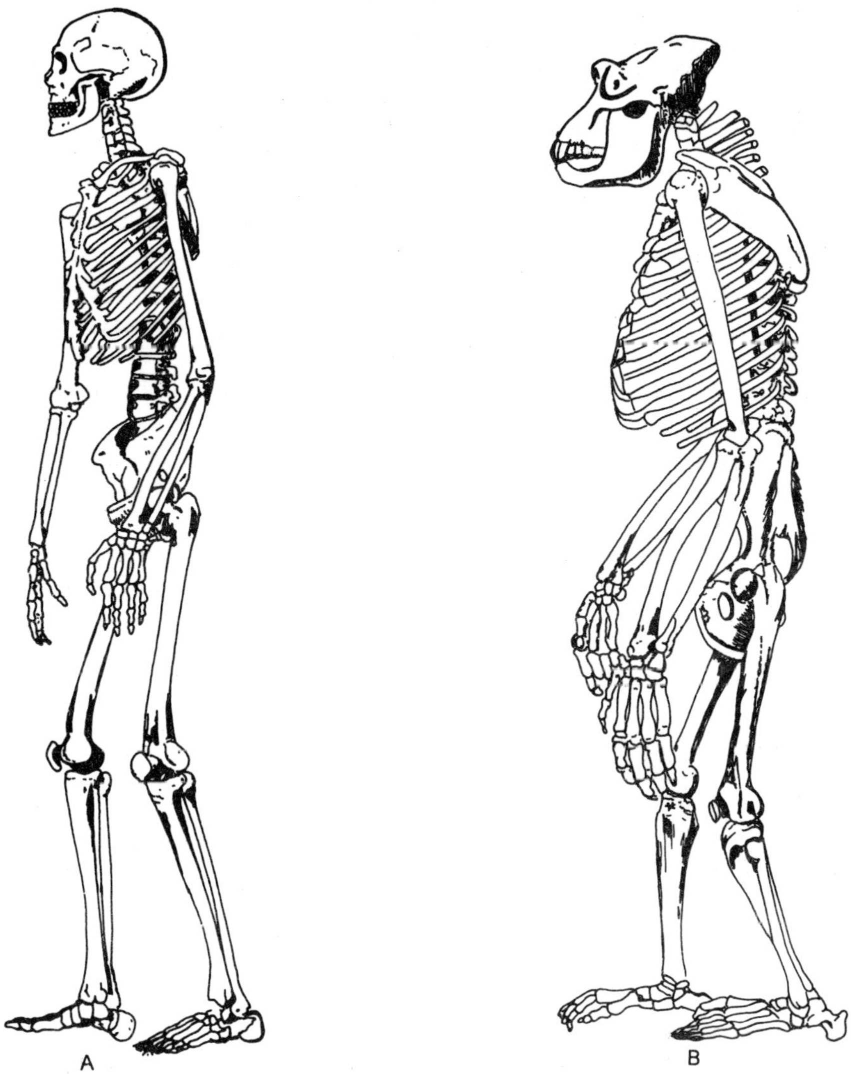

Fig. 12.5: Skeletons of man (A) and gorilla (B).

supporting the viscera while the body is in the erect position. The pelvis of human embryos is at first flattened, apelike, and only gradually assumes the basin form as they approach the time of birth. Third, *relatively short fore limbs,* the hands rarely extending below mid-thigh when the body is held erect. This shows the opposite extreme of a series as compared with the gibbon.

With respect to *foot specialization,* in comparison with the foot of the anthropoid the human foot shows: loss of opposability of the great toe, occasionally offset in some primitive types of man; development of the shock-absorbing arch; tendency toward monodactyly, the axis of the foot running through digit one which thus becomes the "great toe," the others diminishing in size and length. In the fifth digit reduction in the number of phalanges is in progress as the distal ones tend to fuse in a certain percentage of human subjects.

Still another specialization is the *loss of hair* from the body, possibly as a result of the acquisition of artificial clothing. The evidence for this belief cited by Matthew (1915) follows: "(1) It is accompanied by an exceptional and progressive delicacy of skin, quite unsuited to travel in tropical forests. I do not know of any thin-haired or hairless tropical animal whose skin is not more or less thickened for protection against chafing, the attacks of insects, etc. (2) The loss [of hair] is most complete on the back and abdomen. The arms and the legs and, in the male, the chest, retain hair much more persistently. This is just what would naturally happen if the loss of hair were due to the wearing of clothes,—at first and for a long time a skin thrown over the shoulders and tied around the waist. But if the loss of hair were conditioned by climate it should, as it invariably does among animals, disappear first on the under side of the body and the limbs and be retained longest on the back and shoulders." A high specialization is the loss of pigment in the skin of fairer races.

The normal human *dentition* contains the same number of teeth, thirty-two, as that of the other Catarrhini, so that the reduction from the original forty-four is a primate and not a human characteristic. The human teeth, however, are reduced in relative size as compared with those of the anthropoids, the canine no longer exceeds the other teeth in length and it is tending to become incisiform. There is also a loss of the diastemata or spaces between the teeth into which, in the ape, the opposite canines fit.

In the anthropoids the *movement of the lower jaw* is obliquely transverse, ruminant-like; in the human being it is in all directions and partly of a rotary character. This is correlated with the reduction of the interlocking canines. That tooth reduction in humanity is still progressing is shown by the fact that one of the premolars and the second incisors are often reduced and sometimes wanting, and the same is true although to a less extent of the third molars, the so-called wisdom teeth.

The tooth reduction is in turn correlated with the shortening of the muzzle and jaw symphysis—the facial portion of the skull, that concerned with the senses and appetite, in contrast with the enormous expansion and deepening of the brain-case, the seat of mentality. This change in the proportions of face to cranium is expressed in terms of the *facial angle* that is formed between two lines lying in the sagittal plane of the skull, one of which is drawn from the lower margin of the nasal aperture to the ear opening, the other from the forehead to the maxilla. In the higher races of mankind this facial angle approaches a right angle, averaging 85°; in the anthropoid, such as the chimpanzee, it is much less, not more than 45°.

The human *brain* is one of nature's marvels, exceeded in actual size only by that of the elephants and of the greater whales. In relation to bodily weight, man's brain exceeds that of any other creature except some excessively small vertebrates such as the humming-birds and smaller mice. In these forms there has been a dwarfing as a result of evolution, as an increase or decrease in bulk physically seems to be more

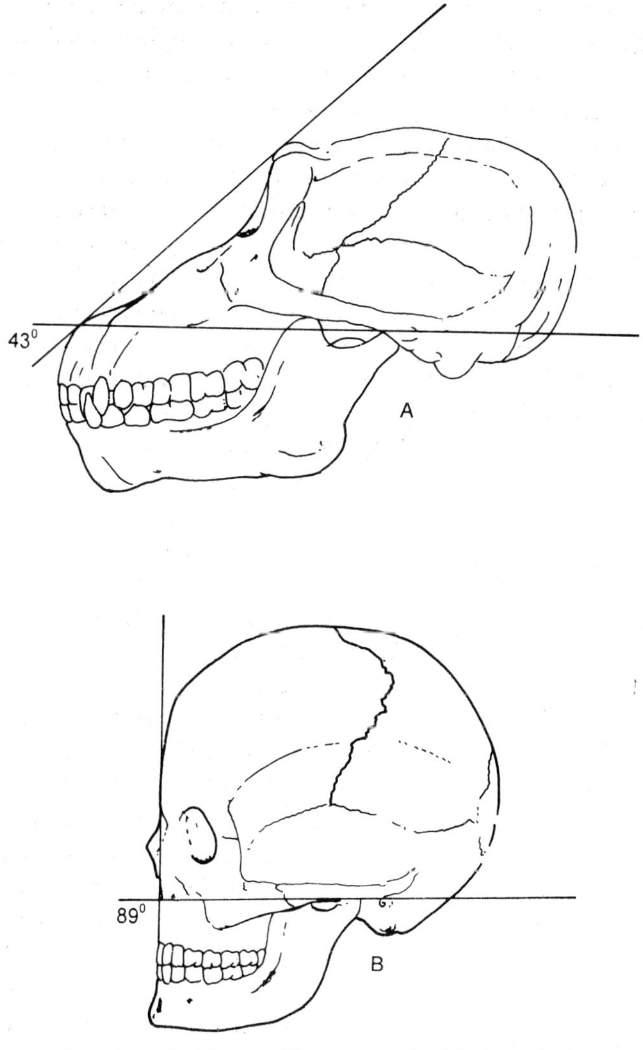

Fig. 12.6: Facial angle of man (B) compared with that of chimpanzee (A).

rapidly attained than a change in brain size in the same animal. Growth of body unaccompanied by equivalent brain growth was seen in the dinosaurs *Stegosaurus* and *Brontosaurus*.

Aside from its mere increase in size, the human brain (see Fig. 12.7) is also of the highest vertebrate type, the huge cerebrum entirely covering the cerebellum, so that the latter is invisible from above. Greater area of the external cortex, the so-called "gray matter" which is the real seat of intellect, is obtained by a deepening and further complication of the depressions between the convolutions. And the great development of the frontal lobes, particularly, gives ample opportunity for the expansion of the higher intellectual faculties. In its subtle fineness of detail, in its ability to record and often to reproduce an almost infinite number of mental perceptions, and in all those other resident facilities which together make up the higher intellectual characteristics of humanity, the human brain stands preëminent as the most complex structure evolution has produced. But a comparison with the brain of an orang shows the self-same fundamental characteristics; the proportions differ, but in a broad way the shapes are similar and the major convolutions are alike, in other words, the two brains differ not in kind but in degree, and that of man is physically merely a relatively larger and more refined example of the same fundamental type. With regard to mental retrogression in the apes Beebe says: "Young orang-utans in their 'talk' as well as in their actions, are the counterparts of human infants. The scream of frantic rage when a banana is offered and jerked away, the wheedling tone when the animal wishes to be comforted on account of pain or bruise, and the sound of perfect contentment and happiness when petted by the keeper whom it learns to love,—all are almost indistinguishable from like utterances of human child. But how pitiless is the inevitable change of the next few years!...Slowly but surely the ape loses all affection for those who take care of it. More and more morose and sullen it becomes until it reaches a stage of unchangeable ferocity and must be doomed to close confinement never again to be handled or caressed." Mr. Beebe adds to this observation in a letter as follows: "I find that while sexual maturity is attained at about six years, the females seem little affected and remain gentle and affectionate. The males, however, begin at about five years to become morose and sullen. This applies both to chimpanzees and orangs. These statements apply only to several animals which have attained their sexual majority in the Park and even among them there is great variation."

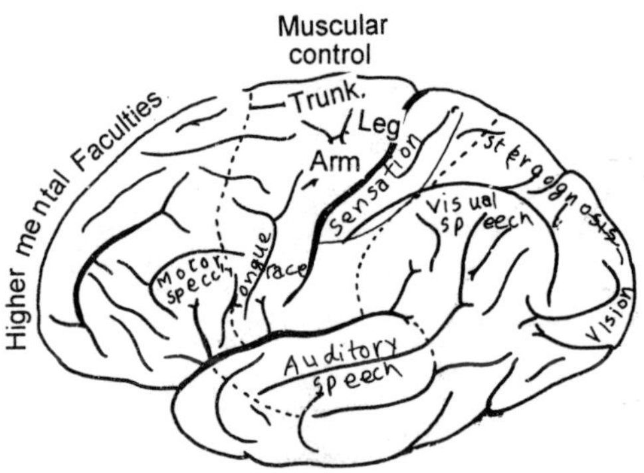

Fig. 12.7: Side view of human brain, cerebrum of high type, showing chief areas of muscular control and of sensory impressions of sight and hearing, also prefrontal area in which higher mental faculties are centered.

The final human characteristic which lifts man high above his fellow creatures is *articulate speech*, the means whereby communication, especially of higher abstract thoughts, is made primarily possible; for the development of a mature written language is clearly the outgrowth of antecedent speech. This human faculty has had great influence in the development of the higher mental traits.

Thus it will be seen that comparative anatomy shows very emphatically our fundamental resemblances to the other anthropoids and that if we would look for differences we must compare details of structure and development rather than distinctions of a larger sort. As Huxley truly said: "The structural differences between man and the man-like apes certainly justify our regarding him as constituting a family apart from them; though, inasmuch as he differs less from them than they do from other families of the same order there can be no justification for placing him in a distinct order...Perhaps no order of mammals presents us with so extraordinary a series of gradations as this—leading us insensibly from the crown and summit of animal creation down to creatures, from which there is but a step, as it seems, to the lowest, smallest, and least intelligent of the placental Mammalia. It is as if nature herself had foreseen the arrogance of man, and with Roman severity had provided that his intellect, by its very triumphs, should call into prominence the slaves, admonishing the conqueror that he is but dust."

Vestigial Organs

Drummond mentions no fewer than seventy such relics which he most appropriately calls the scaffolding left in the body, relics of old-time conditions and needs for which the modern human economy has no further use. They are veritable historical documents enclosed within the limits of each human frame during part or the whole of its existence and may be viewed in no other light. Certain of these features disappear with growth and maturity, others persist during the lifetime of their possessor.

One of these persistent vestigial features is the *direction of hair* on the body. That upon the arms, for instance, slopes from shoulder to elbow and from the wrist upward and outward in such a way that, were the hands clasped above the head with the elbows pointing downward, a posture said to be assumed by the orang, the hair thus arranged sheds the falling rain. In all anthropoids except the gibbon the same direction of hair prevails as with mankind, hence the conclusion of a bygone community of habit between man and these apes is irresistible. Then, too, the absence of hair on the terminal segments of the digits, its scarcity on the second, and greater abundance on the first are true of the anthropoids as well as of man.

The *vermiform appendix* has been mentioned as a diagnostic characteristic of the family Hominidæ and also of the Simiidæ. In man it is not only apparently useless but is sometimes a veritable deathtrap. With herbivorous mammals, on the other hand, its homologue, the cæcum, is large and of high digestive value. Even in man the appendix has the same structure as the large intestine—peritoneum, muscular coat, and mucous layer. In the embryo it has the same caliber as the rest of the bowel but soon ceases to grow and may actually be as long in the new-born babe as in the adult.

The *Darwinian point* to the ear is a little conical projection from the rolled or unrolled margin of the ear, more frequent in the male than in the female, as are all atavistic features. This is in man a relic of the pointed ear found in the human embryo and in lower mammals and is, as Darwin says, a "surviving symbol of the stirring times and dangerous days of his animal youth."

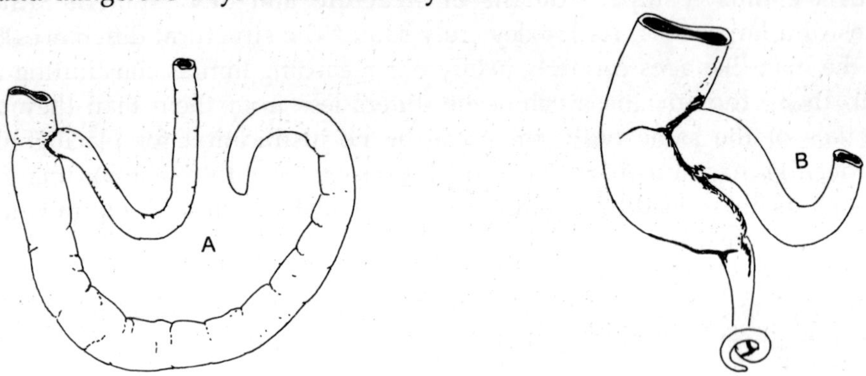

Fig. 12.8: Cæcum of (A) kangaroo and vermiform appendix of (B) human embryo.

Thin bands of *muscle,* formerly of value in moving the shell of the ear to aid "in the appreciation of sound, are still present but usually functionless, as are also the present but involuntary hair-erecting muscles of the scalp. Most of the dermal muscles, in fact, so well developed in lower animals for switching the skin, are retained in the human face only where they are used for the expression of the emotions. Doubtless their retention is in part of defencive significance as they may well have been used to strike terror into the breast of an opponent. In addition to this, they also aided in the expression of the emotions, as of pleasure or pain, and together with the voice formed the first elemental speech.

A further vestige is the *plica semilunaris,* a crescentic fold of membrane in the inner corner of the eye which represents the very efficient third eyelid or nictitating membrane of the eyes of many mammals and of birds. In man it is relatively larger before than after birth. The *pineal body* of the brain is connected in reptiles, notably *Sphenodon,* with a third eye, really the first primordial vertebrate eye. In man this is present as a vestige deep hidden beneath the mass of the cerebrum.

In other mammals that portion of the upper jaw which bears the incisor teeth is separated off by a suture, and is known as the *premaxillary bone.* In man and the chimpanzee this suture is normally obsolete, yet the poet Goethe predicted that some day the separate premaxillary would be found in man, and so it has been. Moreover, the *frontal bone,* single in man, paired in the dog, is paired in an Abyssinian skull in the Yale collection.

ONTOGENETIC EVIDENCES

Ontogenetic Vestiges

Embryology teaches us much of the past life of any race and this is jut as true of humanity as of any other created being; for the unfolding of the miniature man shows precisely the same one-called condition of the ages-remote protozoan ancestry; the same cleavage stages, morula, blastula, and gastrula, as any other metazoan; the gradual assumption of chordate characteristics, of notochord, of hollow nervous system, of gill-clefts; the budding of limbs, at first as ill formed as those of the earliest slime-borne amphibian emergent from the old limiting aquatic environment; the perfected limbs and well-developed tail of an ancient placental mammal, and the ultimate loss of this and other embryonic structures, until a man is born into the world. Thus these wonderful changes, wrought in the dark, reproduce, as in a pageant, the historic changes brought about by the evolutionary process during the long night of the geologic past.

Certain of the ontogenetic features may be more specifically mentioned, such as the *gill-clefts,* of which there are four in the embryo. Occasionally certain of these communicate with the throat so that openings remain on the sides of the neck through which fluids taken in at the mouth can trickle, or white patches on the skin betray their former position. The first gill-cleft, the so-called spiracle of the fish, normally persists and forms the eustachian tube connecting the middle ear with the throat for the purpose of equilizing the air-pressure on either side of the drum. The auricle in fact is developed from this first gill-cleft and the hearing organ may be subsequently repeated down the neck. As Drummond says: "In some human families, where the tendency to retain these special structures is strong, one member sometimes illustrates the abnormality by possessing the clefts alone, another has a cervical ear, while a third has both a cleft and a neck-ear—all of these, of course, in addition to the ordinary ears."

The *tail* is indicated in the human skeleton by the four or five bones at the lower terminus of the spine, coalesced in the adult into the coccygeal bone which is concealed beneath the flesh; but in the embryo not only is it present but is free, movable, and has muscles for wagging it! These are usually reduced later to mere ligaments but may permanently retain their muscular character. The external tail may also persist.

The *lanugo,* or clothing of dark hair which covers the entire body except the palms and soles at the sixth month of prenatal life, usually disappears before birth, but in rare circumstances may persist and give a permanently hairy aspect to both face and body. This fœtal hair is also found in other hairless mammals such as the elephants and whales, and can have but the one historical significance, harking back to the day when hair was a racial necessity and not a superfluity as it is to-day, in all three groups.

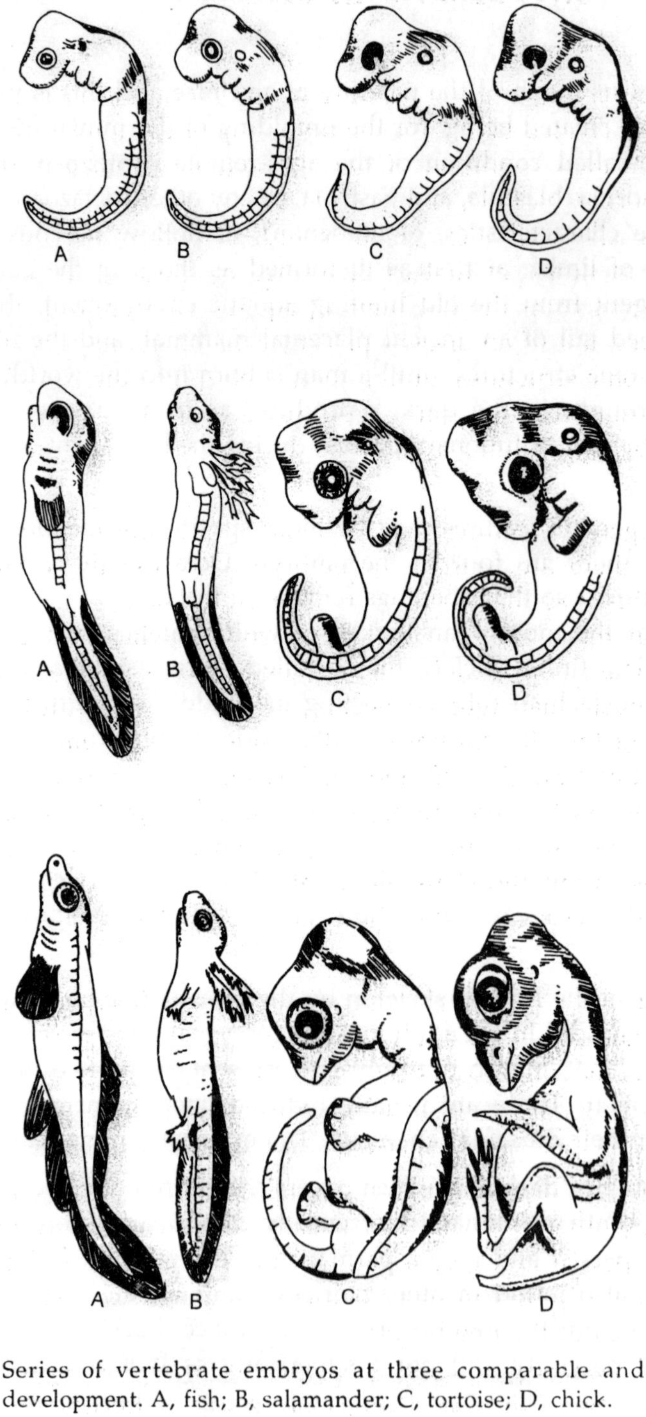

Fig. 12.9: Series of vertebrate embryos at three comparable and progressive stages of development. A, fish; B, salamander; C, tortoise; D, chick.

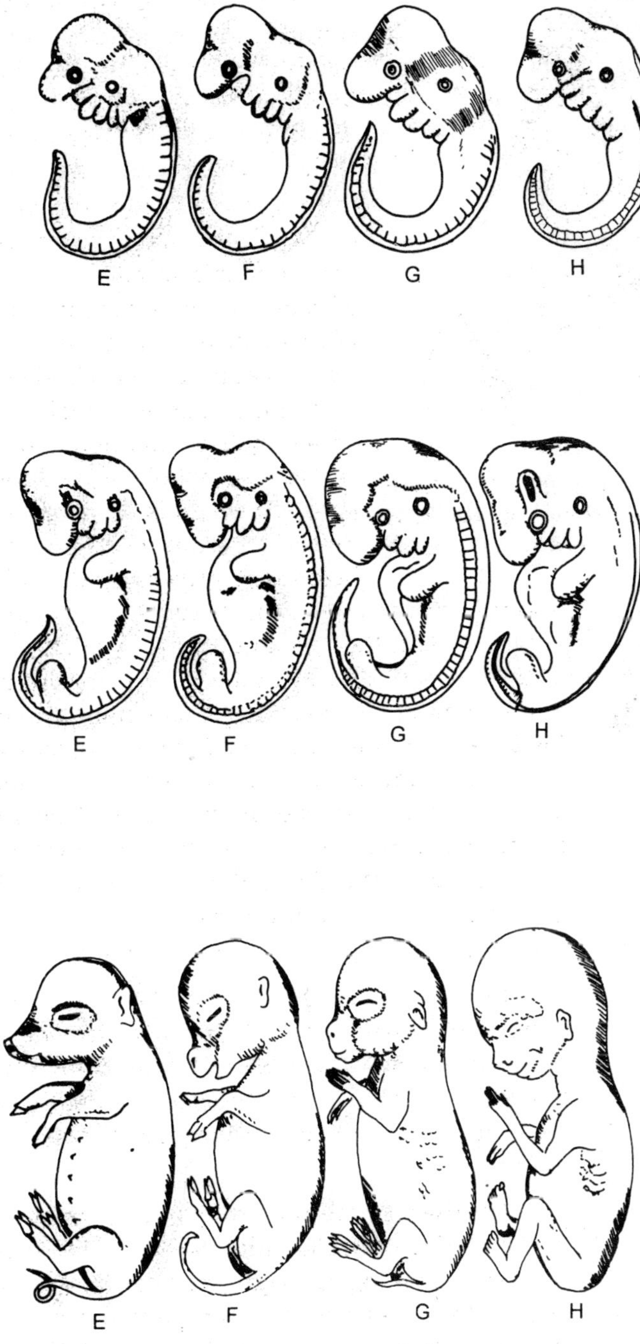

Fig. 12.9: (cont.) Series of vertebrate embryos. E, hog; F, calf; G, rabbit; H, Human.

Fig. 12.10: Russian "dog-man," Adrian Jeftichjew, showing atavistic development of hair.

The "awful grasp of a baby," as one has put it, is also significant, for the *power of grip*, notably great during the first few weeks of its life when it needs the most constant care, sensibly weakens later as experiments have shown. These consisted in the suspending from a stick or from the finger by the power of their hands alone some sixty infants which were under a month old, and in at least half of these the experiment was tried within an hour of birth. "In every instance, with only two exceptions, the child was able to hand on to the finger or a small stick, three-quarters of an inch in diameter, by its hands...and sustain the whole weight of its body for at least ten seconds. In twelve cases, in infants under an hour old, half a minute passed before the grip relaxed, and in three or four nearly a minute. When about four days old...the strength had increased, and nearly all when tried at this age could sustain their weight for half a minute. About a fortnight or three weeks after birth the faculty appeared to have attained its maximum, for several at this period succeeded in hanging for over a minute and a half, two for just over two minutes, and one infant of three weeks old for two minutes thirty-five seconds. Invariably the thighs are bent nearly at right angles to the body, and in no case did the lower limbs hang down and take the attitude of the erect position. Furthermore, the child shows no sign of distress and no cry is uttered until the grasp begins to give way" (Drummond).

Fig. 12.11: Human babe, three weeks old, supporting its own weight for over two minutes. Attitude of lower limbs and feet more simian than human.

This is of course one of the many instances, mainly structural, however, which point to the old-time arboreal life, not perhaps that the infant of that day clung directly to the tree but that the mother did and had to have her hands free for brachiation, hence it was necessary for the infant to cling to her.

Another phenomenon which has received a similar interpretation—that of arboreal life—is the occasional *dreams* one has of falling through space with the violent instinctive effort often undergone to prevent disastrous consequences. And the strange thing about it is that in the dream the fall never ends fatally, for that is an experience which could not be transmitted to offspring, for such would not exist, while that of the fall could. Jack London in his book *Before Adam* makes much of this. Roosevelt says of nightmares, although without necessarily implying an historical interpretation to them: "Civilized man now usually passes his life under conditions which eliminate the intensity of terror felt by his ancestors when death by violence was their normal end and threatened them during every hour of the day and night. It is only in nightmares that the average dweller in civilized countries undergoes the hideous horror which was the regular and frequent portion of his ages-vanished forefathers, and is still an everyday incident in the lives of most wild creatures" But perhaps too much emphasis has been placed upon dreams.

These examples out of many—Wiedersheim says 180—are sufficient to show that the human body cannot be considered as a perfect final work of creation but rather the ultimate product of eons of evolutionary change, resulting in a very imperfect being from the physical point of view—a veritable museum of antiquities!

Pithecanthropus

The Java ape-man, *Pithecanthropus erectus,* was first discovered in Trinil, and elsewhere in central Java, in 1891. The type consists of a calvarium or skull cap, a left thigh bone, and two upper molar teeth. The skull, which has beetling brows or tori, appears relatively long and narrow, a characteristic of most perehistoric crania (except some of the latest), in contrast with those of the apes. In profile the calvarium is extremely low, the highest point being at mid-length, while the rear part is heavy, implying a lack of that delicate balance characteristic of an entirely erect posture. This seems at variance with the evidence offered by the straight thigh bone, which caused by discoverer, Dr. Dubois, to give the specific name *P. erectus* to the form. But heavy neck muscles are indicated, which are correlated with a thrust-forward muzzle and generally bestial appearance. The cranium has been cleared of its contained matrix and a cast of the interior taken which shows the general form and proportions of the brain. While the exterior of the skull is like that of a gigantic gibbon, the brain was essentially human, but with a number of very primitive features. Two or three of these are in the areas which are the last to develop in the brain of a modern child, which throws an interesting side light on the stage of evolution of the *Pithecanthropus* brain.

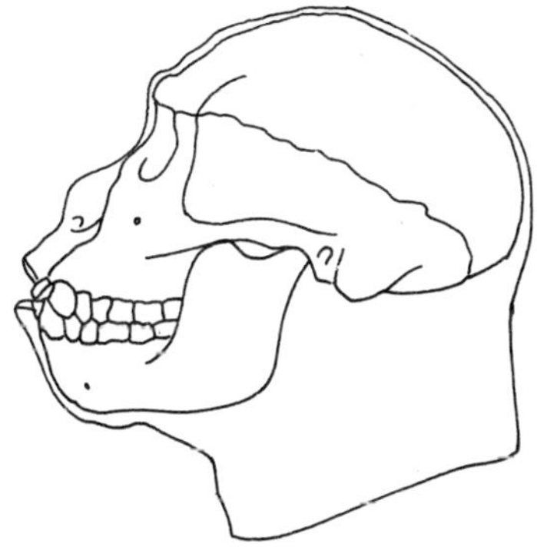

Fig. 12.12: Restored head of Java ape-man (*Pithecanthropus erectus*).

The cranial capacity has been estimated at 940 cubic centimeters, compared with 580 for a very large male gorilla and 1500 for an average European skull. But size is not the only thing, for the undeveloped portions are equally important, and the *Pithecanthrepus* brain is deficient in the so-called prefrontal and parietal areas, the former particularly being the seat of the higher mental faculties. Broca's area, the seat of motor speech, and the auditory speech center are sufficiently developed to show that this primitive type possessed essentially human articulate speech, however, rudimentary it may have been. The restoration of the face has been most carefully done and, while the evidence is far from complete, in all probability represents an approximation of the actual condition. It is based on the calvarium and the teeth and was checked by another very fragmentary specimen found twenty-four miles away.

Sir Arthur Keith estimates the height at 5 feet 8 inches, with a weight of 11 stone or 154 pounds, close to the average of modern man. Thus *Pithecanthropus* lies much nearer to man than the apes but is probably not in the direct line leading to modern man.

Peking Man

Sinathropus pekingensis is thus characterized: cranial walls very thick; tori huge, very low receding forehead, low cranial dome. Mandible chinless with massive canine teeth. Teeth large with both ape-like and human characters. Skeleton unknown, but evidence suggests an erect gait. *Sinanthropus* and *Pithecanthropus* may prove variants of the same human type.

Heidelberg Man

Homo heidelbergensis, the Heidelberg man, represents one of the two oldest recorded European races, geologically speaking. The type was discovered in 1907 in river sands, 79 feet below the surface, at Mauer, near Heidelberg, South Germany. The relic consists of a perfect lower jaw with the dentition. This interesting relic shows a curious combination of characters, for, while the teeth are essentially human, the massive jaw is ape-like, so much so that, were the symphysis alone present, it might be taken for that of a gorilla-like form, while the rear ascending portion resembles a gibbon's, except for size. The chain slopes away as in the Piltdown man, while the area for muscular attachment is extremely powerful. The teeth, relatively small for the jaw, form a continuous series, and the canines do not rise above the level of the other teeth, both of which are human features in contrast with those of the apes. The teeth have short divergent roots, a bulging crown, and a relatively large pulp cavity, which Keith considers as an adaptation to a peculiar sort of diet. He compares the teeth to those of cattle, whose harsh vegetal food is obvious, hence he has coined the word taurodont (ox tooth) by way of a descriptive term. This is a characteristic, unape-like, found nowhere else among men other than in the Neanderthal race, to which Heidelberg man is supposed to be ancestral, and therefore seems to debar both from our own direct lineage.

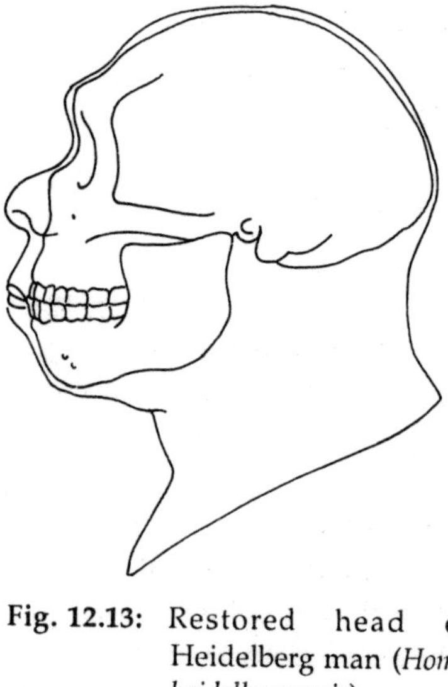

Fig. 12.13: Restored head of Heidelberg man (*Homo heidelbergensis*).

Recently other material has come to light in the Mauer sand pit, some of which may pertain to *Homoheidelbergensis*, but detailed descriptions are not yet available. The material seems, however, to bear out our conception of the brutal appearance of this ancient type.

Associated with the Heidelberg jaw is an extensive warm-climate fauna: straight-tusked elephant (*E. antiquus*), Etruscan rhinoceros, primitive horse, bison, wild cattle (urus), bear, lion, and so on, all of which aid in establishing the date of the jaw as Second Inter-glacial and its age, conservatively estimated, at from 3000,000 to 375,000 years. The cultural evolution of Heidelberg man is indicated by the presence of eoliths, flint implements of the crudest workmanship, if indeed their apparent fashioning is not merely the result of use.

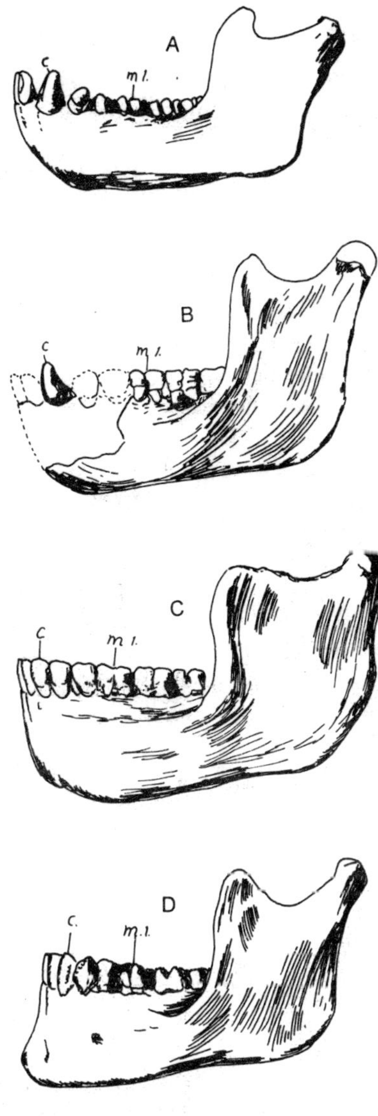

Fig. 12.14: Jaws, left outer aspect, of A, chimpanzee, *Pan*, sp.; B, Piltdown man, *Eoanthropus dawsoni*; C, Heidelberg man, *Homo heidelbergensis*; D, modern man, *H. sapiens*; *c*, canine; *m* 1, first molar.

Neanderthal Man

The specimen of Neanderthal type, *Homo neanderthalensis* or *primigenius*, was discovered in 1856 not far from Düsseldorf in Rhenish Purssia. Here the valley of the Düssel forms the deep Neanderthal ravine, whose limestone walls are penetrated by caverns, in one of which the remains were found. What was doubtless a perfect skeleton at the time of its discovery was so injured by its finders that only a portion of it, which was preserved in the Provincial Museum at Bonn, was saved. Previous to this, however, in 1848, a skull of this race had been found at Gibraltar, but its significance was not realized. Recently another skull has been found in Gibraltar.

The Neanderthal man, a prophet of an unknown race, was for a time utterly without honor though of course the subject of a most heated controversy, being considered as non-human, or as owing its distinctive characters to disease. The sagacity of Huxley threw true light upon the problem, though it was not until the mute testimony of other representatives of the race (the men of Spy) was offered that even Huxley's

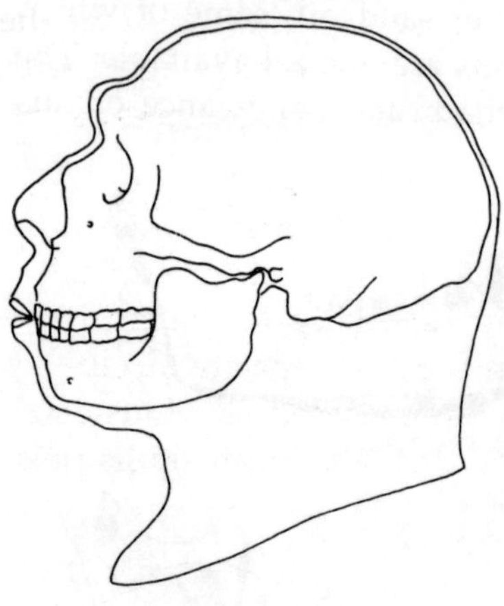

Fig. 12.15: Restored head of Neanderthal man (*Homo neanderthalensis*).

masterful conception of the Neanderthal characters was taken as an accepted fact. Professor Huxley's description of the Neanderthal type is classic. He says:

"The anatomical characters of the skeletons bear out conclusions which are not flattering to the appearance of the owners. They were short of stature but powerfully built, with strong curiously curved thigh bones the lower ends of which are so fashioned that they must have walked with a bend at the knees. Their long depressed skulls had very strong brow-ridges; their lower jaws, of brutal depth and solidity, sloped away from the teeth downwards and backwards, in consequence of the absence of that especially characteristic feature of the higher type of man, the chin prominence."

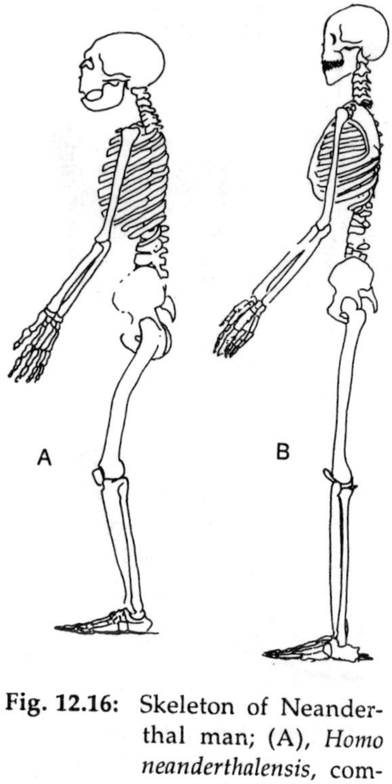

Fig. 12.16: Skeleton of Neanderthal man; (A), *Homo neanderthalensis*, compared with that of a living native Australian; (B), *Homo sapiens*, the latter the lowest existing race.

Subsequently many more specimens have come to light, at Spy in Belgium, at Krapina in Croatia, at Le Moustier, La Chapelleaux-Saints and La Ferrassie in France, and at Mt. Carmel near the Sea of Galilee. These, while differing in various details, effectually serve to establish the race, whose main characteristics are: heavy, overhanging brows, retreating forehead, long upper lip; jaw less powerful than that of Heidelberg man but very thick and massive; chin generally strongly receding but in process of forming; dentition extraordinarily massive in the La Chapelle specimen, whereas in those of Spy the teeth are small. The skull in many characteristics is nearer to the anthropoids than to modern man.

The brain is large and its volume is surely human, but the proportions are again less like those of recent man than like the anthropoids. The chest is large and robust, the shoulders broad, and the hand large, but the fingers are relatively short, the thumb lacking the range of movement seen in modern man. The knee was somewhat bent, the leg powerful, with a short shin and clumsy foot, clearly not of cursorial adaptation. The posture is usually described as not fully erect, but Dr. McGregor says that it was slouching rather than stooping. The average stature was about 5 feet, with a range from 4 feet 8 inches to 5 feet 3 inches, partly sex differences. The body was stocky and powerful with loose jointed limbs, so that the walk must have been an easy, shuffling gait. This, together with the huge head with the face thrust forward to accommodate the heavy jaw and the powerful neck muscles, gives an unlovely picture of this interesting type.

The lower jaw is a derivative of that of Heidelberg, although lighter and with the rudiment of a chin, and the same taurodont specialization of the teeth is indicated. The brain is large but relatively simple in organization, especially in the primitive frontal lobes and Broca's area, implying not only limited intellectual capacity but limited thought transmission as well. Culturally, Neanderthal man is classed as Mousterian, and the presence of a comparable industry in Mongolia and the Gallidean skulls point to Asiatic distribution if not origin.

Neanderthal man lived in Europe from the Third Interglacial stage through the Fourth Glacial, duration of thousands of years, and then became extinct, from twenty to twenty-five millenniums ago. He seems to have been an actual lineal successor of the man of Heidelberg, but was throughout his long career an unprogressive, static race. One of the most remarkable features in connection with this race, however, was the very reverent way in which the dead were buried, with an abundance of ornaments and finely worked flints. This can have but one interpretation, the awakening within this ancient type of the instinctive belief in spiritual immortality!

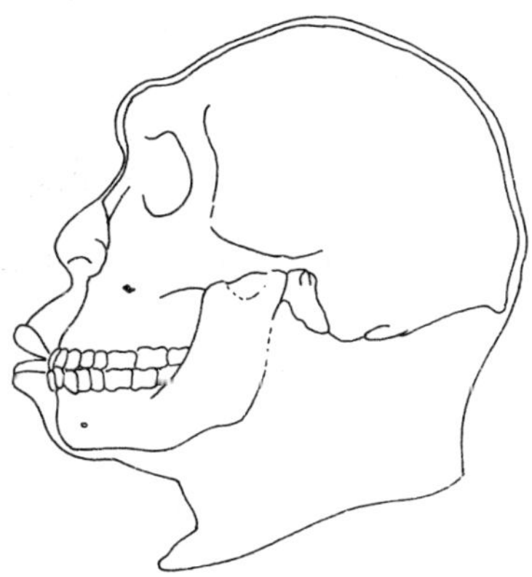

Fig. 12.17: Restored head of Rhodesian man (*Homo rhodesiensis*).

Rhodesian Man

Homo rhodesiensis was discovered in 1921 at Broken Hill in Northern Rhodesia. The remains consist of an almost perfect skull, lacking the lower jaw, together with portions of the limb bones and other skeletal elements. Fragments of a second individual were also found, sufficient to establish the race. Details of the cranium point to a number of simian characters, and the skull, in spite of certain specializations, is of a relatively low type. In the huge beetling brows, the flattened skull top, and other details it shows a strong resemblance to Neanderthal man, an undouted relative. While the face, with its forwardly projecting jaws, is extremely brutal, the brain was large, 1300 cubic centimeters, somewhat less than any known Neanderthal except possibly the first Gibraltar skull, but again deficient in those regions last to be developed in ontogeny. There is a lack of harmony between the relative development of face and brain, as though Rhodesian man had not yet discarded the general utility of the mouth which preceded that of the hands, which in turn depends on brain power.

The teeth are entirely human, set in a finely arched palate and the third molars or wisdom teeth are somewhat smaller than are the others, a modern tendency found in but one other recorded instance among prehistoric men. There is, however, evidence of dental caries or tooth decay, together with other indications of disease in the skull, again an undesirable modern attribute extremely rare in prehistory. The posture of this six-foot man was thought to be erect, but of this there is some doubt.

The dating of this interesting type is difficult. The skull is impregnated with mineral matter, as are the associated animal remains, and Bather thinks that, whatever the age of the latter, that of the skull is appreciably older. But the animals are all of the present African fauna which, however, has a distinct Pleistocene facies as compared with that of Asia. Rhodesian man is therefore of respectable antiquity, surely Pleistocene, although where he should be placed in the European chronology is not apparent.

The human species thus far discussed are all characterized by the huge brow ridges or tori, confluent above the nose, and by a low, flat forehead, inferred of course in the case of Heidelberg man. With the exception of *Pithecanthropus* and *Sinanthropus,* they all belong apparently to one *phylum* the Neanderthalid, in spite of the immense time interval between Heidelberg and the true Nenaderthal men.

Those yet to be discussed, Piltdown and Crô-Magnon men, lack the tori, and the forehead is steep. Otherwise they have little in common, and as a consequence their relationship is not clear.

Piltdown Man

Eoanthropus dawsoni—In 1912 was announced the discovery of a very ancient human relic from the Thames gravels at Piltdown, Sussex, England. When first discovered the skull was in large measure intact but was carelessly broken and thrown on the dump, to be painfully retrieved later. The single fragment saved by the workman was given to Mr. Charles Dawson, who, together with Sir Arthur Smith Woodward of the British Museum of Natural History, secured a number of additional portions, which were then laboriously reconstructed by the latter. As few of the pieces formed actual contacts with one another, the results of several attempts naturally varied slightly, especially as to cranial content, but the general form was clearly established. The parts consisted of parietal, temporal, and a portion of the occipital bones, together with the nasals, a canine tooth, and part of the lower jaw containing several molars. It was not over the essential form of the skull but over the lower jaw that controversy arose, for this bone of contention had so many simian features that many felt it could not possibly belong with the skull, Gerrit Miller going so far as to pronounce the jaw that of an extinct

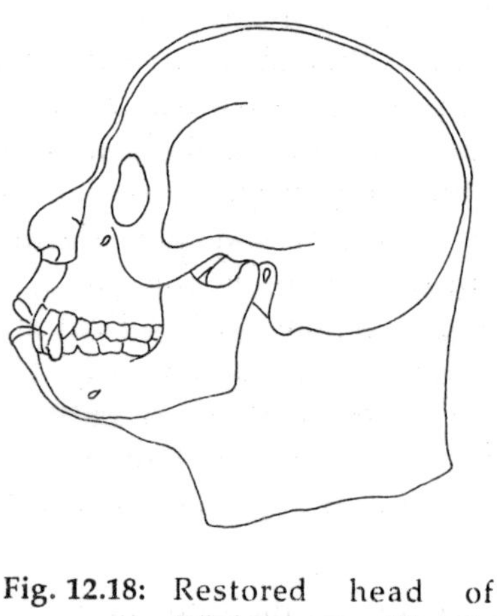

Fig. 12.18: Restored head of Piltdown man (*Eoanthropus dawsoni*).

chimpanzee, to which he gave the name *Pan vetus,* and the association purely accidental. Later opinion practically agrees that, in spite of its lack of harmony with the degree of evolution shown by the cranium, the jaw really may pertain to the same individual, and a second yet more incomplete individual bears out the belief that it is another instance of the evolutionary advance of cranium over face which we have seen in the case of Rhodesian man.

The exterior of the skull, as now restored, is in no sense ape-like, much less so in fact than the far later Neanderthal and Rhodesian men. The tori are lacking, and the forehead is steep, but the relative breadth of the rear portion and the inward slope of the sides, which gradually merge into the crown, do depart toward the ape; in modern man the crown is dome-shaped. The face must have been unusually large. The cranial walls are enormously thick, more so than in any other human skull, which makes the interior comparatively small, 1350 cubic centimetres, compared with 1500 for the average European. Eliot Smith sums up a description of the brain as the most primitive and ape-like human brain hitherto discovered, evidently excluding that of *Pithecanthropus*.

The jaw represents the imperfect right half. Enough remains to show a very retreating chin, but, while the molar teeth are human, the canine is not and must have interlocked with its opposite, as do those of the apes. Thus the jaws seem to have retained a large share of the general utility which goes with the undeveloped brain. The associated flint implements are extremely crude, which is further evidence of the low estate of *Eoanthropus*. Chronologically Piltdown man is not surely dated. He was interglacial, the associated animal remains indicate that, but whether fist or second is in question. He cannot be placed within our genus, *Homo,* since his peculiar morphology excludes him; rather has he been called *Eoanthropus,* the dawn man. *Eoanthropus* may also be barred from the direct ancestry of modern man.

Homo Sapiens

This Linnæan species includes all existing men, of whatever race, and some extinct men. They have a number of distinctive characters in common, although in varying degree. These are, first, the entirely erect posture, with four reversed curves in the spine which act as a shock-absorbing device to protect the nicely poised skull. The limbs are straight, but the segmental proportions vary racially and individually. The skull also varies in size and relative proportions, such, as length to breadth. The forehead is generally steep, and the continuous brow ridge is absent. Thus there is sample space for the fullest mental development. The final distinctive feature is the jutting chin prominence, the result of the reduction of the dental arch. In the lower existing races, Australian and Negro, the teeth are fine and regular and are well spaced; in the higher races, on the other hand, they are apt to be crowded and out of alignment, due to further progressive reduction of the jaws.

Crô-Magnon Man

The original finds of the men of the Cro-Magnon race, *Homo sapiens,* were made at Gower, Wales, and at Aurignac, France. In the latter place seventeen skeletons came to light in 1852, but were buried in the village cemetery and thus lost to science, and not until 1868, when five more skeletons were discovered in a rock shelter at Crô-Magnon, France, was the race established. These individuals, an old man, two young men, a woman and a child, are thus the type Crô-Magnons. This magnificent race is thus characterized:

Skull large but narrow, with a broad face, hence disharmonic. Facial angle equalling the highest type of *Homo sapines*. Jaw thick and strong, with a narrow but very prominent chin. Forehead high and orbital ridges reduced. Brain not only of high type but very large, that of the women exceeding the average male brain of today.

The stature of the old man was 5 feet 11 inches; the average for males being 6 feet 1.5 inches, for women 5 feet 5 inches, a great disparity. The lower segments of the limbs were long, in contrast with the Neanderthal type, hence the men of Crô-Magnon were swift-footed, while those of Neanderthal were slow. Osborn says:

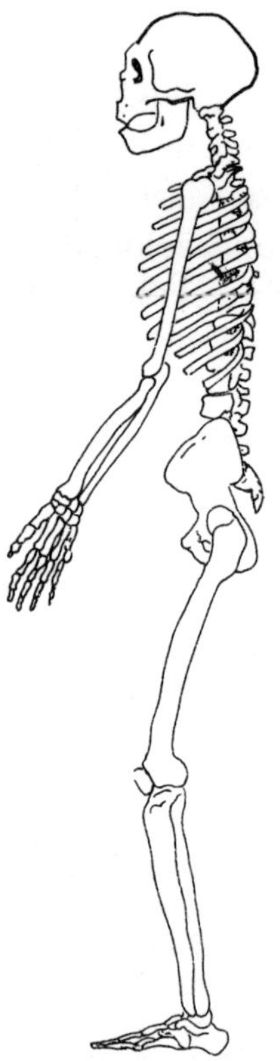

Fig. 12.19: Skeleton of Neanderthal man, *Homo neanderthalensis.*

"The wide, short face, the extremely prominent cheekbones, the spread of the palate and a tendency of the upper cutting teeth and incisors to project forward, and the narrow, pointed chin recall a facial type which is best seen to-day in tribes living in Asia to the north and to the south of the Himalayas. As regards their stature the Crô-Magnon race recall the Sikhs living to the south of the Himalayas. In the disharmonic proportions of the face, that is, the combination of broad cheekbones and narrow skull, they resemble the Eskimo. The sum of the Crô-Magnon characters is certainly Asiatic rather than African, whereas in the Grimaidis [of which specimens have been found in association with Crô-Magnons at the Grotte des Enfants, near Mentone] the sum of the characters is decidedly negroid or African."

The Crô-Magnons again show by their elaborate burial customs how old and well founded is the belief in life after death. They are undoubtedly the people who left in the caverns of France and Spain the marvellous examples of upper Paleolithic art so fully described by Abbé Breuil. They lived for a while contemporaneously with the men of Neanderthal and may have contributed somewhat to the final extinction of the latter. In the course of time, however, they too declined, although to this day survivors of the race may be seen in Dordogne, at Landes near the Garonne in Southern France, and at Lannion in Brittany.

The decline of the Crô-Magnons, with their artistic culture, "may have been partly due to environmental causes and the abandonment of their vigorous nomadic mode of life, or it may be that they had reached the end of a long cycle of psychic development...We know as a parallel that in the history of many civilized races a period of great artistic and industrial development may be followed by a period of stagnation and decline without any apparent environmental cause." (Osborn).

Europe was repopulated after Crô-Magnon decline by later invaders from the Asiatic realm, the so-called Mediterranean narrow-headed and the Alpine broad-headed types, etc., probably differentiated in Asia in early Paleolithic times. The repopulation took place in the Upper Paleolithic. The origin of the Nordic race is in doubt.

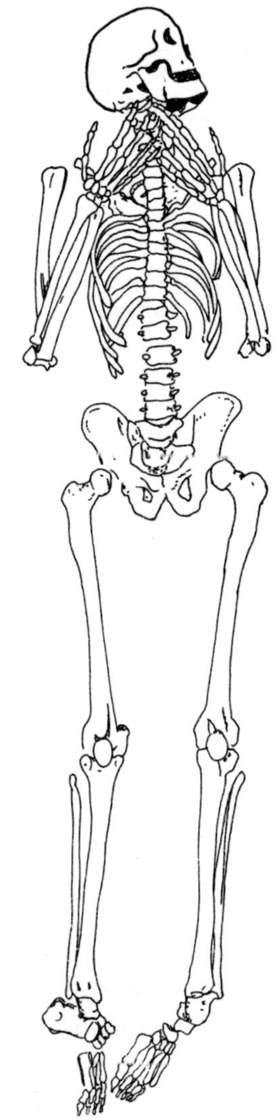

Fig. 12.20: Skeleton of man of Crô-Magnon race, *H. sapiens*, from the Grotte des Enfants, near Mentone.

EVIDENCES OF HUMAN ANTIQUITY

Great Variation

These, briefly summarized, are, first, great variation. If man is monophyletic, that is, derived from a single prehuman species, there is little reason to believe otherwise, he must be old, for while the adaptations to ground-dwelling after the descent from the trees were doubtless relatively rapidly acquired, the differentiation into the various races due perhaps largely to climatic influences rather than to any notable environmental change, must have been slowly attained. As corroborative evidence we have but to point to the mural paintings on Egyptian monuments, dating back several thousand years, in which are depicted the Ethiopian, Caucasian and others, which are in some instances striking likenesses of the present-day Egyptians.

Universal distribution is, in animals, another mark of antiquity; in man, it is probably less so because of his greater intelligence. And yet before transportation had become a science man's spread over land and sea was extremely slow.

High intelligence as compared with that of the anthropoids is also a mark of antiquity, for the brain, especially the type of brain found in the higher human races, must have been very slow of development. Our study of fossil man shows this.

Communal life, division of labour and all of the complicated interactions which it brings about, and the development of law and religions all have taken time. When we realize that Babylonian texts, twice as remote as the

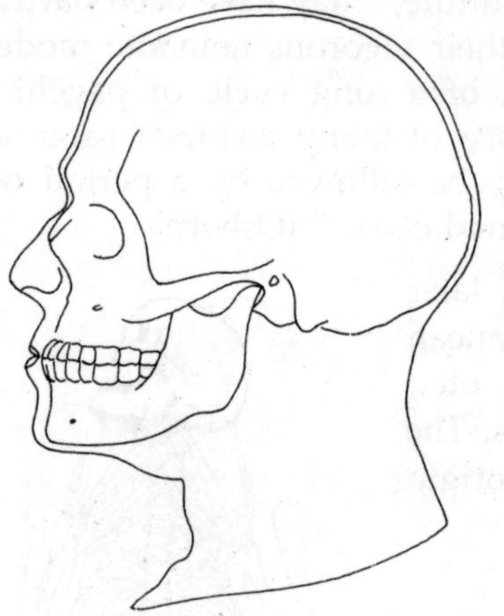

Fig. 12.21: Restored head of Crô-Magnon man (*Homo sapiens*).

patriarch Abraham, give evidence of highly perfect laws and of civilization which must have antedated their production by centuries, we gain another yet more emphatic impression of human antiquity. Add to all this the paleontological evidence of man's association with various genera and numerous successive species of prehistoric animals of which he alone survives, and the evidence is complete.

FUTURE OF HUMANITY

Because of his intelligence, tool using, and communal life man is little subject to the laws which govern the adaptation of animals to their environment. The law of adaptive radiation, which, as we have seen, applies equally well to the insects, reptiles, and mammals, fails in its application of mankind; and yet man has become as thoroughly adapted to speed, flight, to the fossorial and aquatic as they; but his adaptation is artificial and to a very small extent affects only his physical frame, while theirs is natural and the stamp of the environment is deeply impressed upon the organism.

Man's physical evolution has virtually ceased, but in so far as any change is being effected, it is largely retrogressive. Such changes are: reduction of hair and teeth, and of hand skill; and dulling of the senses of sight, smell, and hearing upon which active creatures depend so largely for safety. That sort of charity which fosters the physically, mentally, and morally feeble, and is thus contrary to the law of natural selection, must also in the long run have an adverse effect upon the race, unless offset by an enlightened eugenics.

Man is hardly as yet subject to Malthus' law, for while he is increasing more rapidly than any other mammal, owing largely to the care of the young which makes the expectation of life of the new-born relatively very high, his migratory ability, but above all his intelligence, save him from the application of the law. A single new discovery such as that of electricity may increase his food supply and other life necessities several fold. His future evolution, in so far as it is progressive, will be mental and spiritual rather than physical, and as such will be the logical conclusion of the marvellous results of organic evolution.

Chapter—13

The Gradual Appearance of Man

Biologists probably mostly agree that the question of which fossils are truly those of a man cannot ever be settled. If evolutionary change is gradual it is not simply difficult to solve such questions, it is *impossible* in principle. Apart from the difficulty of the gradualness of change there is the question of definition, which is bound to be to some extent arbitrary. A sudden beginning for man might be imagined as the result of the operation of some creative process, not found elsewhere. The endowment of the race by some outside force, agent, or God with a new entity 'the soul' would be such a process, and of course it is not to be ruled out arbitrarily that this is what happened. Again there might, conceivably, have been a sudden genetic mutation, in one person, leading his descendants to be not only highly successful but also fertile only among themselves. We could then say that he was the first man. A possible suggestion of this sort is that the size of the brain was suddenly doubled by the addition of one more cell division before the end of multiplication of neurons. Such suggestions are not likely to be made by anyone who has studied evolutionary problems closely. He will be much more ready to suppose that some interbreeding but varied population gradually changed until it reached the condition we recognize as that of *Homo sapiens,* itself a collection of very distinct people today.

But the fact that other organisms have been shown to evolve in this way must not prejudice us. What we need is not speculation as to how man might conceivably have arisen, but evidence as to how he did. This can only come from fossils, and the facts we have described already show that up to the stage that may be called pre-man (*Australopithecus*) the changes were gradual. *Ramapithecus* was more like man than were the Dryopithecines, *Australopithecus* was more like a man than *Ramapithecus,* and so on. The series is very incomplete and the existence of sudden small jumps cannot be ruled out, but the evidence is compatible with gradual change. Moreover the human character appears in some features (the gait) before others (size of brain); this is the phenomenon called mosaic evolution.

Perhaps it is at the last stage of all that we should be most prepared to expect the occurrence of some sudden change. Man is indeed very different from all other animals, even from such fossils as those of *Australopithecus*. We feel that we must be in some way widely divergent from them, the product of some special and perhaps sudden process. This feeling is egocentric but cannot be dismissed as wholly irrational. Modern man *is* very different. Can we believe that there has been a gradual evolution of his habits of cooperation, of his language, of his self-consciousness and conscience, of his religion, his art, and his artefacts? The answer in the one case where we have data is *yes*, for the artefacts. There is a continuous series from the first African pebble tools of perhaps 2 million years or more ago, to the flying machines and rockets of today. The development of language is certainly related to that of tools, but we do not know how closely, especially in the early stages.

Man's special features are mostly directed to the habit of co-operation in the business of life, especially perhaps in the use of artefacts. Self-awareness and individuality also have a place in this pattern of co-operation, and we shall have to consider these matters separately later. For the present we are concerned to provide an outline of the time-scale of human origin. It was apparently surprisingly rapid, especially considering the magnitude of the change. Of course there may conceivably have been humans alive during the 20 million years of the Miocene and Pliocene, or even earlier. It is impossible to prove a negative. But the fact that quite numerous ape fossils have been found in that period, but men fossils *only* later makes us conclude provisionally that men have appeared within the last two million years. Assuming rather more than four generations to a century this allows for some 100,000 generations. This certainly gives some margin for gradual change when we think that there have been only perhaps 500 generations since man emerged from the last ice age.

The decisive evidence, however, is that such fossils as have been found show that there have existed creatures who were anatomically intermediate between the Australopithecine types and modern man. The series is not complete and interpretation is difficult, but these early forms are indisputably not quite modern man as we know him today. Taken with the evidence of the evolution of tools the finding of these fossils makes us adopt the hypothesis of continuous steady change. The possibility of sudden jumps will perhaps never be rigorously excluded, but this must not of course be read to mean that the evidence in any way suggests that such jumps occurred.

These intermediate forms are the remains of populations referred to three further taxonomic groups before the appearance of modern man (*Homo sapiens sapiens*). The first of these, the Java and Pekin men, used to be placed in a separate genus *Pithecanthropus erectus*, now usually referred to as a distinct species of the genus *Homo*, *H. erectus*. They lived from perhaps 1,000,000 (Upper Bed II, Olduvai) to 250,000 years ago (Casablanca). The other groups antecedent to modern man include the Neanderthalers, previously also given specific status (*H. neanderthalensis*) but now usually treated as a distinct early subspecies *H. sapiens neanderthalensis*, and two skulls of an earlier date classed as *H. s. steinheimensis*. Besides these three reasonably well-defined forms of hominid there are isolated fossil fragments of some twenty further

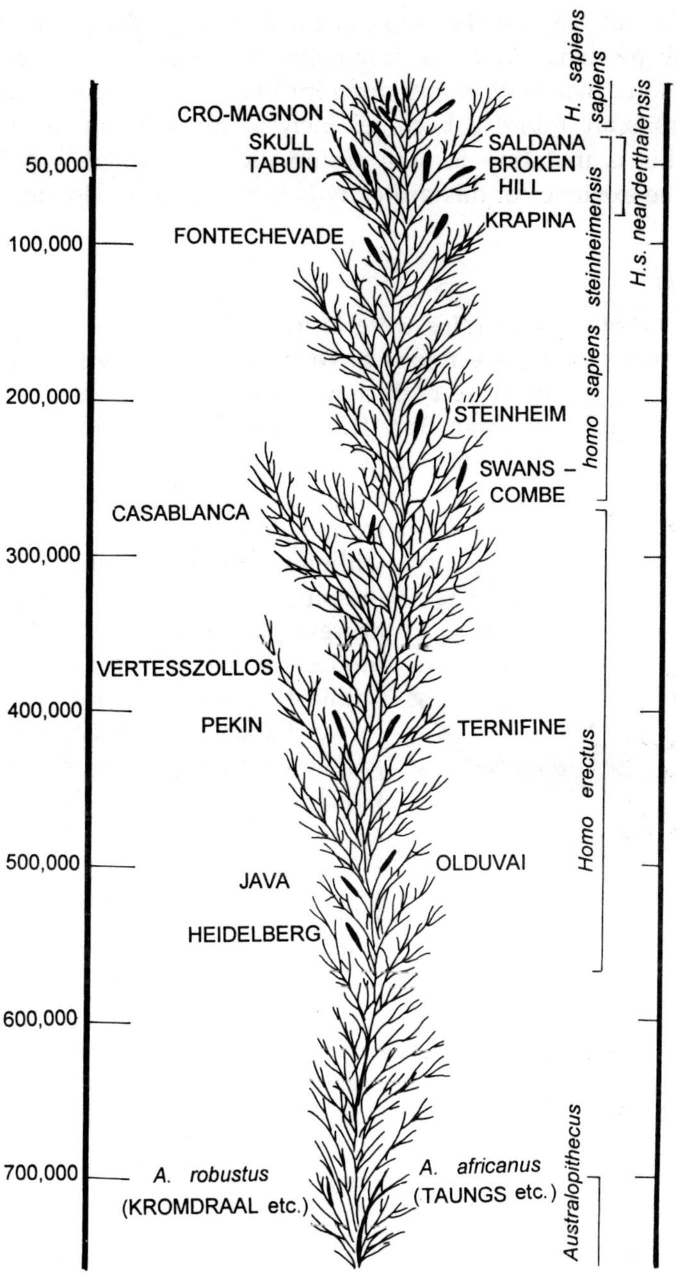

Fig. 13.1: A tentative scheme of human evolution showing the gradual change and that there must have been many different branches and anastomoses of slightly separated lines. Note that there is no main 'trunk'. Some known fossil remains are shown in their relative positions but the dating of the early forms is uncertain. *H. s. steinheimensis* is shown surviving till 45,000 years ago to include such skulls as those of Fontéchevade and Skūhl, but this is doubtfully correct.

types that had not reached the stage of modern man. We shall not describe them in detail but the existence of these fragments strengthens the suggestion that human evolution was gradual, with many different branches. Moreover, as we should expect, as further fossils are found it becomes increasingly difficult to fit them into clear-cut categories. This is irritating for the systematically minded 'typologists', but for the evolutionist the existence of this difficulty is itself the best evidence that gradual change occurred.

Homo Erectus

The first fossils of *H. erectus* were found in Java in 1891 and the species is often called 'Java man'. Other specimens have been found there since and still others in China (*'Sinanthropus'*). More recently remains of similar creatures have been found in East Africa (Olduvai Upper Bed II, hominid 9 'Chellean man'), South Africa (*'Telanthropus'*) and at Ternifine and Casablanca in North Africa (*'Atlanthropus'*). The single jaw of *'H. heidelbergensis'* is a similar type from Germany. Not all are contemporary, the Olduvai specimens being among the earlier and in some respects less advanced. All the remains, however, are of creatures quite close to man (many modern anthropologists wish to include *H. erectus* within *H. sapiens*). The foramen magnum is further forward than in *Australopithecus* and the upright gait is confirmed by remains of several femora, which are very human. No pelves, hands, or feet are known.

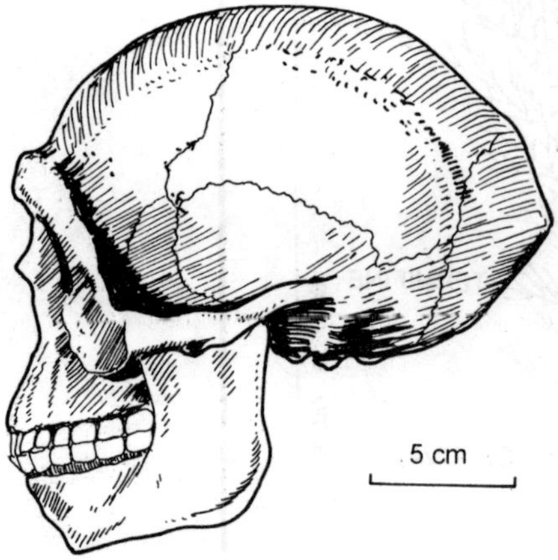

Fig. 13.2: Lateral view of skull of female 'Pekin man' (*H. erectus*). The possessor lived about 400,000 years ago. Note massive jaw and low forehead. From cast in the British Museum (Nat. Hist.).

The skull of *H. erectus* was massive and thick, with cranial capacities from 775 to 1200 ml (see below). The shape of the skull differed from that of *H. sapiens* in the low forehead, perhaps a significant sign that the frontal lobes were poorly developed and hence these were people of not quite human character. On the other hand, the occipital region was well developed. The jaw was much heavier than in modern man, but less so than in *Australopithecus*. Correspondingly there are never special temporal crests (as in *A. robustus*) and the zygoma is thinner than in the Australopithecine fossils but still thicker than in *H. sapiens*. The teeth are of typically human cusp-pattern and the third molar is smaller than the second. Yet, once again the teeth are larger than in most modern men. These people were probably omnivorous.

One of the most important of the many results of investigations by the Leakeys at Olduvai has been to show the possibility of a gradual evolution from *Australopithecus* to *H. erectus*. Specimens that can reasonably be referred to the former genus occur in the lowest beds of the series, now dated up to 2 million years before the present. Maxillae and mandibles found in the lower-middle Bed II at Olduvai have been referred to as a more hominid version of *'H'. habilis'*. These are about 1 million years old and may represent intermediates between *Australopithecus* and *H. erectus*. The latter also occurs at Olduvai, from the top of Bed II, dated at perhaps 500,000 years B.P. (Tobias *et al.* 1965, Hay 1963). These remains are very like those of some of the early Asian *H. erectus*, skulls from Modjokerto and Sangiran in Java, the former in infant. The age of these Javan fossils is uncertain, but is probably greater than 500,000 years and perhaps as much as 700,000. Though fragmentary, they have been judged to have cranial capacities of only 780 ml and 900 ml, while the estimate for the African form is 900-1000 ml. These specimens together thus suggest that by the middle Pleistocene period more nearly human creatures had evolved, perhaps from some Australopithecine stock, possibly represented by *A. africanus*.

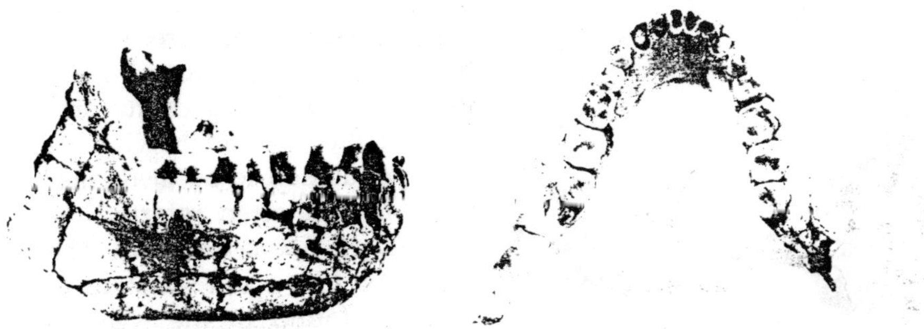

Fig. 13.3: Mandibles of *H. erectus* from Ternifine, Algeria, North Africa. Lived about 400,000 years ago. Note massive jaw and large teeth. Third molar smaller than second.

The remaining specimens of *H. erectus* include the more complete ones from Trinil, in Java, and those from China. They are probably of a rather later date, 550,000-350,000 years B.P., and some had larger brains (range of cranial capacities 915-1225 ml.) It is interesting that the distribution of these finds extends much further north than that of any hominid of earlier date (41° in China, 49° in Germany). Correlated with this there is evidence, from ash heaps and charcoal in the cave in which the remains were found near Peking, that *H. erectus* not only sought shelter in caves but used fire. The date for the cave is about 400,000 years B.P. and this is the earliest evidence of the use of fire. Remains of various mammals were also found, including large carnivores, a deer, and rhinoceros. Cores and flakes of a crude chopper industry were found ('Choukoutienian') and were presumably used to cut up the prey. The *H. erectus* skull at Olduvai is also accompanied by tools, of Chellean type. Fragments of jaws found at Casablanca in Morocco are similar to the other remains of *Homo erectus* and perhaps show that this type survived until 250,000 years ago, but the dating is uncertain. Thus the *Homo erectus* populations, although distinctly different from modern man, had already developed some important features of culture. Their brain size was intermediate between that of *Australopithecus* and modern man. All we know of them suggests that they were truly an intermediate stage of primitive man and they lived throughout Europe, Asia, and Africa from about 1,000,000 to 250,000 years ago.

The Appearance of *Homo Sapiens*

The fact that some modern anthropologists would like to include *H. erectus* within *H. sapines*, though confusing, confirms us in the supposition that our species arose gradually. The earliest fossils referred to *H. sapiens* are the skulls from Swanscombe in the Lower Thames Valley and Steinheim in Germany, dated *c.* 250,000 and *c.* 200,000 years B.P. respectively. They were people we should almost certainly recognize as human, though with thick skulls. Of course we do not know whether they had all the characters we expect in a human today, particularly speech. It is convenient therefore to place them in a distinct subspecies of *H. sapiens steinheimensis*. There is then an unfortunate gap in the record until about 50,000 B.P. After that time, remains are abundant and there is a gratifying confusion of opinions as to how they should be classified. Indeed the next population to be mentioned, Neanderthal man, is still often referred to as a distinct species, *H. neanderthalensis*. The name comes from the first fossil skull to be scientifically studied, found in the Neander valley near Dusseldorf in 1856. Before describing the Neanderthalers we meet again the familiar difficulty that they were not all alike! The 'Classic Neanderthalers' were somewhat different from modern man, the 'Progressive Neanderthalers' much more like him.

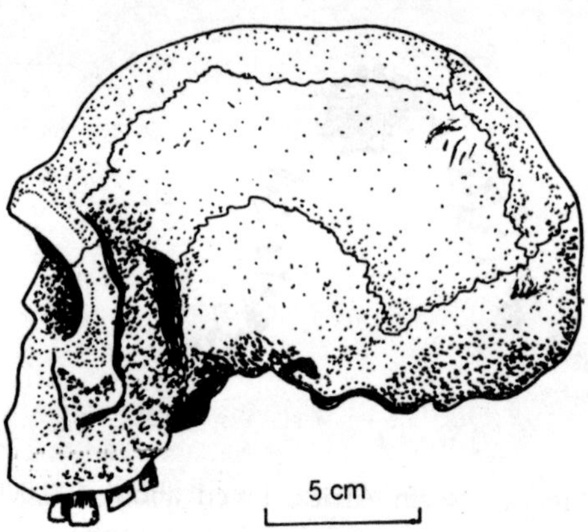

Fig. 13.4: *Homo sapiens steinheimensis*. Age: 200,000 years. From a cast of the Steinheim skull in the British Museum.

In the classic Neanderthal type (*H. s. neanderthalensis*) the face was narrow, long, and prognathous. The brow ridges were quite prominent and formed a continuous supraorbital torus or ridge above the eyes. This character suggests somewhat powerful jaw muscles, and the jaws and teeth were indeed generally heavier than today. The teeth sometimes show a rather enlarged pulp cavity and small roots (taurodontism). The chin was small. The cranial capacity was as large as today, indeed many skulls are very large, about 1600 ml. Characteristically the forehead appeared low, perhaps because the torus was large, and the occipital region was well developed, both being of course also features of *H. erectus*. The postcranial skeleton is well known and shows no major differences from that of today. Neanderthalers are often shown as if walking with a bent back and shuffling gait, but there is no justification for this. These classic Neanderthalers are found at some of the best known sites, including Spy (Belgium), Le Moustier and La Chapelle (France), Neanderthal (Germany), Gibraltar and eastwards to the Crimea, Palestine, and Iraq. There are indications that similar types existed in Africa (at Broken Hill in Zambia, 'Rhodesian Man', and at Saldanha Bay, South Africa). Eleven skulls found at Solo in Java show the existence of 'tropical Neanderthalers'.

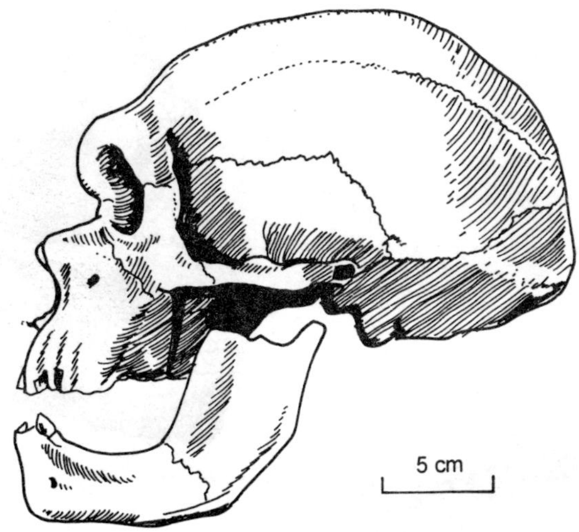

Fig. 13.5: *H. sapiens neanderthalensis.* The 'classic' Neanderthal skull from La Chapelle aux Saints. Age: 40,000 years. From a cast in the British Museum.

In the progressive Neanderthalers living at the same time the brain case was short and narrow, with a high forehead and reduced occiput. These are in fact nearly typical skulls of modern *H. sapiens sapiens.* A difference is in the thickness of the bones of the vault and weight of the jaw. These seem to be the skulls of people like modern man except for the fact that they had to work harder with their jaws to obtain nourishment. Such skulls have been found at many places, Ehringsdorf in Germany, Fontéchevade (France), Saccopastore (Italy), Krapina (Jugoslavia), and at Skūhl (Israel) and in Uzbekistan (U.S.S.R.). With the exception of the last two these progressive Neanderthalers are all from Europe. This may be only an accident of preservation and discovery.

The existence of the two types has led many people to suppose that the classic Neanderthalers were an aberrant line, not leading to modern man, and that they were exterminated by invaders of the progressive type. The incomplete skull at Fontéchevade and a mandible found at Montmaurin in France were associated with bones indicating a warm climate, probably of the Ris-Würm or Last Interglacial period, that is, more than 70,000 years ago. They provide almost the only link between the Swancombe and Steinheim skulls and the progressive neanderthalers, such as those at Skūhl.

On Mount Carmel in Israel there are two nearby caves, one containing classical Neanderthalers (Tabun) and the other those of a more progressive type (Skuhl). Carbon dating gives 45,000±2000 years B.P. for both caves (but Skuhl may be 10,000 years

Fig. 13.6: *H. sapiens steinheimensis.* Progressive Neanderthal skull from Skuhl Age: 45,000 years. From a reconstituted cast in the British Museum.

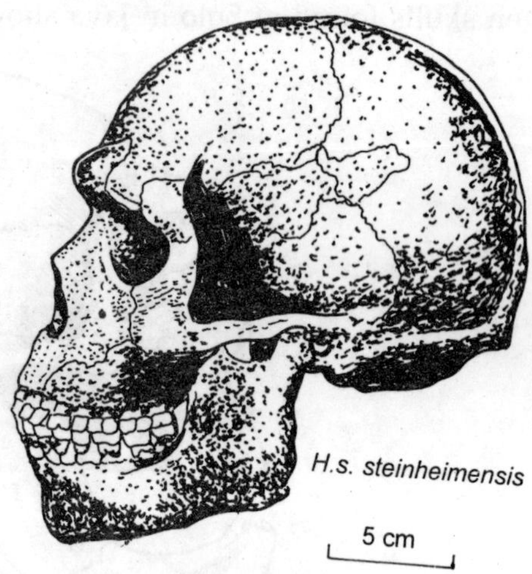

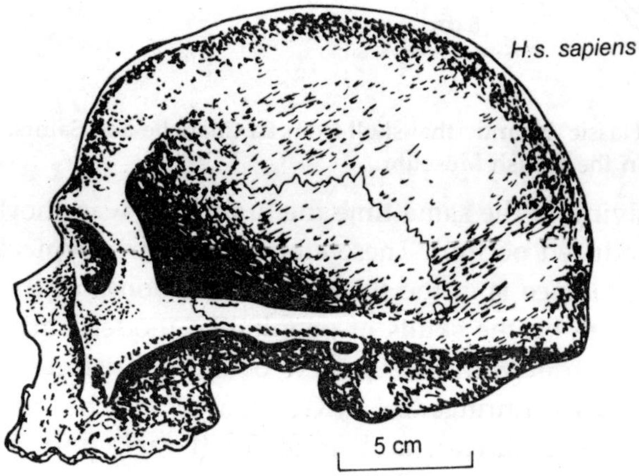

Fig. 13.7: Cro-Magnon skull, an early specimen of *H. sapiens sapiens.* Age: 25,000 years. From photo in Day 1965 and British Museum.

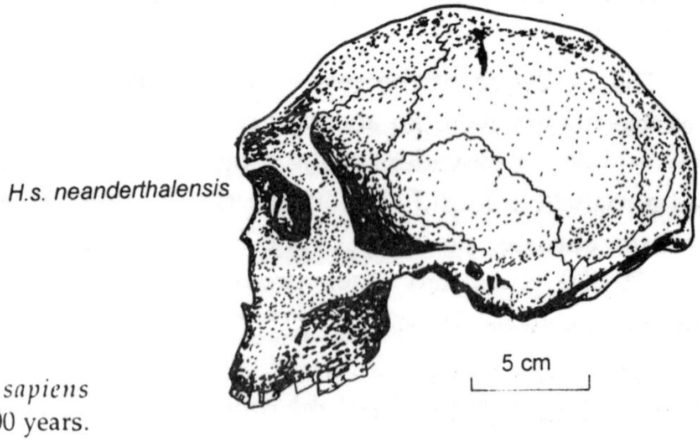

Fig. 13.8: Rhodesian skull (*H. sapiens neanderthalensis*). Age: 40,000 years. From cast in British Museum.

later). These finds show dramatically the problem of analysing the relationships of these remains of the peoples who presumably included our immediate ancestors. From as long ago as 250,000 years there were people with brains fully as large as ours, but with rather heavier jaws and thick skulls. In the classic Neanderthalers living very much later, these primitive features were especially marked and these characteristics remained clear in some populations until 40,000 years ago. But long before that the progressive Neanderthalers were in existence throughout Europe and it was probably from them that undoubtedly human peoples arose (*H. sapiens sapiens*), sometime after 40,000 years ago. For this reason skulls of the progressive Neanderthalers are often said to be of 'presapiens' type. But we really do not know whether they represent a single group persisting over the very long period between the skulls at Swancombe (250,000 B.P.) and Skūhl (45,000 B.P.). At any rate, this last date corresponds approximately in time with the earliest undoubted remains of *H. sapiens sapiens*, found in Borneo (Niah) aged 40,000 years. Fully human remains of somewhat later date have been found in France (Combe-Chapelle, 34,000, and Cro-Magnon, 25,000 years old respectively, and Zambia (Mumbwai, 23,000 years old).

We have to admit, therefore, that in spite of the relatively large number of fossils available the evidence is still not adequate to say much about the subdivisions of these early human populations with any scientific precision. We can make statistical distinctions that will refer specimens to one of the two groups, and methods for estimating the degree of similarity, based upon a number of measurements, have recently been applied to these early human populations. The statistical technique derives from Fisher's discriminant function and it expresses the morphological distance between specimens on a basis of all the characters that can be quantified, thus avoiding the problem of selection of features to measure. Five hundred *H. sapiens sapiens* skulls of Bronze Age were taken as a basis, and the divergence of the Swanscombe, Skūhl and La Chapelle skulls. Different figures are given according to whether the Swanscombe skull is considered to be male or female. In addition the position of eighteen skulls taken from all the major modern 'races' of man are shown. The values express only the degree of divergence from the reference population, not the direction of divergence. It is clear that the 'progressive Neanderthalers' differed strongly from the classical ones and that they also differ from all modern populations, but not by very much. It seems probable that all the Neanderthalers were variants within one interbreeding population, from some part or parts of which modern man evolved.

No doubt there was a great deal of variation in the human population then, as there is today. Probably there were cultural differences, some associated with physical differences. Both the physique and the culture of each group was presumably suited to particular conditions—as also today. There is, however, no more justification for dividing these populations of our ancestors into distinct races, if by that is meant groups that could not mix either genetically or culturally, than there is for man today.

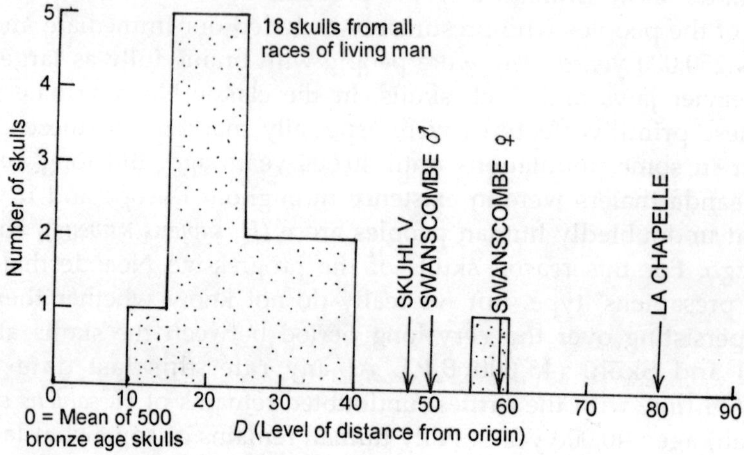

Fig. 13.9: Measurements of a discriminant function ('D'-values') based on seventeen measurements of parietal and occipital bones of modern and fossil men. The ordinate shows the number of skulls that show each level of departure from the mean dimensions of a modern population (at the origin). The Swanscombe and Skūhl skulls are nearer to the modern condition than is a 'classical' Neanderthaler. Two values for Swanscombe are given, one assuming it is a male, the other female.

No more justification and no less. Difference in physique and culture has meant different ways of life, in the past as today. Mixing of the genes and the cultures seems to go on extensively at present and there is no reason to think that it has not done so for a very long time. Nevertheless the variation remains. The fact that it was already present among our ancestors 50,000 or more years ago suggests that variation may be one of the main secrets of the success of man, as it is of many other species. Polymorphic hereditary characteristics only remain side by side in a population if each of them has some special advantage to confer, at least under certain circumstances. Such polymorphism allows the species to maintain itself in a number of slightly different habitats and it is correspondingly ready to become changed if there are large changes in the environment.

It is an interesting question whether human evolution has proceeded especially fast. Using Haldane's (1949) unit of increase or decrease of a character by 1/1000 per 1000 years (=1 darwin) it is calculated from the shape of the skull that from 'Pekin man' to modern man the rate has been 620 millidarwins (md). This is fast compared with maxima of 78 md for the teeth of horses throughout the Tertiary and 60 md for size of dinosaurs. Brain volume seems to have increased very rapidly before and at the time of appearance of modern man, but little since. Indeed some Neanderthalers had larger brains than ourselves. It will be seen that the estimated rate differs somewhat with the character used (620 md for skull shape, 732 md for cranial capacity).

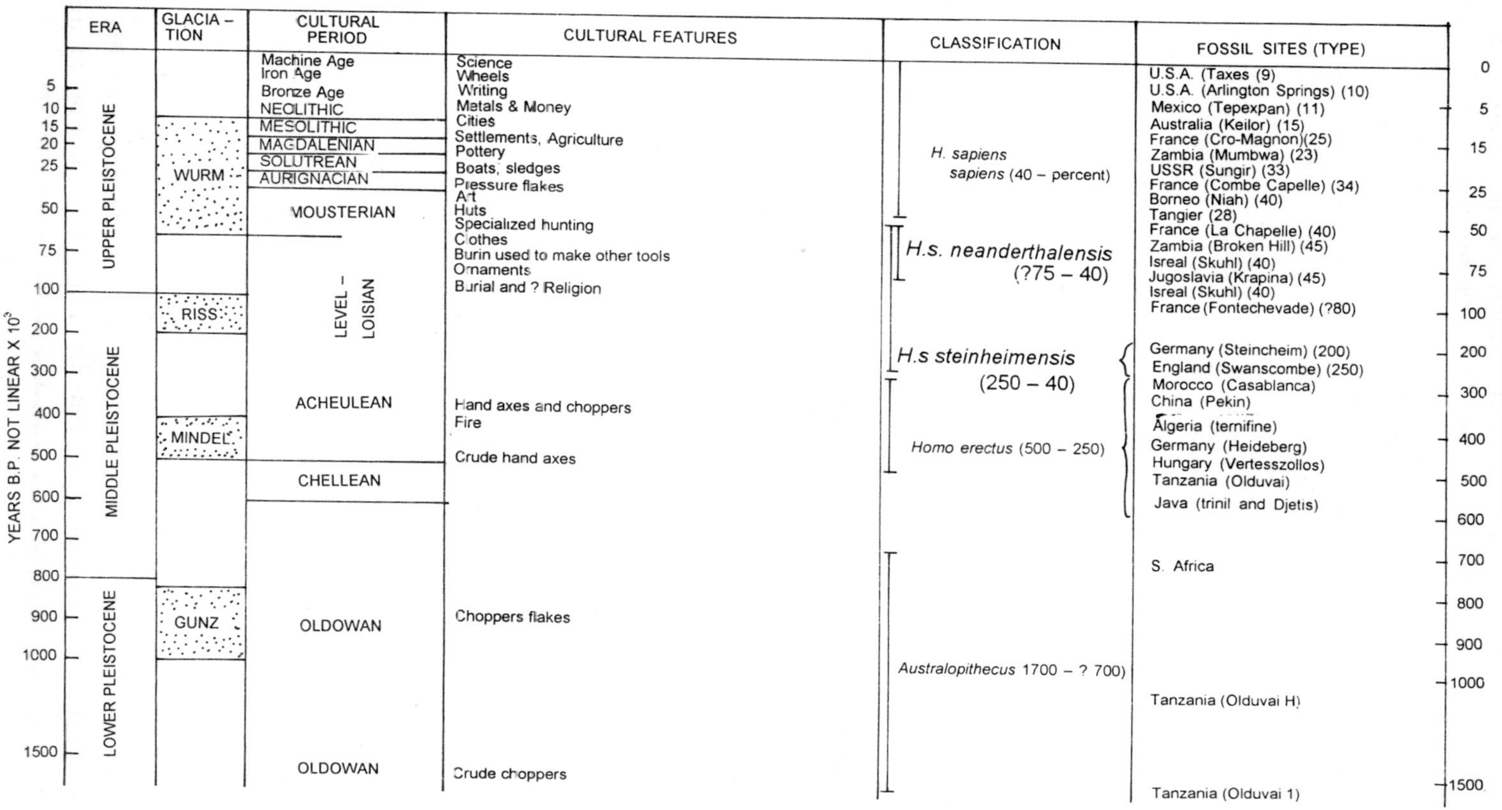

Fig. 13.10: Scheme to show fossils of early man and the various approximate dates and associated cultures. The scale is not linear. (*H. s. steinheimensis* is shown surviving to 45,000 years ago.).

Table 13.1: Rate of evolution of man. Change of 1/1000 part per 1000 years = 1 darwin. Some of the dates used in this table differ from those adopted elsewhere in this chapter.

	Date B. P. X10^6	Cranial capacity (ml)	Rate in md
Australopithecus	1.2	500	..
to			
H. erectus (Java)	0.5	900	280
to			
H. erectus (Pekin)	0.4	1000	351
to			
H. sapiens steinheimensis	0.15	1325	375
to (Swanscombe)			
H. sapiens sapiens (modern)	0.0	1375	6

THE CAUSES OF EVOLUTION

Before trying to find what has 'caused' man to emerge we should have to inquire into the meaning of 'causation' and the question whether evolution can be said to have causes. Evolution is our name for the series of changes that occur as populations adapt to their changing surroundings. Mankind at any given moment of time is one of the populations that have resulted. In so far as the achievement of that state can be said to have causes, they are to be found in the past history of the populations and of the environmental changes that occurred. The nature of the populations will, of course, include those tendencies to strive to survive, to add to living matter, and to find new environments, that together provide the basis for the purposive activities that man, like other organisms, presents by his aspirations.

The prime factors that have caused our evolution must thus be sought in the natures of our ancestors, that is to say, in primate life in general. This in turn is to be sought in the way of life of mammals, of tetrapods, of vertebrates, of their ancestors, and ultimately in the character of life. These are all technically proximate causes and it is questionable whether we can attach any exact significance to the concept of a final cause. We have seen how all the adaptive changes can be considered as a sequence of selections made among rather limited sets of elements, molecules, organelles, and cell-types. As the selection proceeds, under the pressure of environmental stress and competition, new types of living systems are continually emerging. Our present task is to try to identify the factors that have been especially responsible for *human* life.

The Influence of Tree Life

The basic determinant of the primate system of life has been life in the trees. Therefore a prime determinant of human life was the evolution of trees from the

Permian period onwards. Presumably the development of luxuriant tropical forests in the Eocene and Oligocene gave an especially large premium to the development of monkeys and apes. Tree life is often associated with rapid and agile movement, good senses, and a good brain. All of these were essentials as a basis for the human achievements of manipulating, tool-making, observation, and, finally, speaking, thinking, and reasoning. Vision is especially useful in the trees, where olfactory tracks and signs are more difficult to place and recognize than on the ground (though lemurs use marking with scent). Accordingly the olfactory organ and the part of the brain associated with it becomes relatively smaller passing up the primate series. In the trees communication between individuals by smell is also not very effective. Hence visual and auditory methods of signalling are developed by primates, as they are also

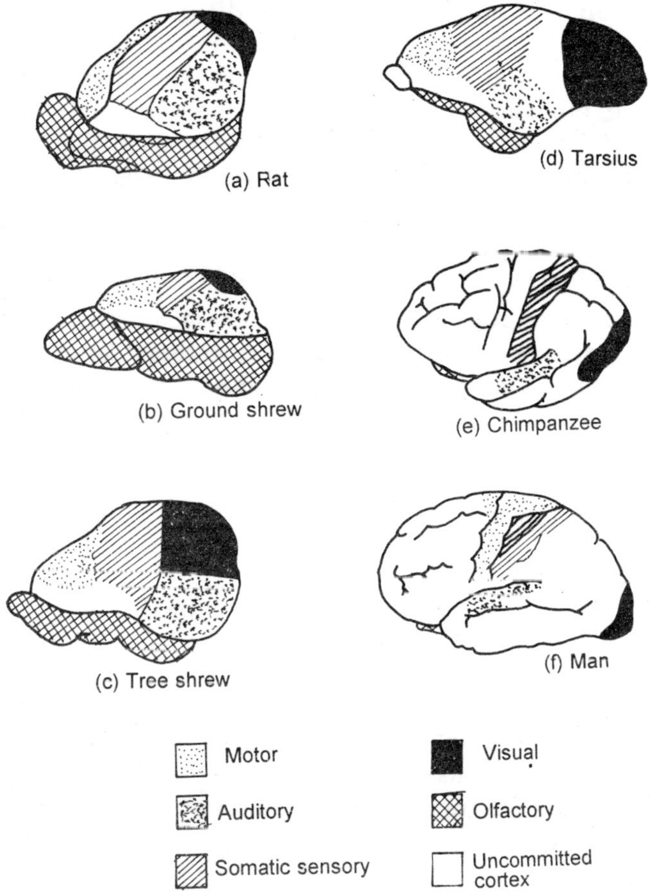

Fig. 13.11: Mammalian brains, to show variations in the proportions devoted to different activities. The comparison can only be rough because important parts are often concealed. The area devoted to olfaction is much reduced a primates. The area marked 'uncommitted cortex' is not related to particular sensory or motor functions. It includes the parietal association areas and much of the frontal and pre-frontal lobes.

in birds. Communication by visual and auditory signals provides the base for much more precise statements than can be achieved by depositing scents. The latter are admirable as signs widely addressed, 'to whom it may concern', by which an animal signals its condition, whether it be the fear substances released by fishes or the sex stimulants of mammals. But scents do not seem to provide an alphabet of signs useful for objective description of outside events (though this may seem so to us only because we do not use them in this way).

Another effect of tree life is that the primate mother can carry only a small number of young and must look after them well. This feature made possible the emergence of the human condition of prolonged childhood and learning.

These are some of the aspects of primates that have provided the background from which human life emerged. Such conditions are sometimes called 'pre-adaptations', but it is not clear what is gained by this label. Adaptation is a process of change to meet a particular existing condition in the environment. It has a future reference, it is true, but this is on the basis that the particular condition will continue to exist. The word 'pre-adaptation' suggests that a particular living state has been developed in anticipation that it will be useful for homeostasis under some conditions that have not yet existed. It is doubtful whether even man can be said to make forecasts in this sense. If we speak of pre-adaptation it should only be in the sense of noting that the existence of one condition made possible the development of some others.

Perhaps the most important of all these characteristics of primate life for the development of man was the tendency to investigate and observe, and then to remember the results and so to plan each individual life, and if necessary to seek out new and diverse environments. These activities were certainly stimulated by tree life and the consequent developments of senses and brain. They led, together, to the appearance of the phenomenon that we call mentality or thinking, dependent upon the use of internal models in the brain.

The Return to the Ground

The particular features of the hominid line presumably emerged when populations endowed with these arboreal gifts were forced to leave the trees. There is much evidence that during the Miocene period many forest areas turned into open plains or tundra. This was the stimulus that led to the evolution of the horses and fast-moving ungulates. Although no undoubted fossil hominids are known from the Miocene there is evidence that types such as *Ramapithecus* were beginning to show human characters, but unfortunately we are ignorant of their method of locomotion.

There has been debate as to whether the bipedal gait emerged from animals that were strictly quadrupedal, or from those that jumped, or from semi-brachiators. The jumping habit produces an upright posture of a sort, and may lead to efficient bipedal locomotion (as seen for example in the Indrid lemurs such as *Propithecus*). There are some similarities between hominids and the jumping tarsioid prosimians. Wood Jones (1943) went so far as to suggest that the human stock arose in this way in the Eocene, and has remained separate from that of monkeys and apes ever since. The evidence that has been reviewed certainly suggests divergence of man from a stock previously related to the tarsioids.

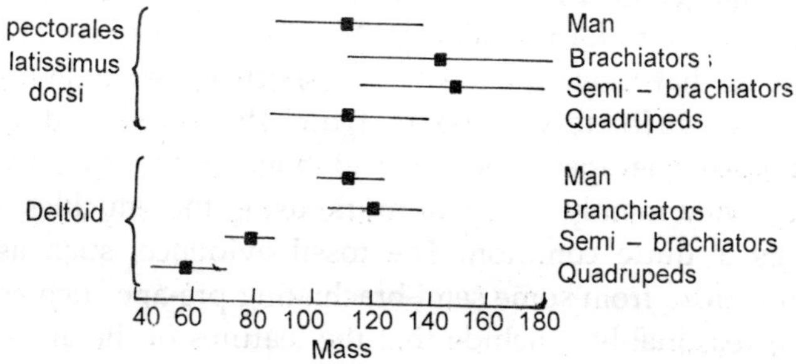

Fig. 13.12: Relative masses of arm muscles in man and various brachiators, semi-brachiators, and quadrupeds among primates. The pectorales and latissimus dorsi are propulsive muscles; the deltoid is an arm-raiser (abductor). Man is nearer to the quadrupeds in respect to the first group but to the branchiators in the second.

It is more likely, however, that the bipedal gait of man emerged from creatures that were at least semi-brachiators, often swinging from their arms, but frequently walking and standing on their hind legs while supported by their arms. The habit of

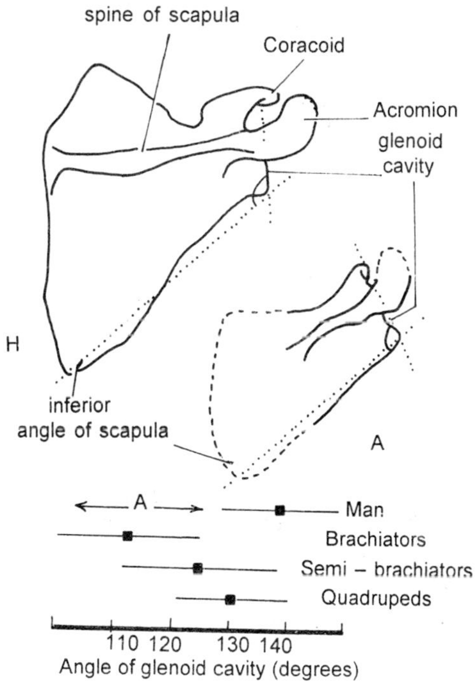

Fig. 13.13: Estimates of the angle of the plane of the glenoid cavity (in degrees) with respect to the scapula (see dotted lines). In this angle man (H) is closer to the quadrupeds than to the brachiators. The angle in *Australopithecus* (A) is much nearer that of the branchiators than of man.

brachiation demands great mobility of the upper limb, especially at the shoulder joint and also by movements of pronation and supination at the forearms and flexion of the long phalanges of the hands. All these features are characteristic of man although he does not use them for brachiation. Man is more similar to the apes than to monkeys in the muscles that move the arms. In respect of the angle of the glenoid cavity of the scapula, man definitely departs from the apes and also, incidentally, from *Australopithecus*. The apes other than gibbons in any case live much of their lives on the ground. They mainly go on all fours, using the knuckles of the hands, but walking on two legs is quite common. The fossil evidence, such as it is, suggests that the hominid line arose from some semi-brachiating pre-ape such as *Dryopithecus* (Proconsul), and we can reasonably conclude that the features of the arms originated in this way.

It is less easy to see signs that the foot of such an animal could readily give rise to the human one. Very drastic changes are required in the whole lower limb to produce the full striding bipedal gait. The gluteus maximus muscle is turned from an abductor into an extensor of the hip, to provide the lift that raises the trunk into the upright position. The propulsive thrust is no longer produced by the hamstring muscles acting as retractors of the hip, as they do in quadrupeds, but by the extensors of the knee

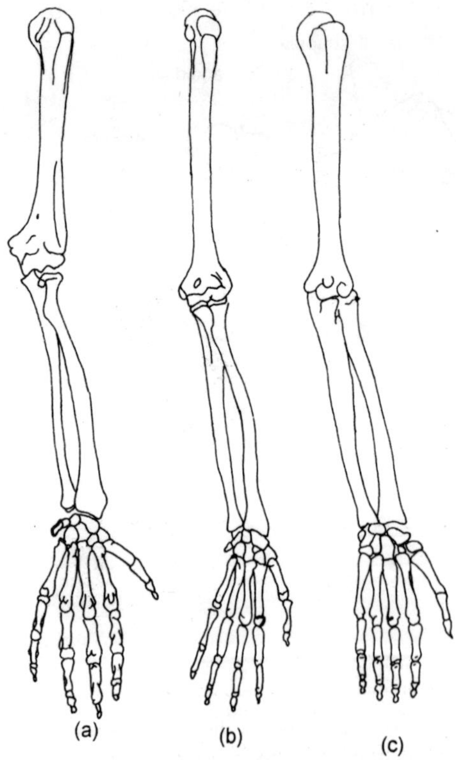

Fig. 13.14: Skeleton of the forelimb of (a) chimpanzee, (b) *Dryopithecus*, and (c) man, reduced to same length. Man resembles the (presumed) semi-brachiator.

(quadriceps) and ankle (calf muscles). In fact the whole rhythmic system for balancing and unbalancing by moving the legs seems to be different in quadrupedal and bipedal gaits. Perhaps the similarities are greater than they appear. The basic neurological mechanism producing the rhythmic movements of locomotion is probably essentially similar throughout the mammals (perhaps with some exceptions, such as whales). Yet these movements themselves differ very greatly. A gallop is very different from a trot. Bipedalism means something quite different according to whether it consists of a short rushing run, or prolonged striding, or standing. Probably the earlier bipedal efforts consisted of short run, which can be achieved by many monkeys and apes, though they are incapable of striding or long standing. In the Australopithecines the ischium was relatively longer than in modern man so that the leg could not have been fully extended backwards in a long stride. Moreover the gluteus medius and minimus muscles, which in man act to prevent the body falling when one foot is off the ground, in Australopithecines probably acted as extensors. Equally large changes were needed to convert the prehensile ape foot into the sprung arches of man, one of his most specialized features.

Bipedal locomotion must have produced some substantial advantage over the older, more stable, quadruped gait. Of course semi-brachiation would become less useful in the absence of trees, but apes often revert to a quadrupedal gait on the knuckles. Bipedalism brings the advantage of a greater range of vision and frees the hands for the use of tools and for carrying food. Presumably it was adopted because it provided a complex of these and other advantages.

From Herbivore to Hunter

Undoubtedly the assumption of a bipedal gait proved a key change in initiating a new form of life. The change was almost certainly gradual, beginning in the Miocene perhaps 20 million years ago. As the hands became freed from locomotion they were available for the using and then the making of tools. We do not know for certain when this began, but the oldest definitely known stone culture, the Oldowan, is associated with *Australopithecus* in Africa, dating from perhaps 2 million years ago or more. Evidence for the use of tools in much earlier times is claimed from the finding of skulls of Miocene animals with depressed fractures. Stones found near by are said to show signs that they were the weapons or tools that were used, perhaps by a *Dryopithecus*-like creature. Presumably, even before this, suitable stones were used for throwing or hitting.

The descent to the plains may well have been associated with a marked change of diet, from mainly herbivorous, as are the apes, to truly omnivorous, with killing first of small animals, lizards, and birds, and later of the increasingly large mammals of the plains. The bones associated with the remains of *Australopithecus* at Olduvai include small animals, birds, lizards, and rodents, as well as large ones. Presumably *A. africanus* ate some of these animals a well as vegetable food.

Some Physical Requirements for Hunting

Co-operative hunting requires many special characteristics both of the individual and of society. In the following sections we shall consider some of these, beginning with the more obvious physical features and proceeding to the social, sexual, and intellectual characters that resulted from less obvious physical changes in brain and glands.

The capture of large prey no doubt depended greatly upon cunning, but swiftness, endurance, and power were probably also important. Bipedal walking and running do not provide the speed of the ungulates or the sudden dash of the cats, but were perhaps suitable for sustained tracking particularly over rough and varied terrain. The foot is one of man's most remarkable organs, allowing sustained locomotion over irregular ground. It seems to have been already evolved by the Australopithecine stage, and the freedom of movement allowed would have been a necessary feature for capturing large ungulates.

In primates the hand, instead of bearing claws for tearing the prey, has developed nails, allowing fine movements and ultimately the holding of tools. The lengthening and opposability of the thumb had perhaps preceded the hunting habit, but were essential for it. Napier has distinguished between a power grip, with the whole hand, and a precision grip, by the thumb and fingers. Selection for hands of the appropriate structure, and brains to use them, would have proceeded if it gave advantage to the individual and to his group. No doubt precision was of special importance, but the load-carrying capacities of the arm and hand are also among man's special features. The great mobility at the shoulder joint is perhaps an inheritance from partial brachiation. It allows the hand to be brought into almost any position in relation to the body. The joints at the elbow ensure that the hand is held away from the body by a 'carrying angle' during walking. Other relevant features of the fore limb are the

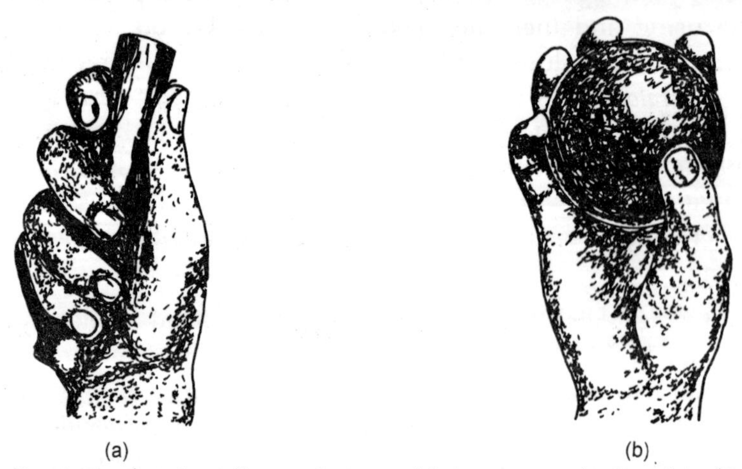

(a) (b)

Fig. 13.15: Demonstrating the difference between (a) the power grip, largely with the long flexors, and (b) the precision grip using also the small muscles within the hand.

great capacity for pronation and supination and the arrangement at the wrist that allows the hand to be deviated through wide angles in both the radial and ulnar directions. All these, together with the structures of the finger and thumb, make possible the working hand of man and all have their beginning in adaptation to arboreal life. The great power of the deltoid muscle allows man to lift and carry weights that are large in relation to himself. In fact, besides his intellect and skill, man is outstanding among all creatures as a great load-carrier and worker. These characteristics must have been developed for some special purposes. We do not know for certain what they were, but it is reasonable to suppose that they included trapping and killing large prey and carrying it home for the women and children.

Tools must have been important from an early stage for killing the prey and also for skinning it and cutting up the meat. Man's largely vegetarian primate ancestry had left him poorly equipped physically for this purpose. The claws were already reduced to nails, in the interests of gripping and manipulation during arboreal life. The teeth of primates have none of the specializations for tearing and cutting that are so useful for the carnivore both before and after the death of the prey. The large canines of some male apes are not used in this way (their true use is, indeed, unknown; may be for display in social situations) and the canines have become *reduced* during hominid evolution.

Among the physical features characteristic of man few are more striking than his nakedness. It would be satisfactory to find that this condition was also in some way related to the hunting habit. Perhaps the best suggestion is that the exertions involved, in a hot climate, required the maximum development of the cooling system by evaporation of sweat. This is ensured by reduction of the layers of hair, lying outside the epidermis, while the insulation needed at other times against heat loss is provided by the fatty layer of the dermis, which does not interfere with water evaporation. The explanation is not very convincing for such a very striking condition. The only other mammals of comparable nudity are elephants, pigs, a few rodents that live underground, and numerous aquatic forms. Comparison with the latter has led to the suggestion that man passed through an aquatic, or at least a wading phase. Many human features are compatible with this possibility but it is difficult to find direct evidence for it (or indeed against it)!

The need to hunt large animals may have been a main stimulus in the development not only of many physical features but also of social life and language. There is little evidence to go on but it is easy to imagine that the populations that survived contained genes ensuring co-operation among individuals, as well as efficient tool-making and tool-using. Hunting may well have been difficult for females and young and this would have led to the formation of stable social groups, with specialization of functions and sharing the results of the chase. This could be achieved by development of tendencies to co-operate, of which there are considerable signs already in monkeys and apes, though not usually for the purpose of getting food. Co-operation to drive, trap, and kill large prey such as antelopes or even elephants requires a reduction of competitive and aggressive behaviour between individuals. In a sense this could be said to involve in man the continuation throughout life of the characteristic features found among juveniles of monkey and other animal communities. There are indeed several signs that the evolution of man has involved a process of 'juvenilization' or even 'foetalization'. Such a change, technically called neoteny (or paedomorphosis), has in fact occurred often in the course of the evolution of diverse animals. It is not difficult to understand how a delay in the rate of development of the reproductive system relative to the rest of the body could be produced by even a small genetic change; for instance, one affecting the pituitary gland. The result might be a striking alteration in the age structure of the population and hence in the psychology of its members. With a long period before puberty the young would be more readily restrained and taught by their elders, to the benefit of the whole community. Moreover some of the physical determinants of aggressiveness and non-co-operation might well be eliminated from the population altogether by neoteny. We know that stimulation of some regions of the brain will release aggressive behaviour and of others will prevent it. We do not fully understand how these centres are normally controlled, but evidently what we regard as 'emotional' responses have very definite and localized physical determinants. There is every reason to think that hereditary influences must operate to modify the operations of either the centres that increase or those that decrease the tendencies that we call 'aggression'.

How far these changes can be regarded as neoteny is perhaps immaterial. Some alterations of conduct of this sort must have occurred even to allow for the elementary co-operation involved in Palaeolithic hunting life. The later progress of humanity seems to have involved still further co-operation, and it is a good guess that the physical changes in man since the middle Pleistocene have included alterations in glands, brains, and perhaps reproductive functions. Increased co-operation was especially in evidence in the change from Palaeolithic hunting to Neolithic culture and further to agriculture and the development of cities and states. Although we are still cursed by our mutual aggressions today, they are probably gradually becoming less and less manifested as individual fights, for which increasingly people have no stomach. Such changes are, of course, deeply influenced by social customs, but it would not be unreasonable to conclude that physical alterations in the relevant parts of the brains and glands of our populations are continuing day by day. The type of aggressivenss required to make a successful business man, politician, or even soldier is no longer that of the bruiser. Ferocity and loss of temper may pay in the underworld but will land most men in the unskilled labour market or lower. And this is true of more and more parts of the world and its societies, though not all. It will indeed be interesting to see what sort of man will prevail and survive.

This is obviously not to say that the problems of human aggressiveness are well on the way to solution. It is all too easy to find evidence of the persistance of gross insensitivity, brutality, and indeed mass murder on the part of the peoples of almost every country of the world. This tendency to human strife has been widely discussed as a strange biological phenomenon. It is probably best to admit that the biologist does not understand it. He can say that physical aggressive reaction almost certainly involves the excitation of centres in the hypothalamus such as those whose electrical stimulation can produce rage or stop it in an animal. We presumably have such centres, and they can probably be activated to produce rage either by physical stimuli (pain) or by 'psychological' ones, such as insult. They are brought into action in many conflict situations, whether between individuals (as in family quarrels) or between gangs, or armies in the field. In this sense they are involved in some acts of warfare, but it seems difficult to hold that the arousal of rage responses is the 'cause' of war or is even a necessary part of it under modern conditions. The struggles between groups of peoples may have biological roots in whatever were the original factors that led to human differentiation. We know only little about these factors, and certainly not enough to disentangle them as causes of conflict from the economic, political, and social forces that are involved in the origin of wars. The biologist is as interested in this problem as anyone else. He sees that for the first time there is a possibility that men will exterminate themselves, and perhaps all life on earth. If he is wise he will deny all those who bring only simple biological arguments to this most serious of all debates, whether they are introduced to prove that war is inevitable ('nature's pruning-hoot') or as claims to methods of preventing it.

What the biologist can perhaps do is to encourage people to realize that continued maintenance and homeostasis are not necessarily served by pursuit of the immediate ends of separate individuals. Indeed, the equipment man receives from his past history is especially designed to ensure communication and co-operation. We have seen how our growth patterns seem to be planned to these ends and that we are endowed with a brain from which speech springs naturally as we mature. These and many other pieces of evidence suggest that the full realization of man's inherited capabilities is to be found in co-operation with others and that in this way also he will find the fullest personal satisfaction. This is no great new discovery and in any case it needs to be tempered with the recognition that the human process of finding a means of survival by skill in thinking requires the maximum creative effort by every individual brain. The problem of how people are best motivated to optimize plans for survival is indeed a difficult one. The biologist, with present knowledge, can make only some contribution to the discussion; he is in no position to claim that he has any technical knowledge that will provide a solution.

This is not to say that inquiry into the origins of the basic cerebral characteristics of man is without importance. We shall consider further features that may have evolved as man turned into a co-operative hunter of large animals.

Some Social Effects of Co-operative Hunting, Caves, and Fire

If the females and infants were to be left behind during the hunt all sorts of developments of the social system may have been required. In our nearest relatives among the surviving apes, the chimpanzees and gorillas, the social systems are rather loose, with variable associations of a male and a group of females and juveniles. Sometimes there are small family groups; in other situations quite large bands may be formed, presumably affording protection. There are only very rare records among apes of any sort of co-operation in collecting food or of food-sharing. Large changes in behaviour must have been necessary to make greater co-operation effective as man evolved. The association of individuals must have become reasonably stable; some kind of home base, even if temporary, was required, to which the food could be brought back. This, as well as the need for shelter, may have been a factor in the development of cave life. The very fact of carrying food to other individuals was a new development among primates, though of course common in carnivorous mammals and in birds. It required the development of a built-in system for assuring attention to the needs of others as well as oneself. The fact that man has such tendencies to altruism is as much a fact of nature as is his selfishness. We need to find out as much as we can about the origins of the habits and drives that lie behind some of our deepest emotions.

It is perhaps unwise to speculate too much about the influence of the development of a 'home' on the origin of the form of life that we call human. The conception is so powerful among peoples today (though not universal) that any clues as to its origin may be important. That the cave is not only for protection against cold is suggested by the fact of early cave occupation in central Africa. It may be that its first importance was as a central site for the group as well as for protection from other carnivores.

The capacity to obtain protection against the weather must have played a large part in the success of man further north in surviving the glacial and pluvial periods. The first signs of the use of fire come during a cold period (the second glacial (Mindel)) about 450,000 years B.P. and from rather northerly sites, in Hungary and at Pekin. In both places there are hearths associated with remains of *H. erectus*. No traces of fire have been found in African open living-floors or rock-shelters.

Presumably, as with stone tools, some use of natural fire preceded man's discovery of how to make it. If the first use was for warmth its value for cooking may have been discovered by accident. As a process of external digestion cooking has no doubt played an important part in the final stages of the development of the human face by reduction of the jaws. One of the most striking characteristics that differentiate all the early hominid fossils, Australopithecines (particularly *A. robustus*) and *H. erectus*, from modern man is the size of the teeth and of the parts of the skull that bear the muscles that move the jaws. The reduction of the teeth and jaws proceeded rapidly after the discovery of cooking, in the middle Pleistocene, and is still proceeding today. The pre-digestion of the food by cooking may have been one of the factors leading to a greater economy of time, with consequent release of capacity to make not only further tools but also statues and other artistic artefacts. The presence of these in turn may have helped to unify societies, to promote communication, and stimulate ideas for further invention.

The capacity to make and control fire we can rank as man's second great technological jump, the first being the manufacture of stone tools, at least a million years earlier. No doubt fire was put to many uses such as splitting stones or hardening wood. But it was only after a further 400,000 years that it led to the next really great technical revolution with the discovery of the smelting of ores to obtain metal. Meanwhile man had made many other discoveries, above all in language and communication.

The Origins of Human Sexual and Mating Habits

One of the very striking characteristics of modern man is to form pairs who remain together for relatively long periods. Such pairing is said to occur in gibbons but is not general in primates. The organized hunting by middle Pleistocene hominids would obviously have required some adjustments if the sexes were separated for long periods. It can be argued that these separations were at least part of the reason for the socialization of mating as marriage and also the particularly intense human interest in sex, even without relation to reproduction. If the sexes are often to be separated, and yet the mother and offspring cared for, there must be strong bonds between individual males and females. This may be the basis for the various features of heightened sexuality that are characteristic of man. Most important of these is the occurrence in the female of an orgasm comparable to that of the male ejaculation. This is accompanied by (or has produced) an almost uniform reciprocal acceptability of males and females throughout the menstrual cycle. There has been an almost complete disappearance of the changes that occur in many primates at the time of ovulation at the middle of the cycle, which serve, for instance in baboons, to advertise the condition of the female and to excite the male, so that copulation occurs only at that time. Some women are regularly aware of the time of ovulation by pains and even slight bleeding at the middle of the month. There are also characteristic slight rises in temperature but data on frequency of copulation do not show clear signs of greater receptivity or attractiveness at this time.

Mating face to face is another characteristic almost unique to man. In nearly all primates the male mounts the female from behind. In gorillas and chimpanzees the female may lie on her back, with the male squatting between her legs rather than lying on her body. The characteristic human position may well have resulted simply from the resting prostrate position of a terrestrial bipedal creature. No doubt, however, it contributed to the formation of a closer communication between the individuals, with corresponding advantage for the family and for the young. Presumably at the same time many mating signs were transferred from the back of the female to the front. The characteristic sexual sign of primates is the sexual skin, a special area around the genitalia and buttocks, which becomes swollen and often bright red at the time of oestrus (corresponding to the mid-month ovulation of human menstrual cycles). In humans the characteristic features of the female are on the ventral side, especially the delicate features of face and lips, rounded breasts with coloured and sensitive skin around the nipples, and the distribution of the pubic hair. Correspondingly the male has beard and distinct distributions of hair on the body.

These and other physical features make the sexual life of hominids more continuous and richer between the members of a pair than in any other primates. In many animals stimulation of the clitoris by the penis of the male serves to ensure entrance at the moment of maximum female response, although a high blood-pressure may persist for a long time in the female (for instance in the bitch). The human response, after penetration, is waves of high blood-pressure producing the orgasm in the female, as in the male. The physical responses are of course the culmination of processes of love-making of a much more subtle character by which the pair maintain their interest in each other and support the lives of themselves and their family. We shall probably never know at what stage of evolution the family emerged from the promiscuous troop. Apparently in baboons and even chimpanzees and gorillas there is no long-lasting pair formation. However, the gibbon is monogamous and the family holds a pair territory (Schaller 1965). It may be that the earlier Palaeolithic hunters lived in small troops. But it is not unlikely that the really successful co-operation began in those populations where the genital and cerebral aspects of sex were such that pairs stayed together long enough to raise families.

EVOLUTION OF ART AND RELIGION

The Importance of Art and Ideology for Survival

There is no body of facts that yet enables us to understand the origins of aesthetic creation or religious beliefs and practices. Presumably both sorts of activity were somehow of assistance to Palaeolithic man in the business of getting a living. This does not mean that carving, or painting, or offering prayers for the dead were crudely of 'practical' value, for instance, by improving hunting technique. Yet there is a case for saying that creation of new aesthetic forms, including those of worship, has been the most fundamentally productive of all forms of human activity. Whoever creates new artistic conventions has found methods of interchange between people about matters that were incommunicable before. The capacity to do this has been the basis of the whole of human history. It remains so today, and has taken many new shapes, including the creative discoveries of science and the inventions of engineering, as well as new forms of visual, literary, and musical artistic invention. Similarly, fresh and deeper religious, political, and other ideologies continually provide more and more convincing means of helping individuals to live together in effective societies, and also helping them with their separate lives. We are all aware how far from perfect our societies are today. The deficiencies make it hard to see that communication and creation are more advanced and helpful than ever before. Indeed, the very fact that they are so makes the deficiencies more obvious, because so much more publicized.

Man has produced a set of techniques for communication that enables individuals to co-operate to obtain a living. These techniques are highly varied and complicated and we still do not know how each of them contributes to the maintenance of society. How much does the transfer of information depend upon the developments that we call Art? Can there be Law and sufficient conformity of individuals without the beliefs and rituals that we call Religion? Can there be Peace without War?

It is beyond the young science of human biology to provide answers to these questions. But we can begin to see some of the factors that have been and are still at work at determining human conduct. In particular we can survey some of the most characteristic human activities of people in the past. We can even very tentatively relate them to the probable human needs at that time. Apart from the scarcity of evidence there is a serious danger of pomposity in interpreting the evidence that is available, such as by labelling sexy statues as symbols of fertility myths, or wounded animals as signs of hunting cults and so on. At first sight it seems that these activities present no problem; most humans (and animals) are interested in sex and in food. Therefore, it is said, 'Of course people *enjoy* talking about them or making symbols of them. What more explanation do we need for representations?' But chimpanzees do not make sex symbols, either sensuous and aphrodisiac or abstract and symbolic. What we have to ask is whether there is any evidence about when and why human beings began to do such things.

The Art of Making Tools and Weapons

We have to think back to the time, say, 3,000,000 years ago, when our still ape-like ancestors had no art and no religion at all. Their 'law' was probably the product of instinctive attack and threat by the strong, and retreat and submission by the weaker. Their language was perhaps a series of grimaces and sounds expressing the reactions of the individual to each particular situation as it occurred and calling for the appropriate reaction of his fellows.

It is surely significant that the first evidence of what we may call nearly human behaviour is of the making of tools, and in particular weapons. A stone chipped to make a chopper is pre-eminently practical and is directly useful for helping in survival. It is a startling fact that making of tools of stone and bone remained the sum of man's creative achievement that left any material trace for perhaps 2 million years. Of course flint is durable, whereas artefacts in wood or other material are unlikely to survive for such long periods. We have continually to remember that the evidence that we have is biased by its very capacity to survive.

Looking at the stone tools, however, it seems that creative efforts for nearly 2 million years were devoted to relatively small improvements in the techniques of napping. We do not know whether changes were gradual or in jumps, but in the aggregate they were exceedingly slow. Human generations succeeded each other for tens of thousands upon tens of thousands of years, leaving hardly any signs of change in way of life. Of course, there were certainly some changes that have left no trace. Yet stability over such a long time suggests that the early techniques for making and using weapons may have proceeded in an almost stereotyped way, depending little on the vagaries of individual creation. Nevertheless, the very slow changes in methods of flint chipping may have contained the first elements of creative art. To make something for future use already argues a capacity for abstraction from the present. To make it so that it is sharper or otherwise does its job better than before is creative even if not 'artistic'. To chip it in such a way that it shows patterns of symmetry or rhythm may make it a pleasure to see, to hold, and to use, and the result may therefore be *more practical because it is more 'beautiful'*.

Some animals as well as humans are said to show responses to rhythms of visual pattern as well as of action. It is claimed that apes, monkeys, and birds choose visual patterns that show rhythm rather than disorder. Many young animals perform rhythmical jumps and gymnastic feats in their play. Juvenile play may be in itself an education for life and could be the basis, with juvenile curiosity, for feats of creativeness. Capacity to develop new and individual action patterns is characteristic of the brains of mammals, and especially of primates, and is of course most evident in the young. One may not think much of chimpanzee painting, but at least the young ones do it. When they get older they can no longer be bothered to be creative.

Rhythm and balance are features of action patterns, and visual art translates them into symbols. It is not impossible that the chipping of a flint to make it better to handle may have been the earliest manifestation of art, as it certainly was of craft. It may have been the parent of many other arts or, indeed, of all of them.

It is worth speculating in this way because the prevalence of stone tools for such a very long preliminary period of human history is our chief source of data on the subject. Of course, there may have been other artefacts that have perished (there were certainly some of bone) and there may have been rhythmical dances before religion, and simple language long before that, all of which left no enduring remains. But the signs that *were* left were of stone, and stone tools are above all *useful*. They may also have been 'beautiful', and the makers may have enjoyed making them, in the sense at least that their life was much improved when they made good ones, easy on the hand as well as the eye, and more effective in the hunt. Those societies in which the young were good learners would be provided with good craftsmen, able by good language to pass on their skills. The earliest consumers of the products of both art and education may well have been the leaders of the hunt, who would have also been the first art critics!

The Use of Fire

The next innovation of which we have evidence is the use of fire, first found in the lower cave near Pekin, with the remains of *Homo erectus*, perhaps 500,000 years ago, and also at Vértesszöllös in Hungary. Fire was no doubt important in many different ways, for keeping away carnivores as well as for keeping people warm. By assisting digestion cooking may have made more food available (and made it more palatable). One of the advantages of a carnivorous way of life is that the time spent on eating is reduced. Large herbivores must eat for many hours a day. The technology of making and keeping fires no doubt also provided further stimulus to language and co-operation.

The Earliest Burials and Evidence of Religious Beliefs

A further major sign of cultural advance that is left to us is burial, first without additions and then with the accompaniment of decorative ornaments and beads and their arrangement in designed patterns. All these are tremendous advances and the signs of them come only after a long interval, with the Mousterian culture, probably of *H. s. neanderthalensis*. At La Ferrassie in France more than 35,000 years ago, two adults were buried in recesses, one natural and one excavated. There were also two child burials, one in a ditch containing animal bones, which were plausibly left as offerings. The other child was buried under a stone decorated on its lower side with a simple pattern of pairs of pits. Such 'cup marks' are often found in later burials, though their symbolism is obscure. Other Neanderthal burials have been found at La Chapelle (35,000-45,000 years old) and Mount Carmel (*c.* 41,000 years old). They are few and simple. But at both sites there are some signs that offerings had been buried, including at La Chapelle the bones of a whole animal still articulated. At Monte Circeo in Italy there was a ring of stones around the skull, whose base was fractured as if (perhaps) the brain had been eaten.

These evidences of offerings are the best signs that we have of the beginnings of religious practices. The fact of burial itself is often difficult to prove and in any case might have involved simple disposal of the body. More complex burials begin to appear some 10,000 years later, in the flourishing Gravettian period of the warmer times of the Paudorf oscillation, around 25,000 years ago. The woman buried near the carving was crouched and facing west at Dolní Véstonice in Czechoslovakia. With her were placed tools, also the paws and tail of an arctic fox in one hand and its teeth in the other. Her body and head had been painted with red ochre and were covered by two large mammoth scapulae, one scratched with lines. Evidently a whole series of symbols are involved, though we can only guess at their meaning. People with such developed symbolism must have had a language with already a long history and possibly not fundamentally different from language today.

The origins of religious practices may have been very much earlier, and have left no trace. Possibly they may have begun simply as patterns of movement or sound. It is likely that such activities were parts of the system of intercommunication that enabled individuals to co-operate to obtain a living. The enforcement of morality and co-operation is a prominent feature of religions today. It is possible to imagine that those groups of early men in which individuals were united by common ceremonies, and enjoyed satisfying rhythmic experiences, whether of visual pattern or dance, would have been more coherent and successful than less 'cultured' neighbours. The linking of such ceremonies with sexual practices may have had a basis in their aphrodisiac effects. The sexual practices would in any case be subject to restrictions to ensure adequate numbers and care of the young, as they are in all primates and, indeed, other animals.

More abstract conceptions of gods, spirits, and magic may have come later (though still long before the Gravettian burials), with the further development of language and especially with the descriptive analysis of self and non-self in terms of person language. One can speculate indefinitely about possible origins for the elaborate web of beliefs and practices that mankind has constructed. They are compounded of fears and hopes and of methods for obtaining a whole variety of biologically practical results, including the proper distribution of territory and of genes. It would be offensive as well as unscientific to attempt to give any superficial general account of these cherished and often socially central practices. The biologists must recognize the fact of the belief of many people in the existence of a transcendent Spirit. To make such a generalization has been one of the most astonishing feats of human brains. Such beliefs are held and stated as a result of cerebral action. We may therefore hope that better knowledge of the brain will give us still better beliefs, or at least better insight into those that we have. The purpose of the study of man is to *enlarge* his capacities and his understanding, including his faculties of wonder and worship, and his appreciation of beauty. This is the process of growth of culture in the widest sense, which has been going on at least for the last 25,000 years, during which men have certainly held some sort of view of the agents at work in the world. Probably the process began long before that, but if so it left no enduring mark until men began to bury their dead.

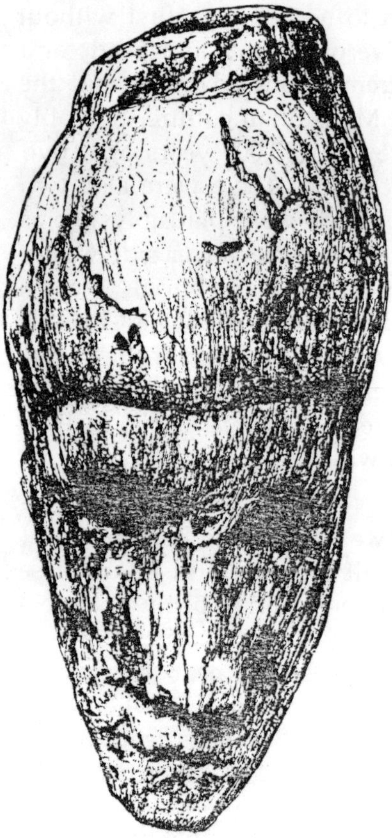

Fig. 13.16: Woman's head carved in ivory from Dolní Véstonice (Czechoslovakia). Gravettian, *c*, 25,000 years B.P.

The recognition of death is probably very ancient. Baboons in zoos give a characteristic bark when surrounding a dead individual. After a chorus of these barks have been heard at night in the wild a dead baboon may be found the next morning. A possible sign of interest in the dead is that the remains of *Australopithecus* in the Makapansgat caves seemed to be placed against the wall, away from the main pile of animal bones.

The solution to the problems of the origins of religion would be of great value, but no one has yet found a way of using the earliest known burials and ornaments to solve them. Burial suggests a developed system of description of the self and its endurance, but other religious forms may have existed before belief in soul and spirit. Rituals and dances, like fears of devils or aspirations towards gods, leave few or no remains.

Apart from burials there are few signs of religious activity until the beginnings of Neolithic life in the Middle East, less than 10,000 years ago. From then onwards shrines and temples and their accompanying images and symbols become increasingly common. There must have been religious gatherings before this, and perhaps particular places were treated as sacred long before temples were built. It may be that these were often hills of striking shape and even that artificial hills were constructed as convenient meeting-places, the first cathedrals. Indeed Canterbury and other cathedrals rest on henge megalithic monuments.

Upper Palaeolithic Carving—the First Symbolic Art Forms

The first evidence that is left of the capacity to make representations of men and animals comes from the Upper Palaeolithic. Aesthetic endeavour may have begun much earlier in improvement of the symmetry of weapons of stone, in carvings of bark or wood, and especially in production of ornaments. In a cave at Brno in Czechoslovakia the figure was buried with a necklace (or head-dress) of hundreds of pieces of mother-of-pearl from the shells of the mollusc *Dentalium*. Ornaments were no doubt used before this, perhaps originally to increase the attractiveness or impressiveness of the individual to his fellow men (or women). The function of trinkets is rather simply to excite attention, even today. The making of representations is a much more complex activity. A carved or painted figure can serve to excite emotion, but it also transmits a message indicating a particular creature *that is not present when the object is seen*. This is the essence of abstract symbolism.

The earliest of these representations that have survived were carved figures, either on cave walls or as small statues. The small figurines may have been carried by the hunters on their wanderings. However, nearly all of them come from huts at Aurignacian or E. Gravettian sites in eastern Europe, from the somewhat warmer period (though the climate was still as hostile as, say, that of much of northern Canada today). There were some signs of settled life in simple houses with clay walls and stone floors. Mammoth and reindeer were main items of food, no doubt along with smaller animals and plants. It seems likely that these people had largely abandoned the nomadic life, which was continued much longer by the Magdalenians of western Europe, but these, although they painted much in caves, left no figurines.

The carvings show a preponderance of animals, but human figures are not rare. Among the oldest is a large statue of a man, carved in mammoth ivory, from a burial

Fig. 13.17: Statue of man carved in mammoth ivory from Brno (Czechoslovakia). Gravettian, *c.* 25,000 years B.P. (Moravian Museum). About half natural size. (After Sandars 1968).

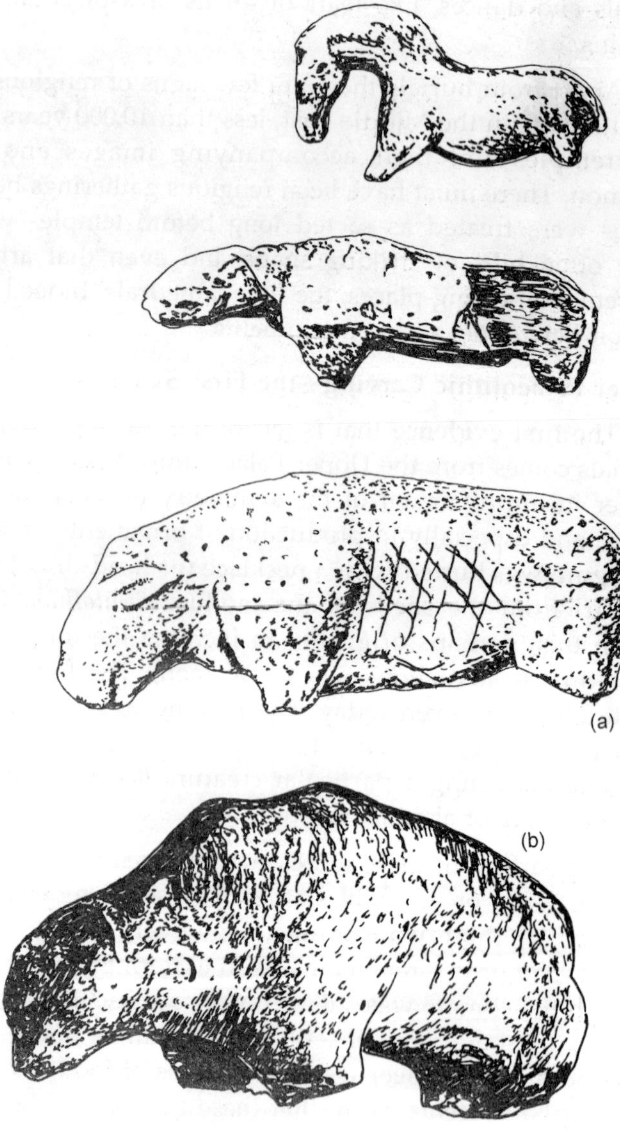

Fig. 13.18: (a) Animals carved in bone and ivory from Württemberg, Germany. Gravettian, *c.* 20,000 years old (Tübingen). (b) Bear in baked clay from Dolni Vestonice. Gravettian, *c.* 23,000 years B.P. (Moravian Museum). (After Sandars 1968).

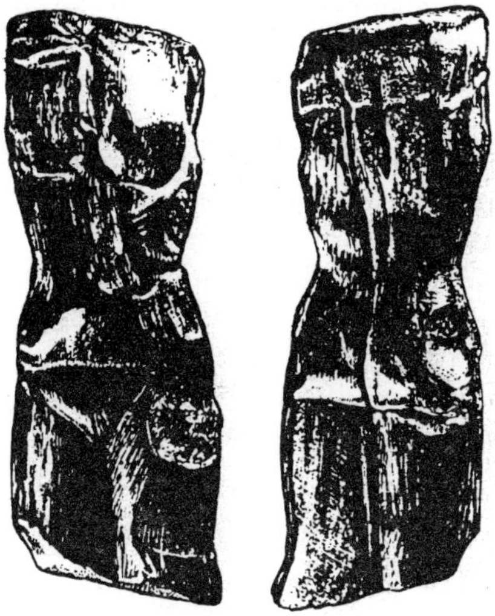

Fig. 13.19: Woman's torso carved in haematite from Ostrava Petrkovice. Gravettian. c. 25,000 years B.P.

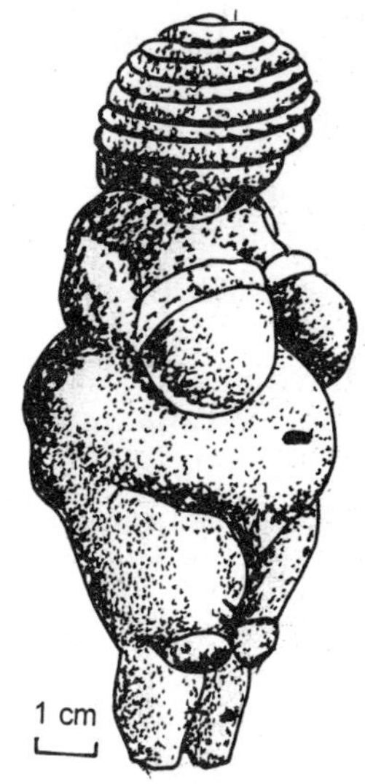

Fig. 13.20: Limestone statue of a woman from Wiliendorf. Late Gravettian-Solutrean, 18,000 years B.P.

in Bron, probably 25,000-30,000 years old. This is a highly naturalistic carving, showing rather heavy brows, as there were indeed also on the skull that was found with the figure. Equally naturalistic are the animals of bone and ivory of about the same date found at Württemberg (Germany) and the remarkable woman's torso from Ostrava Petrkovice (Czechoslovakia).

The Venus of Willendorf, a limestone figure of roughly the same date shows no facial features, but is still not strongly formalized or idealized. We have few samples of this early art of carving to judge from, and dates are approximate, since they are near the limit of the usefulness of the carbon-14 method. But many of the earliest were naturalistic representations of men and animals. Some, indeed, may even be portraits of individuals. The carved face of a crooked mouth and brow such as might perhaps have been due to an injury to the left facial nerve ('Bell's palsy'). Curiously enough there was found near by the skeleton of an old woman of 40, buried with elaborate ritual, and showing damage on the left of the skull. Was she a famous leader 25,000 years ago? Even more curiously there is another face with similar features roughly sketched in a dwelling near by.

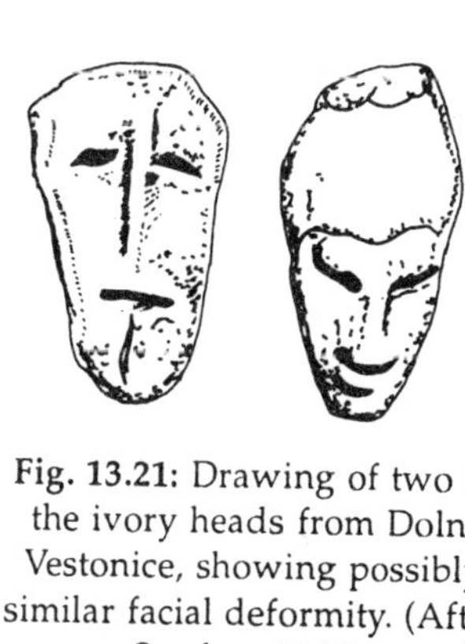

Fig. 13.21: Drawing of two of the ivory heads from Dolni Vestonice, showing possibly similar facial deformity. (After Sandars 1968).

We understand about as much and as little of the significance of these early signs of human artistic creation as we do of the basic social significance of the efforts of artists today. To make the figurines the Palaeolithic sculptors must have required good tools, physical skill, much observation, and also thought, leisure, and some motivation. From the earliest times art involved the double task of making something and of matching it with natural objects. It is interesting that the earliest works of art that have been found include several representations of the human face and figure. This may perhaps be no more than an accident, but it agrees with our suggestion that communication among humans centres on recognition of facial signs and that learning of these is the first task of the child's brain. The earliest communication by symbols may well have included indication of facts about people when they were absent. It is quite possible that the deficiencies of the earliest verbal communications were assisted by the use of visible symbols. Such communications may at first have been only about matters of overwhelming importance, such as the unity and discipline of the group. The name of the leader would be among the verbal signs to demand the greatest attention, and an image would help.

Of course, we know from the prevalence of burials that religion was already well developed much earlier than these signs of art from about 25,000 years ago. The continuation of the name and image of the leader as a symbol of unity may have been as much a cause as a result of the development of religious burial. It is all too easy to speculate—yet perhaps the exercise has some value in making us think about why men act as they do. As we have seen, it is not possible to reconstruct the origins of either art or religion. Yet origins they must have had, in the truest sense, for they are not there in apes. Almost certainly they arose gradually, at a much earlier period, first in some elementary form in which we should hardly recognize them today. They were of value for the powers of co-operation and communication that they conferred on the groups that practised them. Their practices possibly included emphasis on the human face and figure. These may have served as symbols of the dawning consciousness of the individual as an entity able to discourse of himself, others, and the world.

Other pressing subjects of discourse were of course sex, food, and enemies. Besides the human figures the earliest representations include animals and various sex symbols.

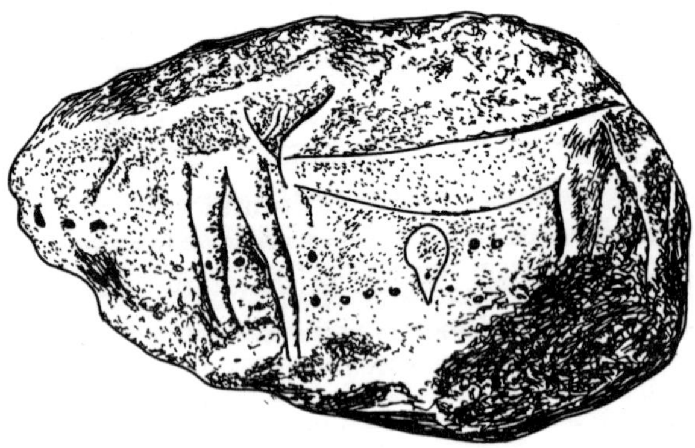

Fig. 13.22: Female signs (vulvae) and an animal engraved on rock at La Ferrassie, Dordogne, France. Aurignacian, *c.* 30,000 years B.P.

Some of these may have served quite simply as salacious objects to arouse or appease sexual desire; pornography has a long and honourable history in the genesis of art. The rich literature of psychological studies today tells us much of the significance of the representations that humans make, and the early propensity to sexual symbolism provides some useful data in forming a realistic view of Man.

Palaeolithic Cave Art

Palaeolithic cave-painting was largely restricted to a limited area in southern France and the Pyreness, with a few outlying examples elsewhere in western Europe. Thus there were evidently great geographical differences between various Upper Palaeolithic cultures. In eastern Europe there was no cave art but many animal and human figurines carved in ivory or stone; in the west there were fewer figurines and fewer representations of man, but many caves with animal paintings. Even allowing for the accidents of survival these differences suggest that from its earliest appearance the making of representations, one of the most distinctive characters of man, was culturally conditioned and used varied conventions. But of course these earliest *surviving* evidences of art forms do not tell us when or how it began. Nor can we tell whether there was any steady progression such as from chipping flints to carving figures, to making of low relief, and so to drawing and painting. Some of the earliest representations are of animal heads engraved on small stones from the early Aurignacian cultures of perhaps more than 30,000 years ago at La Ferrassie. From the later Aurignacian (known as Perigordian) at Labatut there are securely dated painted deer, as well as an engraved horse. The well-known bas-relief figures at Laussel also probably belong to this period.

Few cave-paintings can be referred with certainty to the Solutrean and early Magdalenian periods, but this does not prove that none then existed. The great majority of the work is usually referred to the latter part of the Magdalenian, say 15,000-11,000 years B.P., but much of it cannot be securely dated and may be from a long time earlier.

Techniques of Cave-painting

Painting was practised in a variety of situations, sometimes in caves that were occupied, but more often deep within long tortuous passages, far from the light. In some places the remains of the torches used have been found. When the paintings are high up, as at Lascaux, they must have required either a pole or else some kind of scaffolding, and evidence of the remains of this has been claimed. Bas-relief sculpting, engraving, and painting were often combined, especially the last two. The paints were earth pigments of various shades of red, brown, and yellow, obtained from ochre, native or burnt. Manganese oxides provided violet and black, which were also made from soot and burnt animal substances. No blues or greens are known.

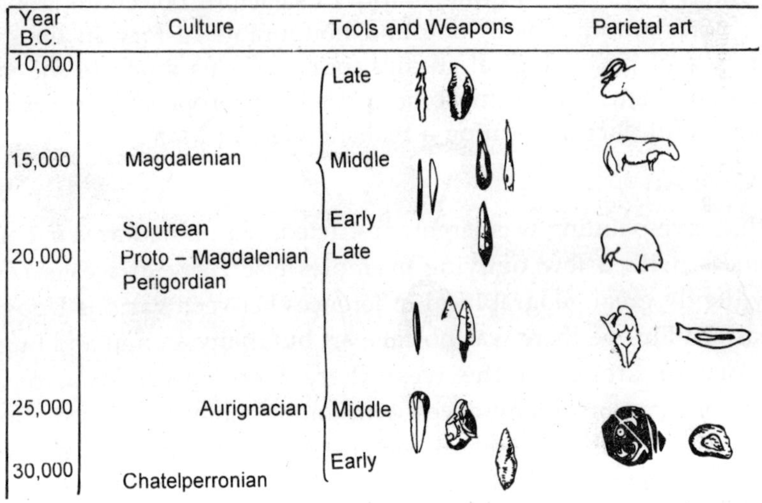

Fig. 13.23: Time chart of the Upper Palaeolithic in western Europe, with carbon-14 determinations and some typical dated examples of parietal (wall) art.

Fig. 13.24: Drawing of black-painted deer from Labatus (Perigordian, *c.* 24,000 years B.P.).

Fig. 13.25: Relief of a woman carved in stone. From Laussel, Dordogne, Gravettian, *c.* 20,000 years B.P. (Musée d'Aquitaine).

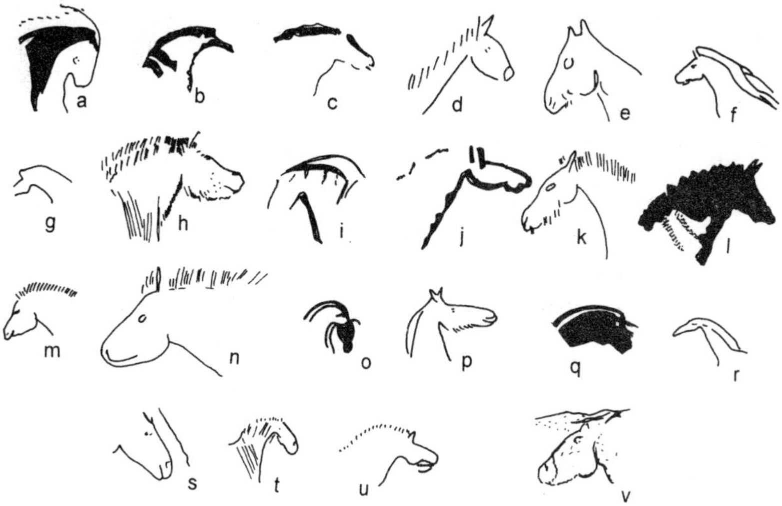

Fig. 13.26: The variety of styles used by Palaeolithic artists to show horses' heads.

The animals are shown by a great variety of clever abstractions and techniques, used to show characteristic details. The styles are so varied that it is hard to identify any specific conventions characteristic of 'Palaeolithic art' as a whole. Moreover the attempt to trace the evolution of styles has so far ended mainly in controversy. Using securely dated works Laming (1959) has modified the original scheme produced by Breuil. Roughly there can be recognized an early Aurignacian-Perigordian period in which there are stencils of hands, rows of dots, large red signs, and meandering lines

Fig. 13.27: Figure of bison engraved in floor with natural hollows. Niaux, France. Magdalenian *c.* 13,000 years B.P.

of clay and simple outlines of animals—in one colour. These are naturalistic but with either stiff legs or none, the horns shown full face in twisted perspective.

In the second 'stage' there are many low-relief sculptures, black animal outlines in blotted or dotted lines, and flat-work figures in red, black, or sepia, sometimes bichromes. The twisted perspective is more elaborately used.

In Laming's third stage we have the fully developed Magdalenian polychrome paintings and black engraved animals, shown in true perspective.

These attempts at classification are very attractive but must be looked at cautiously since they make many assumptions, especially that the styles were the same everywhere at any one time, whereas variety is one of the characteristics of Palaeolithic art, as it is of art today (Ucko and Rosenfeld 1967).

Fig. 13.28: Engravings of bison and bison man and numerous other animals. Trois Frères, Ariège, France.? Magdalenian, *c.* 15,000 years B.P.

The animals shown are mostly horses, cattle, mammoth, woolly rhinoceros, cave lion, red deer, and reindeer. The individual animals are often drawn with striking skill, though with stereotyped techniques. Sometimes the paintings are of groups but there is little composition as we know it, which would indeed be difficult with the limitations of the caves. The figures may be superposed on each other, perhaps to represent herds, perhaps simply because drawn by successive visitors. Human figures bearing animal heads sometimes appear. In other 'compositions' there are strange collections of figures presumably symbolic—but of what?

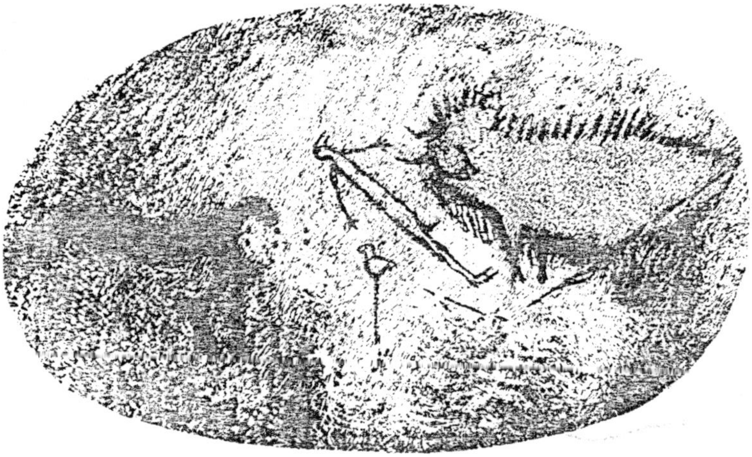

Fig. 13.29: Wall painting said to include a wounded bison, a dead man with a bird's head, a bird on a stick, and a rhinoceros. The scene is in a shaft of the cave and is hard to view. Lascaux, Dordogne, France. Early Magdalenian *c.* 17,000 years B.P.

Composition and Placing of Paintings

Recently, discussion about interpretation of cave-painting has centred around the fact that the representations of various animals are not scattered at random but are grouped more or less systematically. There is often a rather regular sequence, proceeding from the entrance of the cave. First are signs classed by some archaeologists as male and with them animals said to be symbolic of masculinity, horses, deer, and ibex. There follow signs alleged to be female and animals supposed symbolic of that sex—cattle, bison, and mammoth—often forming the main pictures at the centre of the system. Still further in there may be other male symbols.

If the arrangement of the figures and signs is correctly interpreted, it seems to have some sexual symbolic significance. But Leroi-Gourhan, who has analysed the paintings extensively, seems not to be prepared to say much more about the significance than that it was in some way religious. Moreover, the identification and classification of the figures is by no means easy or certain.

The 'Meaning' of Cave-Paintings

The fact is that in spite of all the analysis we do not know why the Upper

Palaeolithic peoples did these paintings. Their reasons may, indeed, have been as varied as those that are given by Australian aborigines, who also do cave painting.

The paintings may have been connected with a system of fertility magic or with a cult of food magic, the representation supposedly serving to increase the probability of finding game. A few of the animals are apparently pierced with spears and this has also suggested that they are enactments of hunting scenes. They could indeed have served as pictorial adjuncts to verbal communication in something of the way suggested above for early sculpture. Pictures are a great help to communication even today and would have been even more so if talking itself was a laborious and perhaps boring task, as writing is for young children now. But such explanations are much too simple. For one thing the caves were dark and inaccessible—not at all the place to teach the young to hunt.

Probably by the Upper Palaeolithic period of man's history ideas had already developed on to much more subtle planes. The pictures almost certainly have some kind of magical or mystical purpose, directed to influencing events, perhaps with rituals to pacify spirits, postulated as likely to be inimical if not placated. It is a pity that we have not yet been able to interpret them well enough to understand how men spoke and thought in those days. Probably the motivation to paint was complex then, as it is today. The urge to represent seems strong in creatures like humans who are steeped in symbolism. The annual migrations of masses of animals upon which the men depended must have powerfully aroused this tendency to represent the objects of interest, their multiplicity, their movement, and their death at the hands of the hunters. Perhaps it was this very tendency to represent that provided the capacity for communication upon which co-operative hunting depended. It may have spilled over from dance form and ritual into sounds and words, and as these became more abstract,

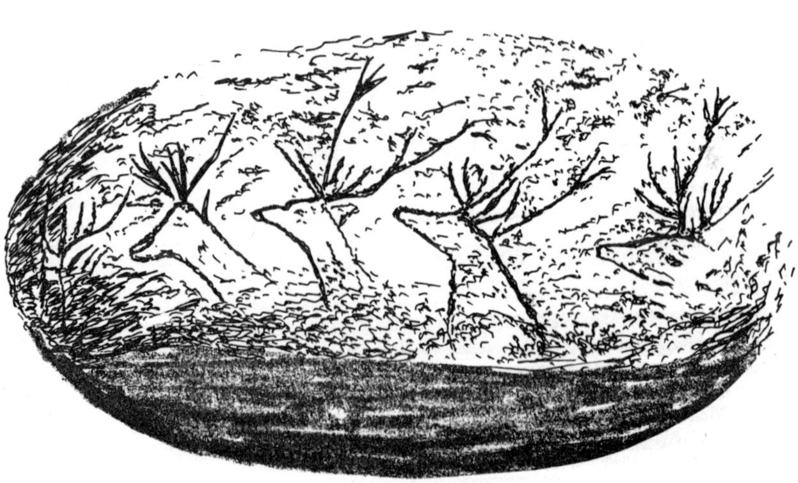

Fig. 13.30: Row of heads of deer tossing as if running or swimming in a herd. Magdalenian, Lascaux. *c*, 15,000 years B.P.

into signs and pictures, conserved deep in the caves and added to by generation after generation of hunters on their regular migrations. The three main theories that have been put forward to 'explain' cave painting may perhaps be summarized as follows.

Art for Art's Sake

The paintings may have been done simply for the pleasure of the painter or to decorate the living space of the group. This motivation cannot be ignored, but neither can it be accepted as the whole answer. Everyday experience tells us that *H. sapiens sapiens* likes drawing and painting. Ethnological studies tell us that food-gathering peoples, such as Australians who decorate caves today, give this as part of their motive. As evidence that the matter is more complex than this is the inconvenience of the placing of many Palaeolithic paintings and the fact that they show only a limited subject-matter—nearly always large food animals, rarely humans, carnivores, or birds, and never plants or inanimate objects.

On the other hand, some paintings are at accessible sites, which were used for occupation; these in fact include more representations of humans than are found in the caves. It may well be that these were done for purposes of decoration or as erotic stimulants, or perhaps just as 'doodles', even by children. Many of the obscure signs may perhaps have no 'meaning' at all; and the superposition of one drawing on another may have no more significance than imitative repetition.

Sympathetic Magic

With all possible allowance for the artistic and casual reasons there remain strong indications that the paintings have an element of symbolism. The choice of subject suggests that this was concerned with the hunt, serving to make more prey available either by magic or instruction, or by a mixture of both these. There is still disagreement about the interpretation of the marks on the animals that are classified as spears or wounds, and not all carry them. One observer who is himself a hunter has suggested that many of the representations seem to be drawings of *dead* animals. The extent of belly revealed, the perspective of the legs, projecting tongue, and prominent jaw suggest that they were drawn from prostrate corpses, not standing animals. Yet another interpretation is that the large bellies sometimes shown indicate pregnant animals, presumably as part of a fertility cult. Difficulties with all such suggestions are that they generalize from a few instances and take insufficient account of the possibility that the features shown are stylistic conventions. The artist may draw animals as if from the side because that was the accepted way to do it; this was the way animals were thought of and that was how you showed them. The concept may have come originally from seeing dead animals, without the artist meaning to represent dead ones in his picture.

The Pictures as Symbols of Religion or Totemism

The Abbé Breuil, who made a comprehensive study of the paintings, insisted that the caves were in essence 'sanctuaries'. There is certainly much to show that they were indeed connected with some kind of ritual. The very fact of difficulty of access points to some secret ceremonial purpose. An obvious suggestion is that the animal signs were somehow connected with totemism, that is the division of society into a number of groups each connected with an animal. The art of modern hunting communities sometimes has some such connection (but by no means always!). There are some suggestions that the caves may have been used for initiation ceremonies; for example, there may be numerous small foot-marks, said to be those of children.

Perhaps the strongest evidence of ritual significance is the fact of the varying proportions of the various animals in different parts of the caves. Counts have been made by Leroi-Gourhan to substantiate this plan, but the identifications of the animals are bound to be difficult and the percentages in different parts of the caves are not very convincing.

Ucko and Rosenfeld conclude that the truth is that we do not understand the paintings. "It is possible...that some representations were the work of children (perhaps some of the floor engravings), that some were used in acts of sympathetic magic (perhaps some of the representations pierced with holes), that some were placed in particular situations in order to please (perhaps some of the open-air low-reliefs), and that some were illustrations of myths and traditions (perhaps those which contain imaginary creatures, anthropomorphs and unexpected combinations of animal species). It is very possible, however, that some and perhaps many Palaeolithic representations were made for reasons which still totally escape the modern observer.'

The Origins of Abstraction

The evidence suggests that already by the Upper Palaeolithic the human brain had begun to see objects in a new way, which we can describe as 'abstract. Men began to 'take-away' parts of descriptions of things from their context and to use them in

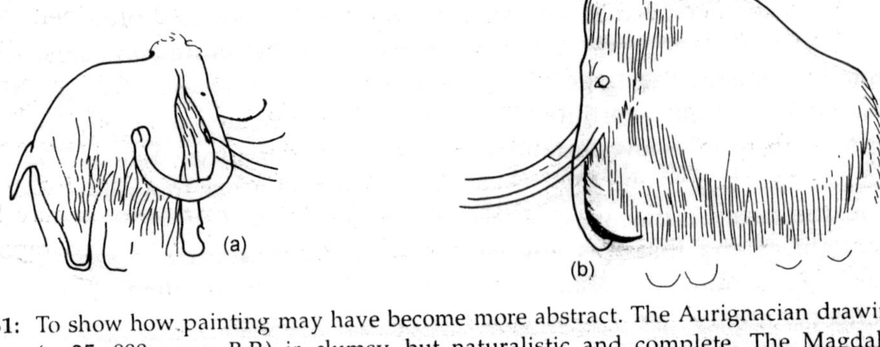

Fig. 13.31: To show how painting may have become more abstract. The Aurignacian drawing (a) (*c*. 25, 000 years B.P.) is clumsy, but naturalistic and complete. The Magdalenian mammoth (b) (*c*. 15, 000 years B.P.) concentrates on essentials ('a general impression of tusky woolliness. The eye is conventionalised and out of drawing. The legs and feet...are suggested by incomplete ellipses').

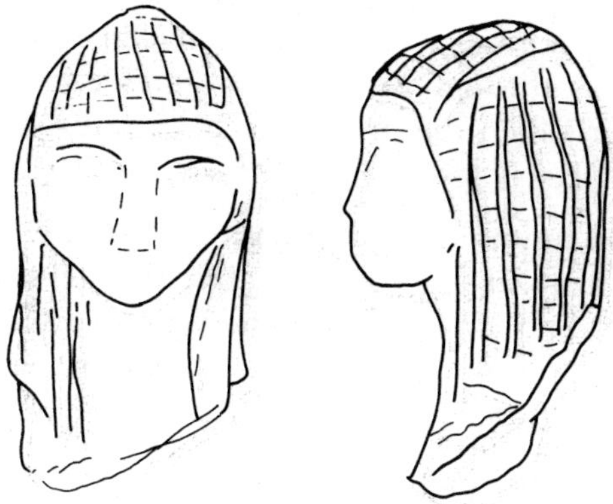

Fig. 13.32: Gravettia-Solutrean sculpture of a woman (*c.* 20, 000 years B P)

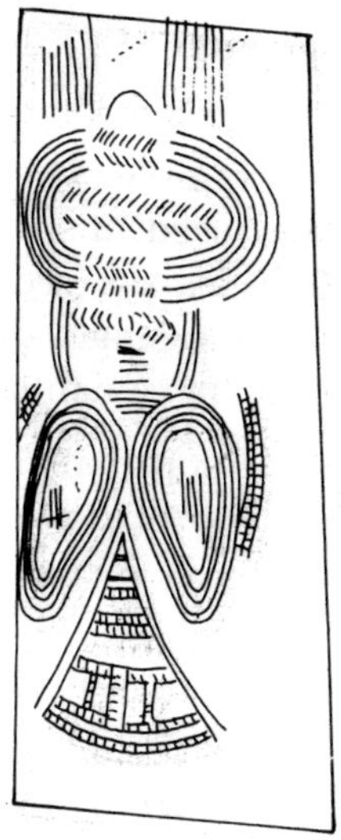

Fig. 13.33: Gravettian engraving representing womanhood,?, 20,000 years B.P.

other connections to make forecasts. Characteristic of this is the use of tools to make other tools, as mentioned already. Again, although many of the figures were naturalistic representations of detail, others were highly abstract. We do not know when this capacity for abstraction began but it may have become progressively more prominent. The early artists showed human figures and faces in full detail very effectively. But in other works and especially later, they tended to emphasize particular features and finally produced purely abstract patterns.

It has been suggested by Pumphrey (1953) that this power of abstraction is related to the development of the power of planning several stages of action in the future, and so to the development of language. He pointed out that sounds constitute an

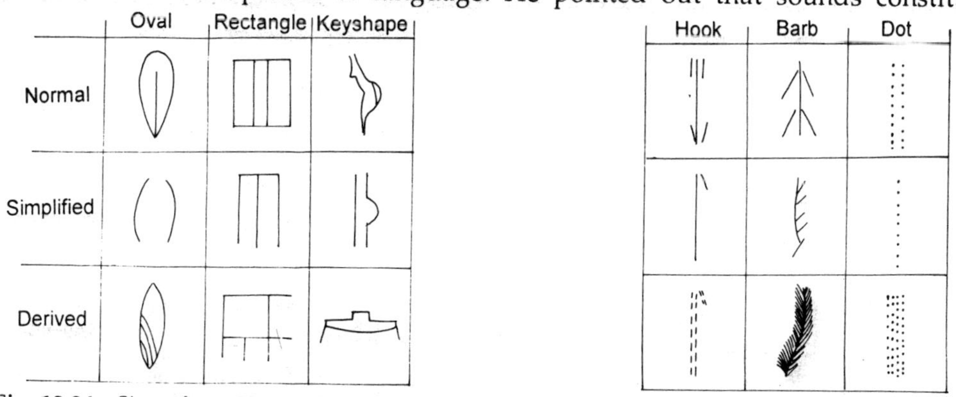

Fig. 13.34: Signs from Upper Paleolithic cave paintings and their suggested interpretations. Those on the left are female; those on the right, male.

exchange both of general information about the state of two or more communicators, and of detailed factual information (which he calls 'intelligence').

For example, studies by telephone engineers of the possibility of limiting the waveband used in speech have shown that if the normal wide band (between 10 and 10, 000 cycles per second) is gated so as to pass through ten narrow gates each 32 cycles wide (using a device known as a 'vocoder') then the 'intelligence' gets through but all the 'personality' and 'emotion' of the speaker is lost.

The lesson that this teaches is that the total amount of information over a wide bandwidth provides us with clues about how to react to situations, but the detailed information is carried in a much more abstract and limited *code.* Indeed, to discard the emotional information is an *advantage.* As Pumphrey pointed out, if the word 'Ion' began as some kind of imitative expression and it at first aroused all the responses

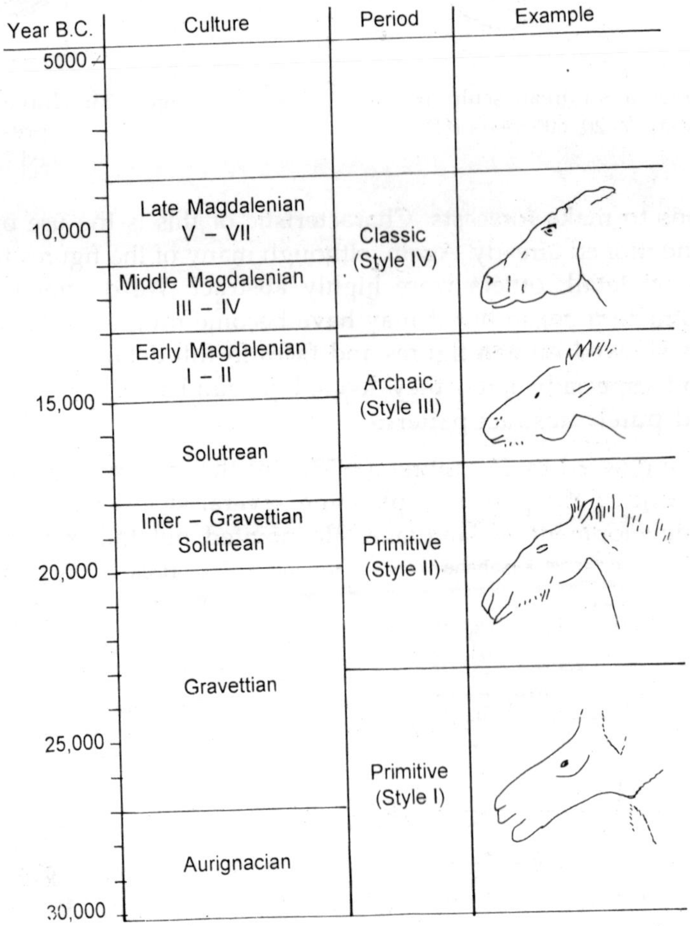

Fig. 13.35: Styles of Upper Palaeolithic art during the 20,000 years from its appearance with the Aurignacians about 30, 000 B. P.

appropriate to the animal, then its use would have been limited if all the hearers gave the right escape reactions! Speech becomes really useful only when the brain learns to attach a core of meaning to abstract conventional symbols used to talk about the animal or object when it is not there, and so to be able to combine word signals to communicate detailed forecasts about the future. We shall probably never know exactly when this first became possible (as Pumphrey said, 'Nothing leaves less trace in history than sound waves'). The early existence of abstract art suggests that it may have occurred already by the Upper Palaeolithic. No doubt long before that time a complex vocabulary of useful sounds had been built up. But it was only when they became abstract, and hence could become brief, that language acquired its flexibility. Even today we see proceeding a continuing tendency to simplify the formal structure of language and to reduce redundancy.

Unfortunately our understanding of Upper Palaeolithic art is still too imperfect to allow us to read from it much about the evolution of brain processes. The interpretation given by Pumphrey would be challenged by others, who have seen in the sequence of styles increasing skill in naturalistic reproduction. No doubt with further analysis we shall learn how to judge these styles better. They provide one of our best hopes of solving the great problem of the origins of our own culture.

Certainly the power of abstraction was highly developed even in the Upper Palaeolithic. The more naturalistic figures were often accompanied by abstract signs, even in the earliest Aurignacian times. Many, perhaps most, of the signs are interpreted as human figures—either female if 'wide' or male if 'narrow'. The tendency to draw representations of human sexual organs upon walls is strong even today, and perhaps we understand the significance of modern graffiti little better than we do those of the Aurignacians.